Current Topics in Microbiology and Immunology

Volume 415

More information about this series at http://www.springer.com/series/82

Ben Adler
Editor

Spirochete Biology: The Post Genomic Era

Springer

Editor
Ben Adler
Department of Microbiology
Monash University
Clayton, VIC, Australia

ISSN 0070-217X ISSN 2196-9965 (electronic)
Current Topics in Microbiology and Immunology
ISBN 978-3-030-07821-8 ISBN 978-3-319-89638-0 (eBook)
https://doi.org/10.1007/978-3-319-89638-0

This Springer imprint is published by the registered company Springer Nature Switzerland AG
The registered company address is: Gewerbestrasse 11, 6330 Cham, Switzerland

Preface

Spirochetes form a fascinating group of bacteria that are usually considered together because of their common helical cellular morphology. Despite this similarity, their biology is very diverse and they occupy a large array of ecological niches. Some members are free living saprophytes, others exist both in the environment and intimately associated with animal hosts, while some have developed such a close relationship with their animal hosts that they are unable to survive independently. Some spirochetes have simple lifestyles, while others infect vertebrate and invertebrate hosts as part of their life cycle. Accordingly, many spirochetes are pathogens and cause a variety of serious human and animal diseases, including syphilis, leptospirosis, Lyme borreliosis, relapsing fever borreliosis, periodontal diseases, digital dermatitis and dysentery.

The understanding of the biology of these bacteria has lagged some way behind what is known for most other species of bacteria of medical, veterinary, agricultural or environmental importance. This gap in our knowledge was due to largely to the fact that until the last couple of decades, genetic manipulation of spirochetes was not possible. While still relatively inefficient, directed mutation of many species of spirochetes is now much easier and in some cases routine.

A further important advance has been the availability, and now routine determination, of whole genome sequences. Currently many hundreds of spirochete genome sequences are in the public domain. These and other developments have resulted in major improvements in the understanding of spirochete biology in the last 10 years.

This volume is not intended to be a comprehensive treatise on spirochete biology. Clearly that would be impossible in the space available. Rather, some key aspects of the interaction of pathogenic spirochetes with their hosts are presented by a group of international experts in the field. There is of necessity some overlap between chapters. That is unavoidable, but also desirable, in that each chapter can be read on a stand-alone basis, with reference to other chapters where appropriate.

Finally, I would like to express my appreciation to all of the contributors to this volume. You are all valued members of my second family, the spirochete family.

Melbourne, Australia Ben Adler

Contents

The *Treponema pallidum* Outer Membrane

Justin D. Radolf and Sanjiv Kumar

Abstract The outer membrane (OM) of *Treponema pallidum*, the uncultivatable agent of venereal syphilis, has long been the subject of misconceptions and controversy. Decades ago, researchers postulated that *T. pallidum*'s poor surface antigenicity is the basis for its ability to cause persistent infection, but they mistakenly attributed this enigmatic property to the presence of a protective outer coat of serum proteins and mucopolysaccharides. Subsequent studies revealed that the OM is the barrier to antibody binding, that it contains a paucity of integral membrane proteins, and that the preponderance of the spirochete's immunogenic lipoproteins is periplasmic. Since the advent of recombinant DNA technology, the fragility of the OM, its low protein content, and the lack of sequence relatedness between *T. pallidum* and Gram-negative outer membrane proteins (OMPs) have complicated efforts to characterize molecules residing at the host–pathogen interface. We have overcome these hurdles using the genomic sequence in concert with computational tools to identify proteins predicted to form β-barrels, the hallmark conformation of OMPs in double-membrane organisms and evolutionarily related eukaryotic organelles. We also have employed diverse methodologies to confirm that some candidate OMPs do, in fact, form amphiphilic β-barrels and are surface-exposed in *T. pallidum*. These studies have led to a structural homology model for BamA and established the bipartite topology of the *T. pallidum* repeat (Tpr) family of proteins. Recent bioinformatics has identified several structural orthologs for well-characterized Gram-negative OMPs, suggesting that the *T. pallidum* OMP repertoire is more Gram-negative-like than previously supposed. Lipoprotein adhesins and proteases on the spirochete surface also may contribute to disease pathogenesis and protective immunity.

J.D. Radolf (✉)
Departments of Medicine, Pediatrics, Molecular Biology and Biophysics, Genetics and Genomic Sciences, and Immunology, UConn Health, Farmington CT 06030-3715, USA
e-mail: JRadolf@uchc.edu

S. Kumar
Department of Medicine, UConn Health, Farmington CT 06030-3715, USA
e-mail: SKumar2@uchc.edu

Current Topics in Microbiology and Immunology (2018) 415:1–38
DOI 10.1007/82_2017_44

Published Online: 29 August 2017

Contents

1 Molecular Architecture of the *T. pallidum* Cell Envelope

1.1 Experimental Obstacles

Venereal syphilis is a sexually transmitted infection renowned for its protean clinical manifestations and protracted natural history (Radolf et al. 2014), both of which reflect the extraordinary invasiveness and immunoevasiveness of its etiologic agent, *Treponema pallidum* subsp. *pallidum* (hereafter referred to as *T. pallidum*) (Lafond and Lukehart 2006; Radolf et al. 2016). It is also a disease that well into the genomics era presents extraordinary challenges to investigators attempting to unravel its many enigmas (Ho and Lukehart 2011; Radolf et al. 2016). *T. pallidum* is one of the few major bacterial pathogens of humans that cannot be propagated continuously in artificial medium (Ho and Lukehart 2011; Norris et al. 2001; Radolf et al. 2016). As they have for decades, investigators must employ intratesticular inoculation of rabbits to isolate and propagate the spirochete (Lukehart and Marra 2007). Because *T. pallidum* cannot be genetically manipulated, experimentalists are restricted primarily to protein-based methodologies to confirm findings and evaluate hypotheses originating from genetic and genomic data. Since the advent of recombinant DNA technology, the fragility of the *T. pallidum* outer membrane (OM) and its low protein content have served as twin confounders of efforts to characterize molecules residing at the host–pathogen interface (Cameron 2006; Radolf 1995; Radolf et al. 2016).

1.2 Historical Misconceptions

Researchers have long appreciated the importance of the *T. pallidum* surface in determining the waxing and waning course of syphilis (Radolf et al. 2006). They also have had to overcome a number of misconceptions to clarify its role in disease pathogenesis. Decades ago, investigators recognized that live (i.e., motile) spirochetes react poorly with the antibodies in patient sera (Nelson and Mayer 1949), and they assumed that this property relates to the pathogen's capacity for immune evasion and persistence (Hardy and Nell 1957; Turner and Hollander 1957). To explain the spirochete's poor surface antigenicity, the notion evolved that the bacterium acquires a protective coat of serum proteins and host-derived mucopolysaccharides (Alderete and Baseman 1979; Christiansen 1963; Fitzgerald and Johnson 1979). In 1973, the existence of the OM was established unequivocally by transmission electron microscopy (TEM) of ultra-thin sectioned, plastic-embedded organisms (Johnson et al. 1973). The ability to express *T. pallidum* antigens in *Escherichia coli* (Norgard and Miller 1983; Stamm et al. 1982; Walfield et al. 1982), the major breakthrough of the early 1980s, attracted to the field many talented molecular biologists intent upon using this powerful new technology to develop a syphilis vaccine. However, these investigators naively assumed that the physical properties and protein content of the syphilis spirochete's OM are similar to those of *E. coli* (Radolf et al. 2006). They also incorrectly assumed that treponemal proteins strongly recognized by the human or rabbit syphilitic sera used to screen recombinant libraries were likely to be surface-exposed in *T. pallidum* (Radolf et al. 2006). The result was the discovery of many notable treponemal antigens, mostly lipoproteins of unknown function at the time, but no OMPs (Cameron 2006; Radolf et al. 2006, 2016).

1.3 The Outer Membrane Hypothesis

As our work with recombinant *T. pallidum* proteins progressed during the 1980s and early 1990s, several observations led us to question prevailing views about the existence of the outer coat and the nature of the spirochete's OM. First, while clumps of testicular debris often were observed in proximity to organisms in negatively stained preparations viewed by TEM, a continuous outer coat or layer was not discernible (Hovind-Hougen 1983; Radolf et al. 1986). Subsequently, we confirmed these findings by radioimmunoassay of freshly harvested treponemes collected onto low-protein-binding polycarbonate filters; only negligible amounts of surface-adsorbed immunoglobulins or serum proteins were detected (Cox et al. 1992). Second, it was noted by routine negative staining that the OM was easily disrupted by routine experimental manipulations, such as centrifugation and suspension, or exposure to low concentrations of non-ionic detergents (Cox et al. 1992; Radolf et al. 1988), conditions that do not perturb the OMs of Gram-negative bacteria. Third, removal of OMs using low concentrations of the non-ionic detergent Triton X-114 (TX114) did not result in an appreciable loss of major membrane immunogens detected by immunoblot analysis with syphilitic serum (Fig. 1a) (Radolf et al. 1988). Finally, organisms lacking OMs showed markedly greater reactivity with syphilitic sera than intact treponemes (Cox et al. 1992, 1995; Radolf et al. 1988).

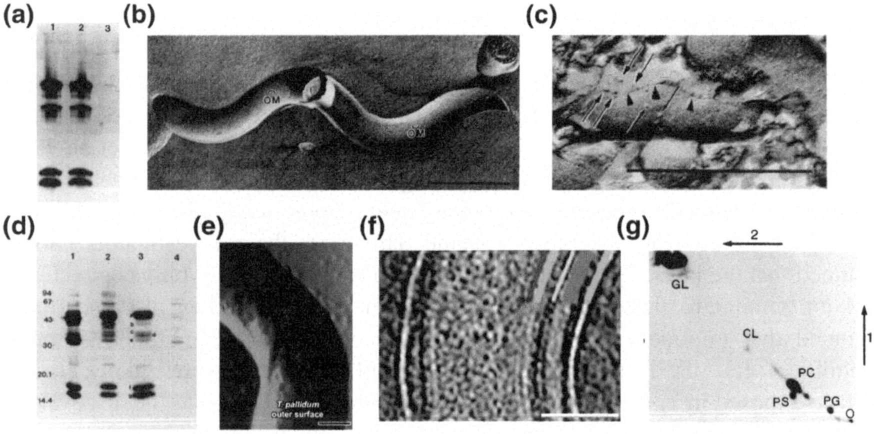

Fig. 1 The *T. pallidum* cell envelope. **a** *T. pallidum*'s major immunogens are associated with the protoplasmic cylinder, not the outer membrane. Reactivity with human syphilitic serum of proteins extracted with Triton X-114 from whole *T. pallidum* cells (*lane 1*), protoplasmic cylinders (*lane 2*), and solubilized outer membranes (*lane 3*); reproduced from reference (Radolf et al. 1988). **b** Freeze-fracture EM reveals scarce intramembranous particles (IMPs) within the *T. pallidum* OM. Convex and concave leaflets of the OM are indicated. Bar, 0.5 μM. Reproduced from reference (Radolf et al. 1994). **c** Deep etching reveals that OM intramembranous particles are surface-exposed. *Arrowheads* indicate the boundaries separating the bacterial surface from the convex fracture face. Particles on the convex fracture face and the treponemal surface are indicated by thin and medium-thickness arrows, respectively. Bar, 0.5 μM. Reproduced from reference (Bourell et al. 1994). **d** TX-114 phase partitioning reveals that the syphilis spirochete's major immunogens (based on reactivity with human syphilitic serum) possess hydrophobic character. Lanes: 1. Percoll-purified *T. pallidum*. 2. TX-114-insoluble material. 3. TX114 detergent-enriched phase proteins. 4. aqueous phase proteins. Reproduced from Reference (Radolf et al. 1988). **e** Scanning probe microscopy reveals rare particles on the *T. pallidum* surface; reproduced with permission from reference (Liu et al. 2010). **f** Cryoelectron microscopy (*longitudinal slice*) showing, from the inside out, cytoplasmic filaments (*red line*), cytoplasmic membrane (*green line*), lipoprotein layer (*purple circles*), peptidoglycan layer (*tan line*), flagellar filament (*thick blue line*), and outer membrane (*green line*). Bar, 50 nM. Reproduced with permission from reference (Liu et al. 2010). **g** [^{3}H]palmitate-labeled lipids were extracted from isolated *T. pallidum* outer membranes and separated by two-dimensional thin layer chromatography. *GL* glycolipids; *CL* cardiolipin; *PC* phosphatidylcholine; *PS* phosphatidylserine; *PG* phosphatidylglycerol; *O* origin. Reproduced from reference (Radolf et al. 1995b)

1.4 Rare Outer Membrane Proteins

Collectively, the above findings led us to hypothesize that the spirochete's fragile OM, not an outer coat, serves as the barrier to antibody binding. Of course, to do so, it would need to have a much lower protein content than conventional Gram-negative bacterial OMs. The question, then, was how to prove this unorthodox idea given how little was known at the time about the molecular architecture and composition of the *T. pallidum* cell envelope. Freeze-fracture EM provided part of the solution. This 'OMP-agnostic' technique revealed that the density of integral membrane proteins (visualized as intramembranous particles, IMPs) in the *T. pallidum* OM is ~100-fold less than that of *E. coli* OMs (Fig. 1b) (Radolf et al. 1989b; Walker et al. 1989). A variant of the freeze-fracture technique, deep etching, showed that these low-abundance particles protrude from the spirochete's surface (Fig. 1c) (Bourell et al. 1994; Radolf et al. 1989b) and, therefore, can interact directly with host cells, tissue components, and circulating molecules, including antibodies. Subsequent efforts to molecularly characterize these morphological entities became known as "the quest for *T. pallidum* outer membrane proteins" (Radolf 1995).

1.5 Lipoprotein Immunogens

TX114 phase partitioning led to the other major piece of the surface antigenicity riddle, which continues today—the identification and localization of the syphilis spirochete's lipoprotein immunogens (Chamberlain et al. 1989a). This technique, developed by Bordier in the 1980s for isolating membrane-associated proteins (Bordier 1981), exploits the relatively low cloud point (~20 °C) of TX114 (Brusca and Radolf 1994). Above the cloud point, TX114 micelles become too large to remain in suspension and can be collected by centrifugation. Membrane proteins incorporated into TX114 micelles at the low temperatures used for solubilization will pellet with the heavier detergent-enriched phase after warming, leaving water-soluble proteins behind in the lighter, aqueous phase. This simple but extremely powerful method revealed that the syphilis spirochete's major immunogens, as determined by reactivity with syphilitic sera, possess hydrophobic character (*i.e.,* they were recovered in the detergent-enriched phase) (Fig. 1d) (Chamberlain et al. 1989a; Radolf et al. 1988). Within the next several years, DNA sequencing determined that these highly immunogenic membrane proteins are synthesized with signal peptides terminated by lipid modification motifs (Akins et al. 1993; Becker et al. 1994; Purcell et al. 1990; Swancutt et al. 1990; Weigel et al. 1992), now referred to as "lipoboxes" (Setubal et al. 2006). These genetic findings were corroborated at the protein level (i) by radiolabeling of polypeptides in *T. pallidum* and/or in *E. coli* with [^{14}C] or [^{3}H]palmitate (Akins et al. 1993; Chamberlain et al. 1989a, b; Purcell et al. 1990; Swancutt et al. 1990); (ii) in some cases, by recovery of radiolabeled fatty acids in the expected 2:1 (ester-to-amide) ratio following sequential alkaline and acid hydrolysis (Chamberlain et al. 1989a; Swancutt et al. 1990); and/or (iii) by showing that processing of native lipoproteins in *T. pallidum* or lipoproteins expressed in *E. coli* was prevented by globomycin (Purcell et al. 1990; Swancutt et al. 1990), a specific inhibitor of signal peptidase II, the enzyme that cleaves the signal peptides of lipoproteins at the lipid-modified cysteine residue (Tokunaga et al. 1984). Importantly, in contrast to the lipid-modified proteins, recombinant lipoproteins without their N-terminal acylation signals partitioned into the TX114 aqueous phase (Akins et al. 1993; Chamberlain et al. 1989b; Purcell et al. 1990; Swancutt et al. 1990), demonstrating that the hydrophobic character and membrane association of native lipoproteins were due to their lipid moieties. Parallel freeze-fracture EM experiments showed that, in contrast to proteins with transmembrane domains (e.g., bacteriorhodopsin and bovine rhodopsin), lipoproteins incorporated into liposomes do not form IMPs (Jones et al. 1995). Together, these results indicated that the protein moieties of lipoproteins are extrinsic to the lipid bilayer and, by extrapolation, that the particles observed in freeze-fractured OMs could not be lipoproteins. Immunoelectron microscopy (IEM) and immunofluorescence analysis (IFA) using antisera generated against numerous recombinant lipoproteins (Tpp47 [TP0574], Tpp15 [TP0171], Tpp17 [TP0435], Tpp34 [TP0971], GlpQ [TP0257]) revealed that none of these immunogens could be detected on the spirochete's surface and, instead, were localized to the periplasmic compartment (Cox et al. 1992, 1995; Deka et al. 2007; Shevchenko et al. 1999). Crucial for these localization experiments was our development of the gel microdroplet method (see below) as a means of maintaining the integrity of the fragile *T. pallidum* OM throughout the labeling process (Cox et al. 1995; Luthra et al. 2015b).

1.6 A Model for the T. pallidum *Cell Envelope*

By the mid-1990s, it was possible to integrate the above information into a model for the *T. pallidum* cell envelope that explains the spirochete's poor surface antigenicity in vitro and its stealth pathogenicity in vivo (Cox et al. 1992; Radolf 1995). The model has two basic components: (i) the OM contains a paucity of integral membrane proteins and surface-exposed lipoproteins and (ii) the preponderance of the spirochete's major membrane immunogens are lipoproteins, with most tethered by their N-terminal lipids to the periplasmic leaflet of the CM. Over the past two decades, data obtained using electron microscopy, biochemistry, and structural biology have supported the model's validity. Consistent with the freeze-etch results, scanning probe microscopy of *T. pallidum* directly visualized sporadic particles on an otherwise smooth bacterial surface (Fig. 1e) (Liu et al. 2010). Cryoelectron microscopy (CryoEM) visualized the native *T. pallidum* OM as a simple lipid bilayer (Fig. 1f) (Izard et al. 2009; Liu et al. 2010), quite unlike that of *Borrelia burgdorferi*, whose external surface possesses an easily discernible proteinaceous layer (Liu et al. 2009). In accord with the notion of a dense array of lipoproteins tethered to the CM's periplasmic leaflet, cryoEM revealed protein "studs" aligned above the CM and below the PG layer (Liu et al. 2010) (Fig. 1f). Tp47, the first *T. pallidum* protein shown to be lipid-modified (Chamberlain et al. 1989b), was found to be a penicillin-binding protein with DD-carboxypeptidase activity involved in PG remodeling (Deka et al. 2002; Weigel et al. 1994), whereas other lipoprotein immunogens are proven substrate-binding proteins (SBPs) for ABC transporters that shuttle a variety of nutrients across the CM (Becker et al. 1994; Brautigam et al. 2016; Deka et al. 2004a, b, 2006, 2013; Machius et al. 2007; Porcella et al. 1996).

2 The Quest for *T. pallidum* Outer Membrane Proteins

2.1 Isolation of Outer Membranes

Prior to the availability of the genomic sequence, isolation of *T. pallidum* OMs seemed the most straightforward approach to identifying rare OMPs (Blanco et al. 1994; Radolf et al. 1995b). The underlying assumption was that rare OMPs enriched in the OM fraction could be identified by SDS-PAGE in combination with peptide sequencing or mass spectrometry and subsequently cloned. Though rational in concept, the results were profoundly disappointing. The most abundant protein in the OM preparations, originally designated *T. pallidum* rare outer membrane protein 1 (Tromp1) (Blanco et al. 1995), was shown by metal analysis and X-ray crystallography to be the SBP for a transition metal ABC transporter (Deka et al. 1999; Lee et al. 1999, 2002) and, therefore, could not be an OMP. Other OM-enriched proteins were obvious periplasmic contaminants (Shevchenko et al. 1997). Isolation of OMs did, however, yield one valuable dividend—determination of the membrane's lipid composition. The *T. pallidum* OM consists principally of phosphatidylcholine, phosphatidylglycerol, phosphatidylserine, and an uncharacterized, poorly immunogenic glycolipid (Fig. 1g) (Radolf et al. 1995b). This lipid profile differs greatly from those of the *E. coli* (Silhavy et al. 2010) and *B. burgdorferi* OMs (Radolf et al. 1995a). Notably absent was lipopolysaccharide (LPS), the highly proinflammatory glycolipid responsible for creating the OM permeability barrier in Gram-negatives (Nikaido 2003), subsequently confirmed by the genomic sequence (Fraser et al. 1998). The lack of LPS likely explains the relative permeability of the *T. pallidum* OM to long-chain fatty acids (LCFAs) compared to that of *E. coli* (Cox and Radolf 2001). It also helps to explain why Toll-like receptor (TLR)-based innate immune surveillance mechanisms (Kawai and Akira 2010) fail to detect hematogenously disseminating bacteria early during the disease (Radolf et al. 2006) as well as the absence of sepsis-like symptomatology in demonstrably spirochetemic secondary syphilis patients (Cruz et al. 2010; Radolf et al. 2014).

2.2 *β-Barrel Predictions*

As the 1990s ended, the need for a new line of attack became obvious. The genomic sequence (Fraser et al. 1998) provided the means for this renewed assault, but not without some twists. One surprise was that the spirochete's genome did not encode orthologs for any well-studied OMPs. Equally unexpected was the finding that it encodes a 12-member paralogous family, designated the *T. pallidum* repeat (Tpr) family, whose members have sequence homology to the major outer sheath protein (MOSP) of *Treponema denticola* (Centurion-Lara et al. 1999; Fraser et al. 1998), a known pore-forming protein and adhesin (Anand et al. 2013; Egli et al. 1993; Ellen 2006). The question for investigators, then, was how to take advantage of this powerful new tool to solve the OMP problem. Structural biology eventually provided a solution in the form of the β-barrel, the hallmark conformation of OMPs in all organisms with OMs as well as eukaryotic organelles derived from them (e.g., mitochondria and chloroplasts) (Wimley 2003). By the mid-2000s, algorithms for identifying proteins predicted to form β-barrels with acceptable false-discovery rates were available. As an alternative to unproductive genome mining for sequence orthologs, we devised a consensus computational framework that used a battery of subcellular localization and β-barrel structural prediction tools to identify and rank candidate OMPs (Cox et al. 2010). Recently, using additional β-barrel prediction algorithms (Markov Chain Model for Beta Barrels [MCMBB]) (Bagos et al. 2004b) and Transmembrane β-barrel proteome database [TMBB-DB] (Freeman and Wimley 2012)), along with structural modeling (Swiss-Model and I-TASSER) (Biasini et al. 2014; Yang and Zhang 2015), and domain identification tools (Conserved Domain Database [CDD], pfam, and InterProScan) (Jones et al. 2014), we have modified and expanded the predicted OMPeome of *T. pallidum* using the Nichols strain as our reference genome (see Table 1 for our current list). Of note, this list should apply to all other syphilis spirochete strains, given their remarkably low degree of sequence divergence (Arora et al. 2016; Smajs et al. 2012). The candidates fall into two classes: *T. pallidum* repeat proteins (Tprs) and a group of unrelated proteins, most of which are annotated as hypotheticals. Despite their lack of sequence homology, the 'hypotheticals' appear to be structural and, presumably, functional orthologs of well-characterized Gram-negative bacterial OMPs. Collectively, these findings suggest that *T. pallidum* is more Gram-negative-like than previously supposed and that the appearance of Gram-negative-like OMs predated the evolution of proteobacteria.

2.3 *Establishing Authenticity–Biophysical Properties*

Bioinformatics is only the starting point for proving that a candidate is a *bona fide* β-barrel-forming OMP. Establishing authenticity requires demonstrating that a protein has the biophysical properties expected of an OMP and is surface-exposed

Table 1 The predicted *Treponema pallidum* OMPeome

TP_ID	Protein annotation	Conserved domains	Structural similarity (PDB ID)	Proposed function	References
TP0011	TPR protein B (TtprB)	$MOSP^C$ and $MOSP^N$	None	Probable porin	Centurion-Lara et al. (1999, 2013)
TP0117	TPR protein C (TprC)	$MOSP^C$ and $MOSP^N$	None	Porin	Anand et al. (2012, 2015), Centurion-Lara et al. (1999, 2013), Gray et al. (2006), Sun et al. (2004)
TP0131	TPR protein D (TprD)	$MOSP^C$ and $MOSP^N$	None	Porin	Anand et al. (2012, 2015), Centurion-Lara et al. (1999, 2013), Gray et al. (2006), Sun et al. (2004)
TP0313	TPR protein E (TprE)	$MOSP^C$ and $MOSP^N$	None	Probable porin	Centurion-Lara et al. (1999, 2013), Gray et al. (2006), Stamm et al. (1998)
TP0317	TPR protein G (TprG)	$MOSP^C$ and $MOSP^N$	None	Probable porin	Centurion-Lara et al. (1999, 2013), Gray et al. (2006), Stamm et al. (1998)
TP0610	TPR protein H (TprH)	$MOSP^C$ and $MOSP^N$	None	Probable porin	Centurion-Lara et al. (1999, 2013)
TP0620	TPR protein I (TprI)	$MOSP^C$ and $MOSP^N$	None	Porin	Anand et al. (2015), Centurion-Lara et al. (1999, 2013), Gray et al. (2006), Sun et al. (2004)
TP0621	TPR protein J (TprJ)	$MOSP^C$ and $MOSP^N$	None	Probable porin	Centurion-Lara et al. (1999, 2013), Gray et al. (2006); Stamm et al. (1998)
TP0897	TPR protein K (TprK)	$MOSP^C$ and $MOSP^N$	None	Unknown	Centurion-Lara et al. (1999), Cox et al. (2010), Giacani et al. (2012), Hazlett et al. (2001), Pinto et al. (2016)
TP1031	TPR protein L (TprL)	$MOSP^C$ and $MOSP^N$	None	Probable porin	Centurion-Lara et al. (1999, 2013)
TP0126	hypothetical protein	None found	2X27	OmpW-like ion-channel involved in transport of small hydrophobic molecules	Giacani et al. (2015), Hong et al. (2006)
TP0326	outer membrane protein	Surface antigen (Beta-barrel), Polypeptide Transport domains (POTRA)	4K3B	BamA, Outer membrane biogenesis	Cameron et al. (2000), Desrosiers et al. (2011), Luthra et al. (2015a)

(continued)

Table 1 (continued)

TP_ID	Protein annotation	Conserved domains	Structural similarity (PDB ID)	Proposed function	References
TP0515	hypothetical protein	LPS-assembly outer membrane protein LptD, Organic solvent tolerance protein OstA	4Q35	LPS-assembly protein LptD, substrate unknown	This chapter, Botos et al. (2016), Gu et al. (2015)
TP0548	hypothetical protein	Uncharacterized protein family (UPF0164)	3BRY	TbuX/FadL-like, long-chain fatty acid transport protein	Cox et al. (2010), van den Berg et al. (2004)
TP0733	hypothetical protein	None found	2MHL	OprG/OmpW-like ion-channel involved in transport of small hydrophobic molecules	This chapter, Hong et al. (2006)
TP0856	hypothetical protein	Uncharacterized protein family (UPF0164)	3BS0	TodX/FadL-like long-chain fatty acid transporter	This chapter, van den Berg et al. (2004, 2005)
TP0858	hypothetical protein	Uncharacterized protein family (UPF0164)	3DWO	TbuX/FadL-like long-chain fatty acid transporter	Cox et al. (2010), van den Berg et al. (2004), van den Berg (2005)
TP0859	hypothetical protein	Uncharacterized protein family (UPF0164)	3BRZ	FadL-like long-chain fatty acid transport protein	This chapter, van den Berg et al. (2004), van den Berg (2005)
TP0865	hypothetical protein	Uncharacterized protein family (UPF0164)	3BRY	TbuX/FadL-like long-chain fatty acid transport protein	This chapter, van den Berg et al. (2004), van den Berg (2005)
TP0966	hypothetical protein	None found	5AZS	OprJ-like outer membrane efflux protein	This chapter, Yonehara et al. (2016)
TP0967	hypothetical protein	None found	5AZO	OprN-like outer membrane efflux protein	This chapter, Yonehara et al. (2016)
TP0969	hypothetical protein	OEP (outer membrane efflux protein) family	2VDE	TolC-like outer membrane efflux protein	Bavro et al. (2008), Cox et al. (2010)

NA = Not available

Signal peptide and transmembrane helixes were predicted by TOPCONS (Tsirigos et al. 2015), LipoP (Juncker et al. 2003), Phobius (Kall et al. 2007), TMHMM (Krogh et al. 2001), and SignalP (Petersen et al. 2011)

Subcellular localizations were predicted by Cello (Yu et al. 2006) and PsortB (Yu et al. 2010)

Conserved domain identification was done by Conserved Domain Database (CDD) (Marchler-Bauer et al. 2015), pfam (Finn et al. 2016), and InterProScan (Jones et al. 2014)

β-barrel outer membrane protein predictions were performed using TMBETA-RBF (Ou et al. 2008), Markov Chain Model for β-barrels (MCMBB)(Bagos et al. 2004a), and transmembrane β-barrel proteome database (TMBB-DB) (Freeman and Wimley 2012) in *T. pallidum* (Anand et al. 2012, 2013, 2015; Desrosiers et al. 2011; Luthra et al. 2011). The three essential biophysical properties of an OM-spanning β-barrel are (i) amphiphilicity (*i.e.*, ability to insert into a lipid bilayer), (ii) extensive β-sheet secondary structure, and (iii) adoption of a closed conformation (Wimley 2003). To examine amphiphilicity, we use TX114 phase partitioning of native (i.e., immunoblotting from phase-partitioned cell lysates) and folded recombinant proteins, and we assess the ability of the folded recombinant protein to insert into liposomes. For the latter, we typically use liposomes with a phospholipid composition simulating that of the *T. pallidum* OM (Radolf et al. 1995b). Proteins that do insert into liposomes can be examined for porin activity; channel formation is a strong evidence for β-barrel formation as well as functional activity (Zeth and Thein 2010). β-sheet content can be assessed quantitatively by far UV circular dichroism (CD) spectroscopy (Shao et al. 1996). Heat modifiability is a technically simple, but powerful, indicator of β-barrel formation; β-barrels are very stable structures that typically migrate faster by SDS-PAGE without than with boiling in final sample buffer (Conlan and Bayley 2003). TEM is an additional means of demonstrating that the folded recombinant protein forms a closed circular structure (Dorset et al. 1983). Incorporation of the folded protein into nanodiscs (a protein scaffold that encloses a lipid bilayer) (Nath et al. 2007) enables one to simultaneously confirm amphiphilicity and ability to circularize (Anand et al. 2015).

2.4 Establishing Authenticity–Surface Exposure in T. pallidum

Because of the many pitfalls inherent in surface labeling spirochetes, particularly an organism with as fragile an OM as *T. pallidum*, complementary methods always should be used before drawing conclusions about surface-exposure. We typically employ (i) IFA in our gel microdroplet system (see reference (Luthra et al. 2015b) for a detailed description), (ii) proteinase K (PK) accessibility, and (iii) opsonophagocytosis assay using rabbit peritoneal macrophages (Anand et al. 2012, 2013, 2015; Desrosiers et al. 2011; Hazlett et al. 2001, 2005; Lukehart and Miller 1978; Luthra et al. 2011). Each method has strengths and weaknesses and presents its own set of technical challenges. The importance of including control antisera for proteins or protein domains whose locations on the surface or in the periplasm are universally accepted cannot be over-emphasized. Antibodies against the flagellar sheath protein FlaA are often used for this purpose. Opsonophagocytosis is sensitive and, because it uses live organisms, surface-specific. However, background levels of internalization can be high and, as with any complex bioassay, reproducibility can be a problem. With PK accessibility experiments, use of motile organisms is extremely important; we use videomicroscopy to document motility throughout the PK incubation period (Desrosiers et al. 2011). Importantly, opsonophagocytosis assay and PK accessibility only determine whether an antigen is surface-exposed. In addition to being able to detect proteins expressed on the *T. pallidum* surface in low copy numbers, the gel microdroplet assay allows localization of periplasmic proteins or the periplasmic domains of bipartite OMPs following controlled removal of OMs. Moreover, when performed in a double-labeling format with an antibody directed against FlaA or another periplasmic marker, the method enables one to assess the intactness of individual organisms thought to be surface-labeled (Cox et al. 2010; Hazlett et al. 2005). This is important because even under optimal circumstances, a small percentage of organisms (usually ~5%) have disrupted OMs.

3 The Expanding Repertoire of Rare Outer Membrane Proteins

3.1 T. pallidum *Repeat Proteins (Tprs)*

3.1.1 TprC/D, TprI, and TprF

Among the Tprs, our original consensus computational matrix (Cox et al. 2010) identified TprC/D (TprC [TP0117] and TprD [TP0131] are identical in the Nichols strain) and TprI (TP0620) as the strongest candidate OMPs. We verified these predictions for the native and recombinant proteins using the methods described above (Anand et al. 2012, 2015). In *T. pallidum*, native TprC/D and TprI are low abundance (~200 copies each per cell), trimeric, amphiphilic, and surface-exposed, while the folded recombinants form β-sheet-rich, heat-modifiable trimers that partition into the TX114 detergent-enriched phase and insert readily into artificial membranes. As with *T. denticola* MOSP (Anand et al. 2013; Egli et al. 1993), integration of TprC/D and TprI into liposomes results in increases in permeability comparable to those produced by the archetypal porin, *E. coli* OmpF (Anand et al. 2012, 2015; Nikaido 2003). With classical porins, the entire polypeptide forms the β-barrel (Nikaido 2003). However, to our surprise, the NCBI conserved domain database (CDD) revealed that Tpr C/D and TprI contain N- and C-terminal regions related to the corresponding domains of *T. denticola* MOSP (Fig. 2a) (Anand et al. 2012, 2015). When examined separately as recombinant proteins, the $MOSP^C$ domains of TprC and TprI formed amphiphilic β-barrels with porin activity in vitro (Fig. 2b) and were surface-exposed in *T. pallidum* (Fig. 2c). The $MOSP^N$ domains, in contrast, were a mixture of α-helix and β-sheet, lacked amphiphilic character, and were periplasmic in *T. pallidum* (Fig. 2c). Consistent with these results, TprF, a truncated protein which contains only a $MOSP^N$ domain (Fig. 2a), lacked amphiphilicity (Fig. 2b), failed to increase liposome permeability (Fig. 2b) and was found by small-angle X-ray scattering (SAXS) analysis to have an elongated structure (Anand et al. 2015). Because TprF is identical to the $MOSP^N$ domains of TprC and TprI along most of its length, this elongated structure almost certainly applies to the N-terminal halves of TprC and TprI as well (Anand et al. 2015).

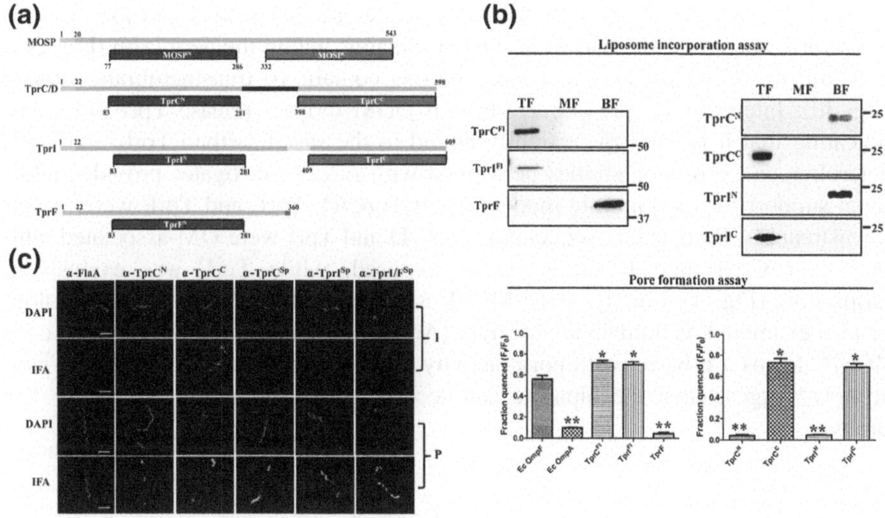

Fig. 2 Bipartite topology of Tpr C/D and I (Nichols strain). **a** Domain architectures of *T. denticola* major outer surface protein (MOSP) and TprC/D/I/F subfamily members. The signal sequences of all three proteins are shown in *blue*. The portions of TprC and TprI colored in *black* and *yellow*, respectively, denote the TprC- and TprI-specific regions of each protein ($TprC^{Sp}$ and $TprI^{Sp}$). The *green* regions in TprI and TprF denote regions present in TprI and TprF but not TprC ($TprI/F^{Sp}$). Reproduced with permission from reference (Anand et al. 2015)). **b** The $MOSP^{C}$ domains of TrpC and TprI are solely responsible for membrane insertion and pore formation by the full-length proteins. Liposomes were reconstituted with folded, full-length recombinant proteins ($TprC^{Fl}$ and $TprI^{Fl}$), TprF, or the $MOSP^{C}$ ($TprC^{C}$ and $TprI^{C}$) or $MOSP^{N}$ ($TprC^{N}$ and $TprI^{N}$) domains of TprC and TprI followed by sucrose density gradient ultracentrifugation and immunoblot analysis. The top fractions (TF) contain liposome-incorporated material, whereas the middle and bottom fractions (MF and BF, respectively) contain unincorporated material. The *bar graphs* show pore formation by the same proteins, along with *E. coli* OmpF (positive) and OmpA (negative) controls, measured by efflux of $Tb(DPA)_3^{3-}$ encapsulated into liposomes (100% efflux = the degree of quenching obtained by detergent lysis). Statistical significance compared with *E. coli* OmpF was assigned according to the following scheme: * $P < 0.05$; ** $P < 0.0001$. Reproduced from Reference (Anand et al. 2015). **c** Bipartite topology of native TprC and TprI in live treponemes. Motile *T. pallidum* were encapsulated in gel microdroplets and probed with 1:100 dilutions of rat antisera against $TprC^{N}$, $TprC^{C}$, or FlaA without (intact I) or with the removal of OMs (Permeabilization P) by pre-incubation with 0.10% Triton X-100. Antibody binding was detected with goat anti-rat Alexa Fluor 488 (*green*) conjugate. Given that $TprC^{C}$ antibodies are highly cross-reactive with $TprI^{C}$, both TprC and TprI are being labeled. Reproduced from reference (Anand et al. 2015)

By conventional TX114 phase partitioning of *T. pallidum*, both native TprC/D and TprI fractionate with the detergent-insoluble material (Fig. 3a), which contains the peptidoglycan sacculus (Fig. 3b) (Radolf et al. 1989a). However, if TprC and TprI were first dissociated from the sacculus by extraction with the detergent n-dodecyl-β-D-maltoside (DDM), they then partitioned into the TX114 detergent-enriched phase (Fig. 3a). These results, in combination with those described above for the recombinant proteins, indicate that, with both Tprs, the

C-terminal β-barrels insert into the OM, while the N-terminal portions extend downward, anchoring the barrels to the PG sacculus within the periplasm (Fig. 3c); structural modeling predicts that the β-barrels contain 10 transmembrane strands (Fig. 3d). Interestingly, pre-extraction with DDM did not release TprF (Fig. 3a), indicating that it is even more tightly bound to the sacculus than TprC and TprI. Heterologous expression studies performed with *E. coli* surrogates provided additional support for the bipartite model. When TprC/D, TprI, and TprF were placed downstream of PelB leader sequences, TprC/D and TprI were OM-associated with only their C-terminal β-barrels surface-exposed, while TprF was exclusively periplasmic (Fig. 4). Finally, since MOSP is considered the parental Tpr ortholog, we also examined its domain architecture (Anand et al. 2013). Our finding that only $MOSP^C$ forms a β-barrel with porin activity and is surface-exposed in *T. denticola* strongly suggests that the bipartite architectural model applies to the entire Tpr family.

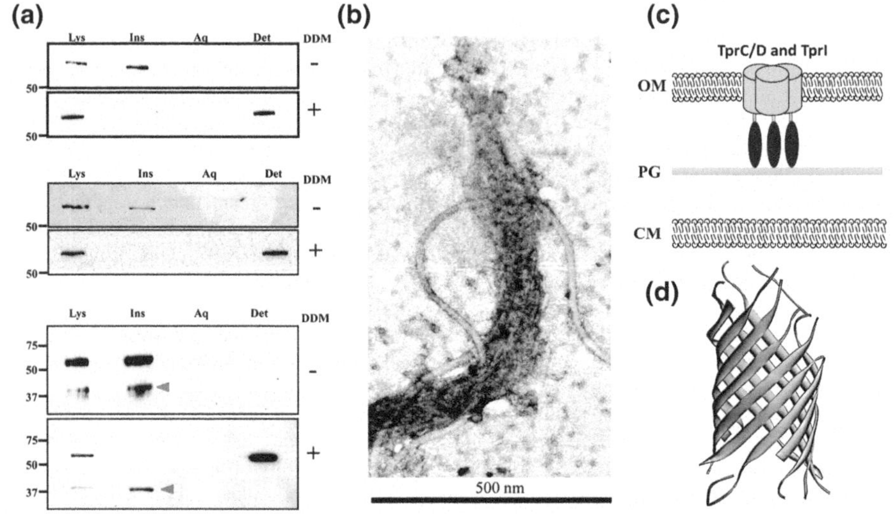

Fig. 3 Native TprC and TprI are amphiphilic but tethered to the peptidoglycan sacculus, whereas TprF is tightly bound to the peptidoglycan sacculus. **a** Triton X-114 phase partitioning of *T. pallidum* lysates without (−) or with (+) pre-solubilization with 2% DDM. Whole cells (WC), Triton X-114-insoluble material (Ins), and aqueous and Triton X-114-enriched phases (Aq and Det, respectively) were separated by SDS-PAGE followed by immunoblotting using antisera specific for TprC (*top*), TprI (*middle*), or TprI and F (*bottom*). *Arrowheads* in bottom panel indicate TprF; TprI is the larger protein. Reproduced from reference (Anand et al. 2015). **b** Extensively washed Triton X-114-insoluble material visualized in negatively stained whole mounts by transmission electron microscopy. Previous studies have shown that this material contains the peptidoglycan sacculus (Radolf et al. 1989a). Reproduced from reference (Anand et al. 2015). **c** Bipartite model for Tpr C/D and TprI. **d** Structural model of TprC (Nichols) generated using TMBpro (Randall et al. 2008) predicts a 10-stranded β-barrel

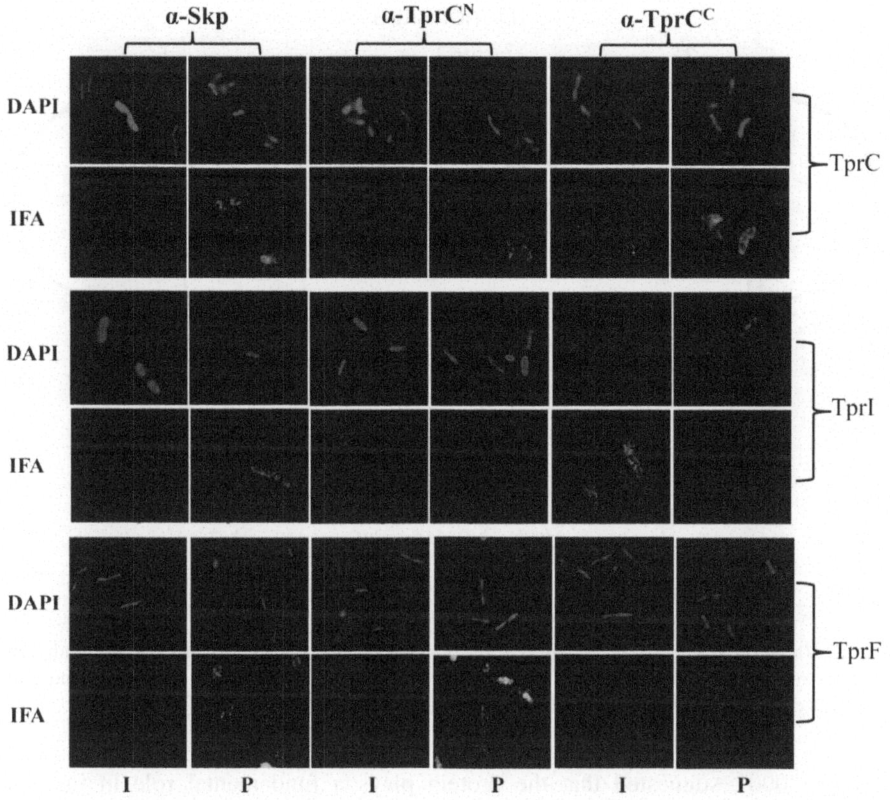

Fig. 4 TprC and TprI, but not TprF, expressed in E. coli with PelB signal sequences display bipartite topology. IFA of intact (I) or permeabilized (P) *E. coli* C41 (DE3) expressing TprC, TprI, or TprF with a PelB signal sequence were probed with rat antisera against $TprC^N$, $TprC^C$, and Skp (periplasmic control). Antibody binding was detected with goat anti-rat Alexa Fluor 488 conjugate. Reproduced from reference (Anand et al. 2015)

3.1.2 TprK

When the *T. pallidum* Nichols genomic sequence became available and the existence of the Tpr family came to light (Fraser et al. 1998), TprK drew immediate attention because of its sequence relatedness to *T. denticola* MOSP and its relatively high level of expression, determined by semi-quantitative qRT-PCR, among *tpr* genes (Centurion-Lara et al. 1999). In their landmark study, Centurion-Lara et al. (1999) reported that antibodies against TprK promote opsonophagocytosis of treponemes by rabbit peritoneal macrophages and that immunization with a large N-terminal fragment of TprK induces partial protection against *T. pallidum* challenge. The subsequent discovery that *TprK* undergoes intra-strain variation, generating subpopulations of organisms with variant *TprK* sequences as infection proceeds (Centurion-Lara et al. 2000; LaFond et al. 2003, 2006a; Stamm and

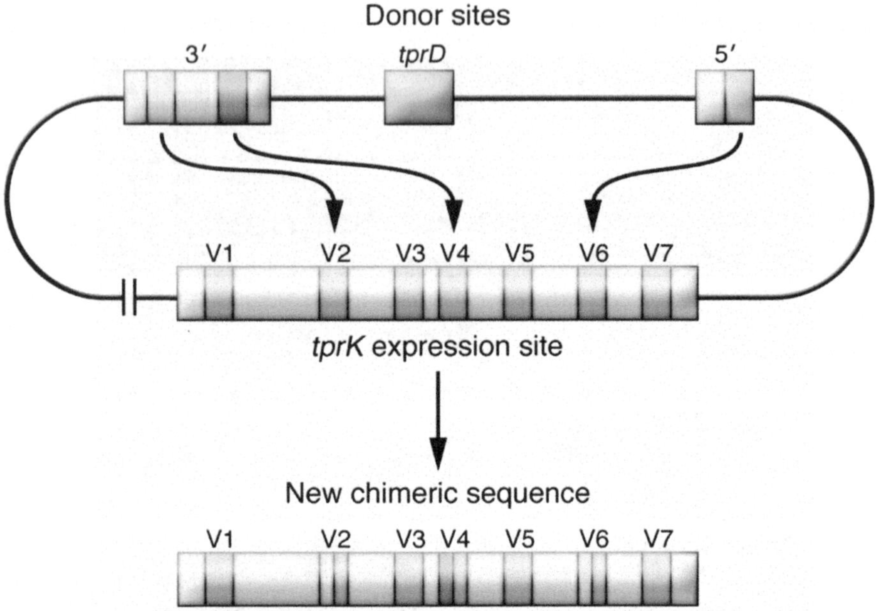

Fig. 5 Variation in TprK is attributed to gene conversion wherein variant DNA segments adjacent to *tprD* recombine with variable regions (V1–V7) of *tprK* to generate new TprK mosaics. Reproduced with permission from reference (Ho and Lukehart 2011)

Bergen 2000), suggested that the protein plays a fundamental role in immune evasion by *T. pallidum* (Deitsch et al. 2009; Ho and Lukehart 2011; Lafond and Lukehart 2006). Importantly, accumulation of TprK sequence variants occurs in syphilis patients as well as infected rabbits (LaFond et al. 2003, 2006a; Myint et al. 2004; Pinto et al. 2016). Indeed, an extraordinary recent study from Portugal in which DNAs for genomic sequencing were 'captured' from genital ulcer swabs described "rampant" sequence variation in *tprK* genes (Pinto et al. 2016). Variability in *tprK*/TprK sequences is not randomly distributed. It occurs in seven discrete variable (V) regions separated by stretches of conserved sequences (Fig. 5), with some V-regions displaying greater sequence diversity than others (Deitsch et al. 2009; LaFond et al. 2003; Pinto et al. 2016). DNA sequence cassettes that correspond to V-region sequences were discovered in an area of the *T. pallidum* chromosome located away from the *tprK* gene (Centurion-Lara et al. 2004). The authors proposed that these cassettes serve as unidirectional donor sites for the generation of variable regions by nonreciprocal gene conversion (Fig. 5) (Deitsch et al. 2009; Ho and Lukehart 2011; Lafond and Lukehart 2006). Generation of TprK variants differs widely among *T. pallidum* strains and, surprisingly, appears to occur at particularly low frequency in the Nichols reference strain (Giacani et al. 2012; LaFond et al. 2006a), perhaps because of extensive passage in rabbits. Sequence variability in the *tprk* donor and expression sites may also explain these strain differences (Giacani et al. 2012). It is important to note that the Lisbon genomic sequences described above were obtained in a relatively confined locale (i.e., Lisbon) and, thus, may be derived from the same *T. pallidum* strain circulating in a circumscribed social network.

The TprK protein elicits both cellular and humoral immunity in infected animals (LaFond et al. 2006b; Morgan et al. 2002a, 2003). TprK antibodies are specifically targeted to the V-regions (Morgan et al. 2002b, 2003), which are thought to be located on extracellular loops (Centurion-Lara et al. 2013); slight changes in the amino acid sequence of a V-region can abrogate antibody binding (LaFond et al. 2006b). Consistent with the notion that immune pressure selects for variants, immunization of rabbits with peptides to V6 resulted in enhanced sequence variability (Giacani et al. 2010). The finding in a rabbit model of secondary syphilis that TprK variability is significantly greater at sites of dissemination compared to the inoculum is further evidence that immune pressure selects for variants and that sequence variation facilitates immune evasion (Reid et al. 2014).

Nevertheless, TprK presents something of a conundrum. While the genetic data collectively constitute a powerful argument that TprK is an authentic OMP, the information available about the protein does not agree with this assertion. In contrast to TprC/D and TprI, both recombinant and native TprK (Nichols) are hydrophilic by TX114 phase partitioning (Cox et al. 2010; Hazlett et al. 2001). In accord with these results, localization data obtained by proteinase K susceptibility and IFA in the Nichols strain place the native protein in the periplasm (Cox et al. 2010; Hazlett et al. 2001). Like other full-length Tprs, TprK is predicted to have a bipartite domain architecture (Anand et al. 2013). Three of the V domains are predicted to be upstream of the $MOSP^C$ domain, a location that would be of no value for immune evasion. Also worth noting is that the portions of the protein reported to confer partial protection (Centurion-Lara et al. 1999; Morgan et al. 2002a) are from the hydrophilic N-terminal half one would expect to be periplasmic based on our bipartite model for the Tprs; indeed, in other hands (Hazlett et al. 2001), this portion of the protein is not protective. However, why a periplasmic protein would undergo such extensive sequence variation is, without question, perplexing. Definitive structural and topologic analyses of intra- and interstrain TprK variants, including localization of epitopes subject to sequence variation, are needed to resolve these discordances.

3.1.3 Why a Family?

Finally, why so many Tprs? One obvious answer is that sequence diversity in the extracellular loops of the β-barrel domains of the full-length Tprs, in conjunction with differential expression, enhances the spirochete's capacity for immune evasion (see below). A second, non-mutually exclusive possibility is physiological. Sequence diversity in the strands that form the walls of the barrel lumen would be expected to affect the conductance properties and substrate specificities of the channel (Nikaido 2003; van den Berg 2012). Indeed, in addition to a role in immune evasion, we propose that the Tprs function collectively as a family of OM transporters whose varied substrate specificities enable the spirochete, an extreme auxotroph (Fraser et al. 1998; Norris et al. 2001), to import the wide range of water-soluble small molecules needed for survival in diverse microenvironments.

3.2 BamA

In 2000, Cameron et al. discovered TP92 (TP0326) using a differential screening strategy to identify *E. coli* clones expressing *T. pallidum* opsonic targets (Cameron et al. 2000). In this seminal report, they noted that TP0326 had sequence similarity to Gram-negative OMPs whose function(s), at that time, were poorly defined. A decade later, TP0326 emerged from our consensus computational framework as one of the most probable β-barrel formers encoded in the *T. pallidum* genome (Cox et al. 2010). By that time, it was apparent that TP0326 belongs to the Omp85 superfamily whose members play a central role in OM biogenesis in double-membrane organisms and eukaryotic organelles (*i.e.*, mitochondria and chloroplasts) derived from them (Heinz and Lithgow 2014). Subsequently, we demonstrated that, like other members of the superfamily, TP0326/BamA has a dual-domain architecture (Fig. 6a) consisting of an amphiphilic, surface-exposed C-terminal β-barrel and a periplasmic region containing five polypeptide transport-associated (POTRA) repeats presumed to associate with other BAM complex subunits and interact with the chaperones that shuttle nascent OMPs across the periplasm (Desrosiers et al. 2011). A homology model based upon the solved structure of *Neisseria gonorrhoeae* BamA (Noinaj et al. 2013) predicted that TP0326 contains a 16-stranded β-barrel with characteristic BamA features (Fig. 6b) (Luthra et al. 2015a). One is the shortened and incompletely hydrogen-bonded β1 and β16 strands that theoretically can separate to allow insertion of nascent OMPs into the OM bilayer. Another is the putative dome formed by three of the barrel's eight extracellular loops (L4, L6, and L7) that occlude the extracellular opening, 'forcing' OMP precursors within the channel to move laterally through the separated barrel strands.

L4 contains a surface-exposed, immunodominant epitope and proven opsonic target (Luthra et al. 2015a). Antibodies directed against this epitope could be responsible for the partial protection against intradermal challenge with *T. pallidum* observed in the initial TP92 report (Cameron et al. 2000). The L4 epitope also provides potential insights into the selective pressures acting on spirochetes within at-risk populations. Some *T. pallidum* strains have BamA β-barrels in which a single amino acid substitution in the immunodominant L4 epitope markedly reduces the reactivity of sera from patients infected with strains containing Nichols-like BamA β-barrels (Fig. 6c) (Luthra et al. 2015a). Petrosova et al. (2012) noted that the β-barrel in the TP0326 of Mexico A is identical to that in subsp. *pertenue* strains. They proposed that the subsp. *pertenue* TP0326 β-barrel was introduced into the subsp. *pallidum* genome as a result of recombination in an

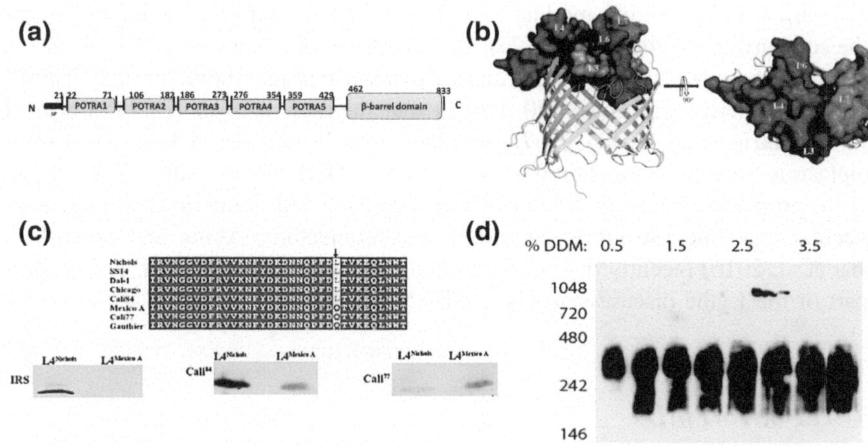

Fig. 6 TP0326/BamA, the central component of the *T. pallidum* β-barrel assembly machine (BAM) complex. **a** Schematic of BamA bipartite topology showing five N-terminal periplasmic POTRA domains and a C-terminal β-barrel. **b** A homology model based upon the solved structure of *Neisseria gonorrhoeae* BamA (Noinaj et al. 2013) predicts that TP0326 contains a 16-stranded β-barrel with characteristic BamA features, including three extracellular loops (L4, L6, and L7) that occlude the barrel's extracellular opening. Reproduced from reference (Luthra et al. 2015a). **c** L4 is an immunodominant surface feature of BamA. Multiple-sequence alignment of the L4 regions of BamAs from geographically diverse *T. pallidum* strains, sequences amplified from skin biopsy specimens from two secondary syphilis patients (Cali-77 and Cali-84) enrolled at our Cali, Colombia, study site, and the Gauthier strain of the *T. pallidum* subsp. *pertenue*. All strains of *T. pallidum* subsp. *pertenue* in the database have same the L4 sequence. Immunoblot relativities of pooled IRS and pooled sera from U.S. (HSS^U) and Colombian (HSS^C) HIV-negative patients with secondary syphilis against L4, L6, and L7 loop peptides and a control peptide L3β6 derived from a portion of the barrel not surface-exposed. NRS and NHS, normal rabbit and normal human serum, respectively. Reproduced from reference (Luthra et al. 2015a). **d** Identification of the BAM complex in *T. pallidum*. Lysates solubilized with graded concentrations of the detergent DDM were separated by Blue-Native PAGE followed by immunoblot analysis with antisera directed against the POTRA arm of TP0326/BamA. Reproduced from reference (Desrosiers et al. 2011)

individual co-infected with yaws and venereal syphilis. Because of its diminished susceptibility to pre-existing Nichols anti-L4 antibodies, the resulting subsp. *pallidum* strain would theoretically be capable of making inroads into populations in which Nichols-like strains predominate.

In genetically manipulable microorganisms, including *Borrelia burgdorferi* (Dunn et al. 2015), BamA was shown to be the central component of a molecular machine whose subunits work cooperatively to catalyze the insertion of newly exported OMPs into the OM (Rollauer et al. 2015). The large size of the BAM complex in *T. pallidum* (~400 kDa vs. ~200 kDa *E. coli*) (Fig. 6d), along with the absence of orthologs for known BAM subunits, strongly suggests that the mechanisms in *T. pallidum* for chaperoning nascent OMPs into the OM differ substantially from those in Gram-negative prototypes (Desrosiers et al. 2011; Luthra et al. 2015a). In addition to the BAM complex, *E. coli* also contains a translocation and assembly module (TAM) required for export of autotransporters to the cell surface (Webb et al. 2012a). The TAM complex consists of TamA, also a member of the OMP85 superfamily, and a cytoplasmic membrane protein, TamB, containing a periplasmic DUF490 domain that interacts with the POTRA arm of TamA (Selkrig et al. 2015). In *T. pallidum, tp0325* encodes a large (108 kDa) cytoplasmic membrane-anchored protein with a DUF490 domain. Webb et al. (2012b) proposed that *T. pallidum* evolved a BAM–TAM chimeric OM biogenesis molecular machine. Strongly supporting this conjecture, Akins and coworkers (Iqbal et al. 2016) recently demonstrated that the TamB ortholog of *B. burgdorferi* is part of the Lyme disease spirochete's BAM complex.

3.3 TP0515/LptD

As noted earlier, the genomic sequence confirmed that *T. pallidum* lacks the LPS biosynthetic pathway (consisting of up to 100 genes) (Fraser et al. 1998; Whitfield and Trent 2014). TP0515 is a structural ortholog of LptD, the OM component of the Gram-negative apparatus that conveys LPS from its site of synthesis at the inner membrane to its ultimate destination in the outer leaflet of the OM (Okuda et al. 2016). Like other LptDs (Gu et al. 2015), TP0515 is predicted to consist of a periplasmic N-terminal domain adopting a β-jellyroll fold and a crenelated 26-stranded β-barrel in which the first and last strands are incompletely hydrogen-bonded, enabling lateral opening of the barrel, as with BamA (Fig. 7a). In Gram-negative bacteria, LptD forms a heterodimeric translocon with LptE, an OM-tethered lipoprotein that resides within the lumen of the LptD barrel (Okuda et al. 2016). Perplexingly, the *T. pallidum* genome lacks a recognizable LptE but encodes orthologs for the other five components of the LPS transport pathway (Fig. 7b). TP0784 (LptC), TP0785 (LptA), and the soluble (i.e., periplasmic) N-terminal portion of LptD are predicted to form a contiguous structure that bridges the inner and outer membranes (Fig. 7b). By analogy with the 'PEZ' model developed for Gram-negative bacteria (Okuda et al. 2016), once the as yet unidentified cargo reaches the outer leaflet of the CM, TP0883 (LptF), TP0884 (LptG), and TP0786 (LptB) form an ATP-dependent machine that pushes it toward LptC and then across the periplasmic bridge. What could be the cargo for this system in an organism lacking LPS? One possibility is the previously identified OM glycolipids (Radolf et al. 1995b). Conceivably, the biochemical differences between these glycolipids and conventional LPS either obviate the need for LptE or, by analogy with the BAM complex, necessitate a markedly different functional ortholog. Interestingly, *Leptospira* species, which do contain true LPS, encode LptE (Haake and Zuckert 2015).

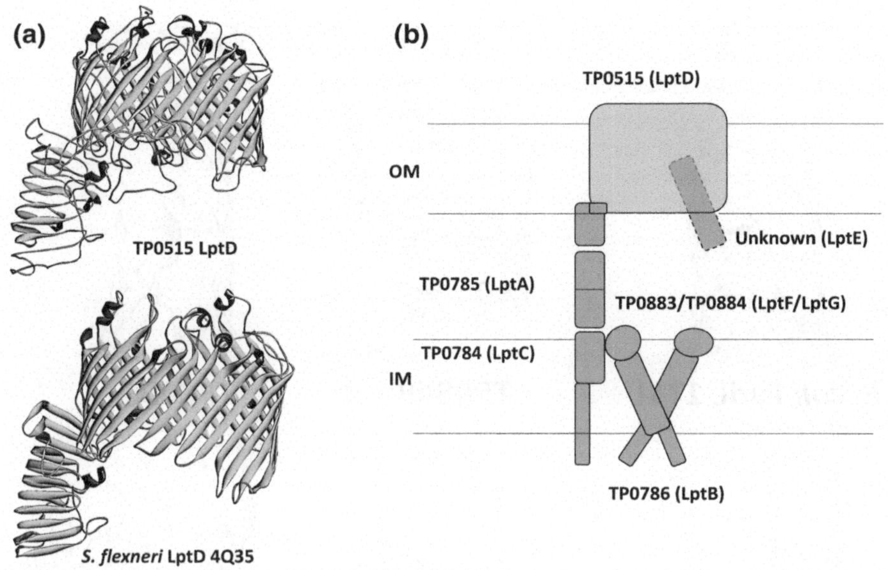

Fig. 7 *T. pallidum's* LptD ortholog. **a** Homology model for TP0515 based upon the solved structure of *Shigella flexneri* LptD (PDB 4Q35). The model and figure were generated using I-TASSER (Yang and Zhang 2015) and Discovery Studio, respectively. **b** *T. pallidum* contains orthologs for all of the components of the *E. coli* LptD complex except LptE

3.4 FadLs

Although biomembranes typically are highly permeable to long-chain fatty acids (LCFAs), Gram-negative OMs are a notable exception because of the permeability barrier created by an outer leaflet composed of LPS (Nikaido 2003). Consequently, transport of LCFAs across the OM of Gram-negative bacteria requires a dedicated transporter, FadL, a crenelated 14-stranded β-barrel that opens laterally to allow diffusion of hydrophobic molecules into the interior of the OM lipid bilayer (Fig. 8) (van den Berg et al. 2004; van den Berg 2005). How *T. pallidum* obtains LCFAs, which it cannot synthesize (Fraser et al. 1998), has been a longstanding question. As noted earlier, the syphilis spirochete's OM is more permeable to LCFAs than that of *E. coli*, suggesting that *T. pallidum* could obtain LCFAs by diffusion (Cox and Radolf 2001). However, the discovery that *T. pallidum* appears to harbor five FadL structural orthologs (TP0548, TP0856, TP0858, TP0859, and TP0865) argues that diffusion of LCFAs alone is likely not sufficient to meet the bacterium's needs. Structural models (Fig. 8) predict that all five proteins possess an N-terminal hatch domain or plug within the barrel lumen, a characteristic FadL feature (van den Berg et al. 2004; van den Berg 2005). At least four are predicted to possess the characteristic FadL kink in the third transmembrane strand for lateral exit of substrate (van den Berg et al. 2004; van den Berg 2005) (Fig. 8). The question, then, is why five orthologs? Members of the FadL family are substrate-specific transporters (Hearn

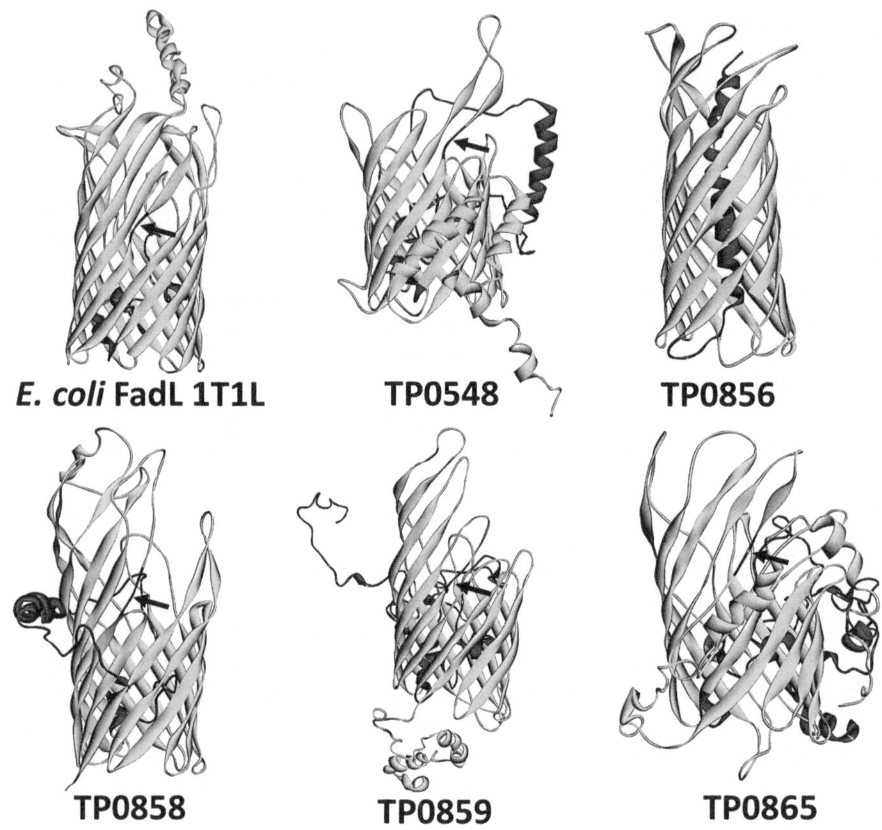

Fig. 8 *T. pallidum* contains five FadL orthologs. Structural models and figures for TP0548, TP0856, TP0858, TP0859, and TP0865 were generated using I-TASSER (Yang and Zhang 2015) and Discovery Studio (BIOVIA 2015), respectively. The N-terminal hatch domains are shown in magenta, while the kinks in the third transmembrane strands of TP0548, TP0858, TP0859, and TP0865 are shown in black and indicated by arrowheads. *E. coli* FadL (PDB 1T1L) is shown for comparison

et al. 2008). If all five *T. pallidum* proteins are FadLs, the implication is that they are required to import distinct classes of essential hydrophobic nutrients. It is also possible that one or more FadLs interact with a "tetratricopeptide repeat" (TPR) protein-associated transporter (TP0956/TatT and TP0957/TatP) postulated to convey hydrophobic substrates across the periplasm (Brautigam et al. 2012; Deka et al. 2012).

3.5 *Efflux Pumps*

Bacteria live in complex communities in which they must protect themselves against toxic compounds, including antibiotics, elaborated by their microbial and

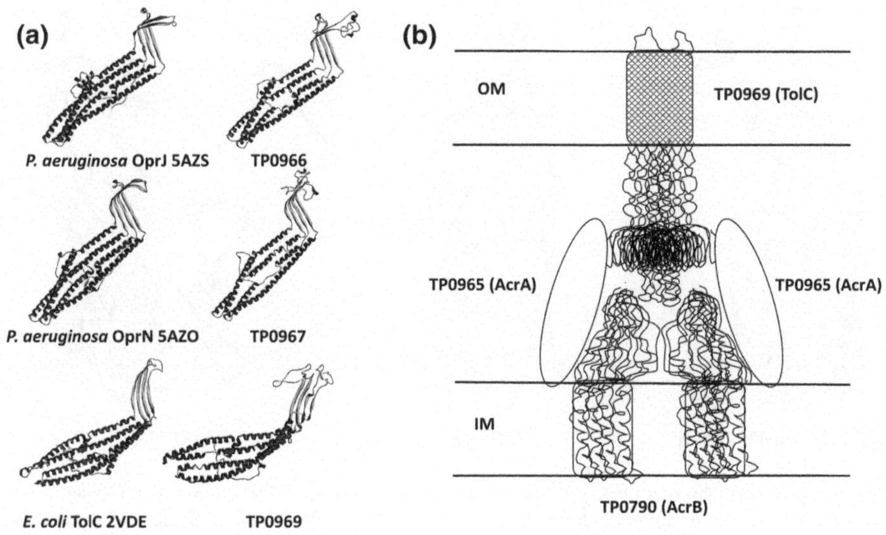

Fig. 9 *T. pallidum* TolC/OprJ/OprN orthologs and putative TolC complex. **a** Structural models for *T. pallidum*'s OprJ (TP0966), OprN (TP0967), and TolC (TP0969) orthologs. The models for TP0966 and TP0967 are based upon *Pseudomonas aeruginosa* OprJ (PDB 5AZS) and OprN (PDB 5AZO), while that for TP0969 is based upon *E. coli* TolC (PDB 2VDE). Structural models and figures were generated using SWISS-MODEL (Biasini et al. 2014) and Discovery Studio (BIOVIA 2015), respectively. **b** Schematic for *T. pallidum*'s putative AcrAB–TolC complex

fungal competitors (Sassone-Corsi and Raffatellu 2015). Among the resistance mechanisms they employ to defend themselves against cytotoxic molecules are tripartite efflux pumps, cell-envelope-spanning machines consisting of an OMP, a cytoplasmic membrane protein, and a "fusion protein" or adaptor that connects the two (Nikaido and Takatsuka 2009). The prototype RND-type efflux OMP, TolC, is a trimer in which each protomer contributes four transmembrane strands of a 12-stranded β-barrel (Nikaido and Takatsuka 2009; Phan et al. 2015). Our analysis (Fig. 9a) reveals that *T. pallidum* contains orthologs for OprJ (TP0966) and OrpN (TP0967) from *Pseudomonas aeruginosa* (Yonehara et al. 2016) as well as *E. coli* TolC (TP0969) (Koronakis et al. 2000). We have also identified orthologs for the AcrA adaptor (TP0965) and AcrB cytoplasmic membrane protein (TP0790) predicted to complex with TP0969 to form a complete TolC-like RND pump (Fig. 9b). Because of its rapid invasiveness, *T. pallidum* is usually conceived of as a 'loner' pathogen that lives in otherwise sterile microenvironments. However, during sexual activity, syphilis spirochetes are inoculated into areas that teem with commensal microorganisms (Radolf et al. 2014). Efflux pumps, hitherto unknown to exist in *T. pallidum*, appear to be part of its strategy for resisting the efforts by commensals to repel the interloper.

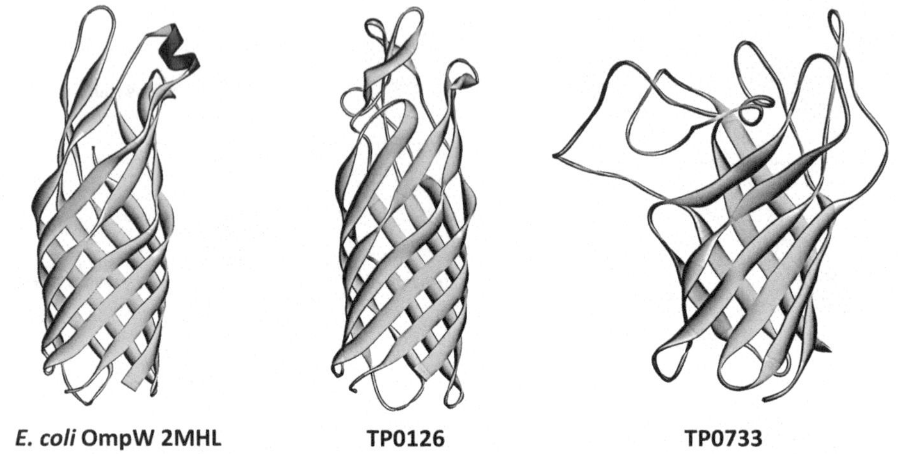

Fig. 10 OmpW orthologs in *T. pallidum*. Structural models for TP0126 and TP0733 are based upon *Pseudomonas aeruginosa* OprG (2X27) and *E. coli* OmpW (2MHL). Structural models and figures were generated using SWISS-MODEL (Biasini et al. 2014) and Discovery Studio (BIOVIA 2015), respectively

3.6 OmpWs

OmpW is a small monomeric OMP commonly found in Gram-negative bacteria (Nandi et al. 2005). The crystal structure of *E. coli* OmpW revealed an 8-stranded β-barrel with a long, narrow hydrophobic channel (Fig. 10); these results, along with the demonstration of ion-channel properties in planar lipid bilayers, suggested that OmpW functions to transport small hydrophobic molecules across the OM (Hong et al. 2006). In subsequent studies, OmpWs have been shown to be functionally versatile proteins. They can protect bacteria against environmental stress (Asakura et al. 2008) and enhance virulence by conferring resistance to phagocytosis (Wu et al. 2013) and killing by the alternative complement pathway (Li et al. 2016). Interestingly, the *T. pallidum* genome encodes two OmpW orthologs, TP0126 and TP0733 (Fig. 10). In a recent report, Giacani and coworkers (Giacani et al. 2015) demonstrated that folded recombinant TP0126 has extensive β-sheet secondary structure and that changes in length of a polyG tract within the promoter may promote phase variation of the protein.

4 Outer Membrane Lipoproteins

In Gram-negative bacteria, mature lipoproteins typically either remain anchored to the periplasmic leaflet of the CM or are transported to the inner leaflet of the OM by the Lol (localization of lipoproteins) system (Okuda and Tokuda 2011; Silhavy et al. 2010). The Lol system in *E. coli* (Okuda and Tokuda 2011) consists of an ABC transporter (LolCDE) that uses the energy from ATP hydrolysis to extract the proteins from the CM and transfer them to a soluble periplasmic carrier, LolA. A hydrophobic pocket in LolA binds the lipoprotein via its hydrophobic N-terminus and delivers it to the OM, where it is passed "mouth-to-mouth" to the OM lipoprotein receptor LolB. Through an unknown mechanism, LolB then catalyzes insertion of the lipoprotein via its N-terminal lipids into the periplasmic leaflet of the OM. Lipoproteins that contain a "Lol avoidance" (or CM "retention") signal, an aspartate at the +2 position in *E. coli*, are not taken up by the Lol transporter and remain CM-associated. Most *E. coli* lipoproteins lack this signal and, consequently, are routed to the OM. Not all Gram-negative bacteria adhere to the +2 rule. In *Pseudomonas aeruginosa*, Lys and Ser at positions 3 and 4, respectively, are the typical CM retention signals (Narita and Tokuda 2007).

T. pallidum contains orthologs for LolCDE and LolA, but no discernible ortholog for LolB (Luthra et al. 2011). Nevertheless, the presence of a Lol system suggests that some *T. pallidum* lipoproteins make their way to the OM. We serendipitously identified one such lipoprotein, TP0453 (p30.5), when we used a hydrophobic, photoactivatable probe, 3-(trifluoromethyl)-3-(m-[125I] iodophenyl-diazarene, in a pre-genomic attempt to identify OM-associated proteins (Hazlett et al. 2005). Although this approach was intended to identify OM-spanning proteins, the only radiolabeled protein was the lipoprotein TP0453. PK accessibility demonstrated that TP0453 is not surface-exposed. Double-label (i.e., FlaA and anti-TP0453 antibodies) localization experiments in the gel microdroplet assay confirmed this, but also revealed that subcritical micellar concentrations of Triton X-100 (i.e., 0.02%), which would be expected to preferentially perturb the outer leaflet of the OM, markedly increased the percentage of treponemes labeled by anti-TP0453 antibodies without a corresponding increase in labeling of FlaA. We concluded that TP0453 is anchored to the inner leaflet of the OM. Subsequent investigations revealed that the TP0453 polypeptide possesses unique physical properties (Hazlett et al. 2005; Luthra et al. 2011). As noted earlier, polypeptide moieties of spirochetal lipoproteins typically are hydrophilic. The polypeptide portion of TP0453, however, is amphiphilic by TX114 phase partitioning and liposome incorporation. The X-ray crystal structure of TP0453 revealed that it exists in closed and open conformations (Luthra et al. 2011). Membrane insertion occurs as a result of lateral movement of amphipathic helices on the surface of the closed conformer; these movements also expose a large hydrophobic pocket. We postulated that TP0453 functions as a carrier of lipids, glycolipids, and/or derivatives during OM biogenesis. We also proposed that the ability of the TP0453 polypeptide to spontaneously insert into membranes obviated the need for LolB.

In Gram-negative bacteria, lipoproteins translocate the OM via an extension of the Sec-dependent general export pathway, which involves 12 to 16 different proteins comprising the type II secretion machine or "secreton" that bridges the CM and OM (Douzi et al. 2012). Importantly, since lipoproteins must engage the secreton from the periplasmic side of the CM, the Lol and type II pathways are considered to be mutually exclusive (Douzi et al. 2012). The *T. pallidum* genome does not encode the Type II pathway (Fraser et al. 1998), and there are no other known mechanisms whereby lipoproteins can reach the surface of a double-membrane organism. However, this, does not, in itself, rule out the possibility that *T. pallidum* can express lipoproteins on its surface. *B. burgdorferi* also lacks a secreton but clearly expresses numerous outer surface lipoproteins (Radolf et al. 2012), and it has been proposed that an OM-situated "flippase" translocates folded lipoproteins delivered by the Lol system (Zuckert 2014). The need for LolA–lipoprotein complexes to engage the flippase from the underside of the OM perhaps explains why neither spirochete has a LolB (Zuckert 2014).

In recent years, three putative *T. pallidum* outer surface lipoproteins have been identified. One is the multifunctional TP0751. First identified as a laminin-binding adhesion (Cameron et al. 2005), TP0751 is also a zinc-dependent metalloprotease (hence, its designation as "pallilysin") that forms a complex with TP0750 (which contains a Von Willebrand factor type A domain) capable of degrading clots and extracellular matrix, activities that could facilitate spirochetal dissemination as well as attachment (Houston et al. 2012, 2014). Unfortunately, expression levels of pallilysin in *T. pallidum* (unpublished observations) or following heterologous expression in *B. burgdorferi* (Parker et al. 2016) are so low that it cannot be detected by IFA or chemiluminescence-based immunoblot analysis. Also noteworthy, the recently solved X-ray structure revealed that pallilysin is a member of the lipocalin superfamily (Parker et al. 2016). In *E. coli*, the lipocalin Blc is a lipoprotein tethered to the inner leaflet of the OM where it binds lysophospholipids (Campanacci et al. 2006). On the other hand, three lines of evidence support its surface location: (i) pallilysin antibodies yielded positive results in opsonophagocytosis assays (Houston et al. 2012); (ii) FLAG-tagged pallilysin expressed in *B. burgdorferi* could be detected by flow cytometry on the surface of intact Lyme disease spirochetes and promoted their attachment to HUVEC monolayers in "gain of function" experiments (Parker et al. 2016); and (iii) most convincing is the recent demonstration that immunization of rabbits with pallilysin protected against dissemination of *T. pallidum* following intradermal challenge (Lithgow et al. 2017).

TP0136, a second putative *T. pallidum* outer surface lipoprotein, has been characterized as a fibronectin-binding lipoprotein adhesin with sequence heterogeneity among *T. pallidum* strains (Arora et al. 2016; Brinkman et al. 2008; Ke et al. 2015). In contrast to IEM data published by Palzkill and coworkers (Brinkman et al. 2008), we found no evidence for surface-exposure of TP0136 in the Nichols strain using our gel microdroplet assay (Cox et al. 2010). TP0435/Tpp17, the third, has been known for years as a major treponemal antigen with great utility for serodiagnosis of syphilis (Sena et al. 2015). Evidence that it functions as a cytadhesin was obtained using *B. burgdorferi* as a surrogate genetic host (Chan et al. 2016). The case for surface-exposure of at least small amounts of native TP0435 in *T. pallidum* was bolstered by IEM and opsonophagocytosis assays with *T. pallidum* (Chan et al. 2016). In considering these findings, one must bear in mind that Tpp17 is a structural ortholog for NlpE (Brautigam et al. 2015), a lipoprotein tethered to the inner leaflet of the OM in Gram-negative bacteria (Hirano et al. 2007) and that, in our localization studies, we did not observe evidence for surface-exposure (Cox et al. 1995).

5 Concluding Remarks

5.1 Immune Evasion

While at one time it was thought that the OM of *T. pallidum* is antigenically inert (Cox et al. 1992; Penn et al. 1985), we now know that this simplistic notion is incorrect and that syphilitic infection does, in fact, induce antibodies capable of promoting bacterial clearance (so-called "functional" antibodies) (Radolf and Lukehart 2006). After many false starts, the field has progressed to the point where we can topologically map humoral responses to *bona fide* OMPs and identify protective B cell epitopes. Besides their obvious relevance to vaccine development, such studies are essential to addressing a central issue of syphilis pathogenesis—why, unlike 'classic' extracellular pathogens, spirochetes not only fail to be rapidly cleared, but actually continue to disseminate in the presence of surface-reactive antibodies (Lafond and Lukehart 2006; Radolf et al. 2006; Salazar et al. 2002).

Immunolabeling and opsonophagocytosis assays have shown that *T. pallidum* populations consist of antibody-binding and non-binding subpopulations and that the minority of organisms that bind antibodies do so in minute amounts and with delayed kinetics (Lukehart et al. 1992; Cox et al. 2010). One can envision a scenario at sites of infection whereby the nonbinders evade host antibodies, replenishing the ranks of spirochetes that do bind sufficient amounts of antibodies to be cleared (Radolf et al. 2016). If correct, then understanding the basis for the heterogeneity of *T. pallidum*'s surface antigenicity becomes critical to unraveling the immune evasion enigma. Now that we have a much clearer understanding of the spirochete's OMP repertoire, a "one size fits all" explanation seems unlikely. The picture emerging appears to be multifactorial, involving (i) poor target availability due to low copy numbers (Anand et al. 2012, 2015; Desrosiers et al. 2011; Luthra et al. 2015a), (ii) limited production of antibodies against surface-exposed epitopes along with skewed production of antibodies against periplasmic domains (Anand et al. 2012, 2015; Desrosiers et al. 2011; Luthra et al. 2015a), (iii) organism-to-organism variation in the levels of expression of OMPs and outer surface lipoproteins (Chan et al. 2016; Giacani et al. 2007, 2015; Pinto et al. 2016), and (iv) in the case of TprK, antigenic variation as a result of intra-genomic recombination (Deitsch et al. 2009; Ho and Lukehart 2011; Pinto et al. 2016). Conceivably, when the antibody repertoire covers a sufficiently broad range of surface epitopes on a population-wide basis, humoral responses eventually drive down spirochete burdens and keep them suppressed (Radolf et al. 2016).

5.2 Functional and Regulatory Considerations

Early on, we recognized that our ultrastructural model raises physiologic- and virulence-related issues that could not be ignored: how does the spirochete meet its nutritional requirements and carry out its complex parasitic program with a minimalist OM (Radolf 1995)? A partial answer may lie with the bacterium's extremely slow ($\sim$30 h) rate of replication (Magnuson et al. 1948), presumably an evolutionary 'compromise' between the density of surface molecules needed for viability and the demands of stealth. Multifunctionality of surface molecules, a well-accepted facet of Gram-negative OMP biology (Galdiero et al. 2012; Smith et al. 2007), is likely another piece of the puzzle. In addition to being nutrient transporters, the OMPs described herein could mediate a multitude of interactions and functions at the host–pathogen interface. Of course, the answers ultimately lie in further detailed characterization of *T. pallidum*'s OMP repertoire. One of the take-home lessons from our recent bioinformatics exercises is that the syphilis spirochete's OMP repertoire seemingly encompasses a much greater degree, as well as greater redundancy, of transport capabilities than previously recognized. One might postulate that this diversity serves the ends of stealth pathogenicity in two ways: by enhancing nutrient acquisition in the myriad niches *T. pallidum* is known to inhabit within its obligate human host (Radolf et al. 2014) and promoting persistence in these niches through differential expression driven by host environmental pressures (Giacani et al. 2009, 2015; Radolf et al. 2016). Given the limitations of animal models for experimental syphilis and ethical issues involved in translational human research, devising strategies to elucidate in vivo expression profiles and underlying regulatory mechanisms is going to be a formidable challenge. Mining of the spirochete genome, the obvious starting point, has revealed that the spirochete's regulatory apparatus is far more intricate than previously suspected. As is true of all bacteria, *T. pallidum* appears to sense its environment to a large extent through the lens of intermediary metabolism (Radolf et al. 2016). OMPs, the bacterium's eyes for sensing nutrient availability, undoubtedly will be part of these regulatory networks.

Acknowledgments Funded in part by R01 AI-26756 from the National Institutes of Health (NIAID) and the Department of Research, Connecticut Children's Medical Center.

References

Akins DR, Purcell BK, Mitra MM, Norgard MV, Radolf JD (1993) Lipid modification of the 17-kilodalton membrane immunogen of *Treponema pallidum* determines macrophage activation as well as amphiphilicity. Infect Immun 61:1202–1210

Alderete JF, Baseman JB (1979) Surface-associated host proteins on virulent *Treponema pallidum*. Infect Immun 26:1048–1056

Anand A, Luthra A, Dunham-Ems S, Caimano MJ, Karanian C, LeDoyt M, Cruz AR, Salazar JC, Radolf JD (2012) TprC/D (Tp0117/131), a trimeric, pore-forming rare outer membrane protein of *Treponema pallidum*, has a bipartite domain structure. J Bacteriol 194:2321–2333

Anand A, Luthra A, Edmond ME, Ledoyt M, Caimano MJ, Radolf JD (2013) The major outer sheath protein (Msp) of *Treponema denticola* has a bipartite domain architecture and exists as periplasmic and outer membrane-spanning conformers. J Bacteriol 195:2060–2071

Anand A, LeDoyt M, Karanian C, Luthra A, Koszelak-Rosenblum M, Malkowski MG, Puthenveetil R, Vinogradova O, Radolf JD (2015) Bipartite topology of *Treponema pallidum* repeat proteins C/D and I: outer membrane insertion, trimerization, and porin function require a C-terminal β-barrel domain. J Biol Chem 290:12313–12331

Arora N, Schuenemann VJ, Jager G, Peltzer A, Seitz A, Herbig A, Strouhal M, Grillova L, Sanchez-Buso L, Kuhnert D, Bos KI, Davis LR, Mikalova L, Bruisten S, Komericki P, French P, Grant PR, Pando MA, Vaulet LG, Fermepin MR, Martinez A, Centurion Lara A, Giacani L, Norris SJ, Smajs D, Bosshard PP, Gonzalez-Candelas F, Nieselt K, Krause J, Bagheri HC (2016) Origin of modern syphilis and emergence of a pandemic *Treponema pallidum* cluster. Nat Microbiol 2:16245

Asakura H, Kawamoto K, Haishima Y, Igimi S, Yamamoto S, Makino S (2008) Differential expression of the outer membrane protein W (OmpW) stress response in enterohemorrhagic *Escherichia coli* O157:H7 corresponds to the viable but non-culturable state. Res Microbiol 159:709–717

Bagos PG, Liakopoulos TD, Hamodrakas SJ (2004a) Finding B-barrel outer membrane proteins with a markov chain model. WSEAS Trans Biol Biomed 2:186–189

Bagos PG, Liakopoulos TD, Spyropoulos IC, Hamodrakas SJ (2004b) A hidden Markov model method, capable of predicting and discriminating beta-barrel outer membrane proteins. BMC Bioinform 5:29

Bavro VN, Pietras Z, Furnham N, Perez-Cano L, Fernandez-Recio J, Pei XY, Misra R, Luisi B (2008) Assembly and channel opening in a bacterial drug efflux machine. Mol Cell 30:114–121

Becker PS, Akins DR, Radolf JD, Norgard MV (1994) Similarity between the 38-kilodalton lipoprotein of *Treponema pallidum* and the glucose/galactose-binding (MglB) protein of Escherichia coli. Infect Immun 62:1381–1391

Biasini M, Bienert S, Waterhouse A, Arnold K, Studer G, Schmidt T, Kiefer F, Gallo Cassarino T, Bertoni M, Bordoli L, Schwede T (2014) SWISS-MODEL: modelling protein tertiary and quaternary structure using evolutionary information. Nucleic Acids Res 42:W252–W258

Biovia DS (2015) Discovery studio modeling environment. San Diego, CA, USA, Dassault Systèmes

Blanco DR, Reimann K, Skare J, Champion CI, Foley D, Exner MM, Hancock RE, Miller JN, Lovett MA (1994) Isolation of the outer membranes from *Treponema pallidum* and *Treponema vincentii*. J Bacteriol 176:6088–6099

Blanco DR, Champion CI, Exner MM, Erdjument-Bromage H, Hancock RE, Tempst P, Miller JN, Lovett MA (1995) Porin activity and sequence analysis of a 31-kilodalton Treponema pallidum subsp. pallidum rare outer membrane protein (Tromp1). J Bacteriol 177:3556–3562

Bordier C (1981) Phase separation of integral membrane proteins in Triton X-114 solution. J Biol Chem 256:1604–1607

Botos I, Majdalani N, Mayclin SJ, McCarthy JG, Lundquist K, Wojtowicz D, Barnard TJ, Gumbart JC, Buchanan SK (2016) Structural and Functional Characterization of the LPS Transporter LptDE from Gram-Negative Pathogens. Structure 24:965–976

Bourell KW, Schulz W, Norgard MV, Radolf JD (1994) *Treponema pallidum* rare outer membrane proteins: analysis of mobility by freeze-fracture electron microscopy. J Bacteriol 176:1598–1608

Brautigam CA, Deka RK, Schuck P, Tomchick DR, Norgard MV (2012) Structural and thermodynamic characterization of the interaction between two periplasmic *Treponema pallidum* lipoproteins that are components of a TPR-protein-associated TRAP transporter (TPAT). J Mol Biol 420:70–86

Brautigam CA, Deka RK, Liu WZ, Norgard MV (2015) Insights into the potential function and membrane organization of the TP0435 (Tp17) lipoprotein from *Treponema pallidum* derived from structural and biophysical analyses. Protein Sci 24:11–19

Brautigam CA, Deka RK, Liu WZ, Norgard MV (2016) The Tp0684 (MglB-2) lipoprotein of *Treponema pallidum*: a glucose-binding protein with divergent topology. PLoS ONE 11: e0161022

Brinkman MB, McGill MA, Pettersson J, Rogers A, Matejkova P, Smajs D, Weinstock GM, Norris SJ, Palzkill T (2008) A novel *Treponema pallidum* antigen, TP0136, is an outer membrane protein that binds human fibronectin. Infect Immun 76:1848–1857

Brusca JS, Radolf JD (1994) Isolation of integral membrane proteins by phase partitioning with Triton X-114. Methods Enzymol 228:182–193

Cameron CE, Lukehart SA, Castro C, Molini B, Godornes C, Van Voorhis WC (2000) Opsonic potential, protective capacity, and sequence conservation of the *Treponema pallidum* subspecies *pallidum* Tp92. J Infect Dis 181:1401–1413

Cameron CE, Brouwer NL, Tisch LM, Kurowa JM (2005) Defining the interation of the *Treponema pallidum* adhesin TP0751 with laminin. Infect Immun 73:7485–7494

Cameron CE (2006) The *T. pallidum* outer membrane and outer membrane proteins. In: Radolf JD, Lukehart SA (ed) Pathogenic treponema: molecular and cellular biology, Caister Academic Press, Norwich, UK, pp. 237–266

Campanacci V, Bishop RE, Blangy S, Tegoni M, Cambillau C (2006) The membrane bound bacterial lipocalin Blc is a functional dimer with binding preference for lysophospholipids. FEBS Lett 580:4877–4883

Centurion-Lara A, Castro C, Barrett L, Cameron C, Mostowfi M, Van Voorhis WC, Lukehart SA (1999) *Treponema pallidum* major sheath protein homologue Tpr K is a target of opsonic antibody and the protective immune response. J Exp Med 189:647–656

Centurion-Lara A, Godornes C, Castro C, Van Voorhis WC, Lukehart SA (2000) The tprK gene is heterogeneous among *Treponema pallidum* strains and has multiple alleles. Infect Immun 68:824–831

Centurion-Lara A, LaFond RE, Hevner K, Godornes C, Molini BJ, Van Voorhis WC, Lukehart SA (2004) Gene conversion: a mechanism for generation of heterogeneity in the *tprK* gene of *Treponema pallidum* during infection. Mol Microbiol 52:1579–1596

Centurion-Lara A, Giacani L, Godornes C, Molini BJ, Brinck Reid T, Lukehart SA (2013) Fine analysis of genetic diversity of the *tpr* gene family among treponemal species, subspecies and strains. PLoS Negl Trop Dis 7:e2222

Chamberlain NR, Brandt ME, Erwin AL, Radolf JD, Norgard MV (1989a) Major integral membrane protein immunogens of *Treponema pallidum* are proteolipids. Infect Immun 57:2872–2877

Chamberlain NR, DeOgny L, Slaughter C, Radolf JD, Norgard MV (1989b) Acylation of the 47-kilodalton major membrane immunogen of Treponema pallidum determines its hydrophobicity. Infect Immun 57:2878–2885

Chan K, Nasereddin T, Alter L, Centurion-Lara A, Giacani L, Parveen N (2016) *Treponema pallidum* Lipoprotein TP0435 Expressed in *Borrelia burgdorferi* produces multiple surface/periplasmic isoforms and mediates adherence. Sci Rep 6:25593

Christiansen S (1963) Protective layer covering pathogenic treponemata. Lancet 1:423–425

Conlan S, Bayley H (2003) Folding of a monomeric porin, OmpG, in detergent solution. Biochemistry 42:9453–9465

Cox DL, Chang P, McDowall AW, Radolf JD (1992) The outer membrane, not a coat of host proteins, limits antigenicity of virulent *Treponema pallidum*. Infect Immun 60:1076–1083

Cox DL, Akins DR, Porcella SF, Norgard MV, Radolf JD (1995) *Treponema pallidum* in gel microdroplets: a novel strategy for investigation of treponemal molecular architecture. Mol Microbiol 15:1151–1164

Cox DL, Radolf JD (2001) Insertion of fluorescent fatty acid probes into the outer membranes of the pathogenic spirochaetes *Treponema pallidum* and *Borrelia burgdorferi*. Microbiology 147:1161–1169

Cox DL, Luthra A, Dunham-Ems S, Desrosiers DC, Salazar JC, Caimano MJ, Radolf JD (2010) Surface immunolabeling and consensus computational framework to identify candidate rare outer membrane proteins of *Treponema pallidum*. Infect Immun 78:5178–5194

Cruz AR, Pillay A, Zuluaga AV, Ramirez LG, Duque JE, Aristizabal GE, Fiel-Gan MD, Jaramillo R, Trujillo R, Valencia C, Jagodzinski L, Cox DL, Radolf JD, Salazar JC (2010) Secondary syphilis in Cali, Colombia: new concepts in disease pathogenesis. PLoS Negl Trop Dis 4:e690

Deitsch KW, Lukehart SA, Stringer JR (2009) Common strategies for antigenic variation by bacterial, fungal and protozoan pathogens. Nat Rev Microbiol 7:493–503

Deka RK, Lee YH, Hagman KE, Shevchenko D, Lingwood CA, Hasemann CA, Norgard MV, Radolf JD (1999) Physicochemical evidence that *Treponema pallidum* TroA is a zinc-containing metalloprotein that lacks porin-like structure. J Bacteriol 181:4420–4423

Deka RK, Machius M, Norgard MV, Tomchick DR (2002) Crystal structure of the 47-kDa lipoprotein of *Treponema pallidum* reveals a novel penicillin-binding protein. J Biol Chem 277:41857–41864

Deka RK, Goldberg MS, Hagman KE, Norgard MV (2004a) The Tp38 (TpMglB-2) lipoprotein binds glucose in a manner consistent with receptor function in *Treponema pallidum*. J Bacteriol 186:2303–2308

Deka RK, Neil L, Hagman KE, Machius M, Tomchick DR, Brautigam CA, Norgard MV (2004b) Structural evidence that the 32-kilodalton lipoprotein (Tp32) of *Treponema pallidum* is an L-methionine-binding protein. J Biol Chem 279:55644–55650

Deka RK, Brautigam CA, Yang XF, Blevins JS, Machius M, Tomchick DR, Norgard MV (2006) The PnrA (Tp0319; TmpC) lipoprotein represents a new family of bacterial purine nucleoside receptor encoded within an ATP-binding cassette (ABC)-like operon in *Treponema pallidum*. J Biol Chem 281:8072–8081

Deka RK, Brautigam CA, Tomson FL, Lumpkins SB, Tomchick DR, Machius M, Norgard MV (2007) Crystal structure of the Tp34 (TP0971) lipoprotein of *Treponema pallidum*: implications of its metal-bound state and affinity for human lactoferrin. J Biol Chem 282:5944–5958

Deka RK, Brautigam CA, Goldberg M, Schuck P, Tomchick DR, Norgard MV (2012) Structural, bioinformatic, and in vivo analyses of two *Treponema pallidum* lipoproteins reveal a unique TRAP transporter. J Mol Biol 416:678–696

Deka RK, Brautigam CA, Liu WZ, Tomchick DR, Norgard MV (2013) The TP0796 lipoprotein of *Treponema pallidum* is a bimetal-dependent FAD pyrophosphatase with a potential role in flavin homeostasis. J Biol Chem 288:11106–11121

Desrosiers DC, Anand A, Luthra A, Dunham-Ems SM, LeDoyt M, Cummings MA, Eshghi A, Cameron CE, Cruz AR, Salazar JC, Caimano MJ, Radolf JD (2011) TP0326, a *Treponema pallidum* beta-barrel assembly machinery A (BamA) orthologue and rare outer membrane protein. Mol Microbiol 80:1496–1515

Dorset DL, Engel A, Haner M, Massalski A, Rosenbusch JP (1983) Two-dimensional crystal packing of matrix porin. A channel forming protein in *Escherichia coli* outer membranes. J Mol Biol 165:701–710

Douzi B, Filloux A, Voulhoux R (2012) On the path to uncover the bacterial type II secretion system. Philos Trans R Soc Lond B Biol Sci 367:1059–1072

Dunn JP, Kenedy MR, Iqbal H, Akins DR (2015) Characterization of the β-barrel assembly machine accessory lipoproteins from *Borrelia burgdorferi*. BMC Microbiol 15:70

Egli C, Leung WK, Muller KH, Hancock RE, McBride BC (1993) Pore-forming properties of the major 53-kilodalton surface antigen from the outer sheath of *Treponema denticola*. Infect Immun 61:1694–1699

Ellen RP (2006) Virulence determinants of oral treponemes. In: Radolf JD, Lukehart SA (ed) Pathogenic treponema molecular and cellular biology, Norwich, UK, Caister Academic Press, pp. 357–386

Finn RD, Coggill P, Eberhardt RY, Eddy SR, Mistry J, Mitchell AL, Potter SC, Punta M, Qureshi M, Sangrador-Vegas A, Salazar GA, Tate J, Bateman A (2016) The Pfam protein families database: towards a more sustainable future. Nucleic Acids Res 44:D279–D285

Fitzgerald TJ, Johnson RC (1979) Surface mucopolysaccharides of *Treponema pallidum*. Infect Immun 24:244–251

Fraser CM, Norris SJ, Weinstock GM, White O, Sutton GG, Dodson R, Gwinn M, Hickey EK, Clayton R, Ketchum KA, Sodergren E, Hardham JM, McLeod MP, Salzberg S, Peterson J, Khalak H, Richardson D, Howell JK, Chidambaram M, Utterback T, McDonald L, Artiach P, Bowman C, Cotton MD, Fujii C, Garland S, Hatch B, Horst K, Roberts K, Sandusky M, Weidman J, Smith HO, Venter JC (1998) Complete genome sequence of *Treponema pallidum*, the syphilis spirochete. Science 281:375–388

Freeman TC Jr, Wimley WC (2012) TMBB-DB: a transmembrane β-barrel proteome database. Bioinformatics 28:2425–2430

Galdiero S, Falanga A, Cantisani M, Tarallo R, Della Pepa ME, D'Oriano V, Galdiero M (2012) Microbe-host interactions: structure and role of Gram-negative bacterial porins. Curr Protein Pept Sci 13:843–854

Giacani L, Lukehart S, Centurion-Lara A (2007) Length of guanosine homopolymeric repeats modulates promoter activity of subfamily II tpr genes of *Treponema pallidum* ssp. *pallidum*. FEMS Immunol Med Microbiol 51:289–301

Giacani L, Godornes C, Puray-Chavez M, Guerra-Giraldez C, Tompa M, Lukehart SA, Centurion-Lara A (2009) TP0262 is a modulator of promoter activity of tpr Subfamily II genes of *Treponema pallidum* ssp. pallidum. Mol Microbiol 72:1087–1099

Giacani L, Molini BJ, Kim EY, Godornes BC, Leader BT, Tantalo LC, Centurion-Lara A, Lukehart SA (2010) Antigenic variation in *Treponema pallidum*: TprK sequence diversity accumulates in response to immune pressure during experimental syphilis. J Immunol 184:3822–3829

Giacani L, Brandt SL, Puray-Chavez M, Reid TB, Godornes C, Molini BJ, Benzler M, Hartig JS, Lukehart SA, Centurion-Lara A (2012) Comparative investigation of the genomic regions involved in antigenic variation of the TprK antigen among treponemal species, subspecies, and strains. J Bacteriol 194:4208–4225

Giacani L, Brandt SL, Ke W, Reid TB, Molini BJ, Iverson-Cabral S, Ciccarese G, Drago F, Lukehart SA, Centurion-Lara A (2015) Transcription of TP0126, *Treponema pallidum* putative OmpW homolog, is regulated by the length of a homopolymeric guanosine repeat. Infect Immun 83:2275–2289

Gray RR, Mulligan CJ, Molini BJ, Sun ES, Giacani L, Godornes C, Kitchen A, Lukehart SA, Centurion-Lara A (2006) Molecular evolution of the tprC, D, I, K, G, and J genes in the pathogenic genus *Treponema*. Mol Biol Evol 23:2220–2233

Gu Y, Stansfeld PJ, Zeng Y, Dong H, Wang W, Dong C (2015) Lipopolysaccharide is inserted into the outer membrane through an intramembrane hole, a lumen gate, and the lateral opening of LptD. Structure 23:496–504

Haake DA, Zuckert WR (2015) The leptospiral outer membrane. Curr Top Microbiol Immunol 387:187–221

Hardy PH Jr, Nell EE (1957) Study of the antigenic structure of *Treponema pallidum* by specific agglutination. Am J Hyg 66:160–172

Hazlett KR, Sellati TJ, Nguyen TT, Cox DL, Clawson ML, Caimano MJ, Radolf JD (2001) The TprK protein of *Treponema pallidum* is periplasmic and is not a target of opsonic antibody or protective immunity. J Exp Med 193:1015–1026

Hazlett KR, Cox DL, Decaffmeyer M, Bennett MP, Desrosiers DC, La Vake CJ, La Vake ME, Bourell KW, Robinson EJ, Brasseur R, Radolf JD (2005) TP0453, a concealed outer membrane protein of *Treponema pallidum*, enhances membrane permeability. J Bacteriol 187:6499–6508

Hearn EM, Patel DR, van den Berg B (2008) Outer-membrane transport of aromatic hydrocarbons as a first step in biodegradation. Proc Natl Acad Sci U S A 105:8601–8606

Heinz E, Lithgow T (2014) A comprehensive analysis of the Omp85/TpsB protein superfamily structural diversity, taxonomic occurrence, and evolution. Front Microbiol 5:370

Hirano Y, Hossain MM, Takeda K, Tokuda H, Miki K (2007) Structural studies of the Cpx pathway activator NlpE on the outer membrane of *Escherichia coli*. Structure 15:963–976

Ho EL, Lukehart SA (2011) Syphilis: using modern approaches to understand an old disease. J Clin Invest 121:4584–4592

Hong H, Patel DR, Tamm LK, van den Berg B (2006) The outer membrane protein OmpW forms an eight-stranded b-barrel with a hydrophobic channel. J Biol Chem 281:7568–7577

Houston S, Hof R, Honeyman L, Hassler J, Cameron CE (2012) Activation and proteolytic activity of the *Treponema pallidum* metalloprotease, pallilysin. PLoS Pathog 8:e1002822

Houston S, Russell S, Hof R, Roberts AK, Cullen P, Irvine K, Smith DS, Borchers CH, Tonkin ML, Boulanger MJ, Cameron CE (2014) The multifunctional role of the pallilysin-associated *Treponema pallidum* protein, Tp0750, in promoting fibrinolysis and extracellular matrix component degradation. Mol Microbiol 91:618–634

Hovind-Hougen K (1983) Morphology. In Schell RF, Musher DM (ed) Pathogenesis and immunology of treponemal infection, Marcel Dekker, New York, pp. 3–28

Iqbal H, Kenedy MR, Lybecker M, Akins DR (2016) The TamB ortholog of *Borrelia burgdorferi* interacts with the beta-barrel assembly machine (BAM) complex protein BamA. Mol Microbiol

Izard J, Renken C, Hsieh CE, Desrosiers DC, Dunham-Ems S, La Vake C, Gebhardt LL, Limberger RJ, Cox DL, Marko M, Radolf JD (2009) Cryo-electron tomography elucidates the molecular architecture of *Treponema pallidum*, the syphilis spirochete. J Bacteriol 191:7566–7580

Johnson RC, Ritzi DM, Livermore BP (1973) Outer envelope of virulent *Treponema pallidum*. Infect Immun 8:291–295

Jones JD, Bourell KW, Norgard MV, Radolf JD (1995) Membrane topology of *Borrelia burgdorferi* and *Treponema pallidum* lipoproteins. Infect Immun 63:2424–2434

Jones P, Binns D, Chang HY, Fraser M, Li W, McAnulla C, McWilliam H, Maslen J, Mitchell A, Nuka G, Pesseat S, Quinn AF, Sangrador-Vegas A, Scheremetjew M, Yong SY, Lopez R, Hunter S (2014) InterProScan 5: genome-scale protein function classification. Bioinformatics 30:1236–1240

Juncker AS, Willenbrock H, Von Heijne G, Brunak S, Nielsen H, Krogh A (2003) Prediction of lipoprotein signal peptides in Gram-negative bacteria. Protein Sci 12:1652–1662

Kall L, Krogh A, Sonnhammer EL (2007) Advantages of combined transmembrane topology and signal peptide prediction–the Phobius web server. Nucleic Acids Res 35:W429–W432

Kawai T, Akira S (2010) The role of pattern-recognition receptors in innate immunity: update on Toll-like receptors. Nat Immunol 11:373–384

Ke W, Molini BJ, Lukehart SA, Giacani L (2015) *Treponema pallidum* subsp. pallidum TP0136 protein is heterogeneous among isolates and binds cellular and plasma fibronectin via its NH2-terminal end. PLoS Negl Trop Dis 9:e0003662

Koronakis V, Sharff A, Koronakis E, Luisi B, Hughes C (2000) Crystal structure of the bacterial membrane protein TolC central to multidrug efflux and protein export. Nature 405:914–919

Krogh A, Larsson B, von Heijne G, Sonnhammer EL (2001) Predicting transmembrane protein topology with a hidden Markov model: application to complete genomes. J Mol Biol 305:567–580

LaFond RE, Centurion-Lara A, Godornes C, Rompalo AM, Van Voorhis WC, Lukehart SA (2003) Sequence diversity of *Treponema pallidum* subsp. *pallidum tprK* in human syphilis lesions and rabbit-propagated isolates. J Bacteriol 185:6262–6268

LaFond RE, Centurion-Lara A, Godornes C, Van Voorhis WC, Lukehart SA (2006a) TprK sequence diversity accumulates during infection of rabbits with *Treponema pallidum* subsp. *pallidum* Nichols strain. Infect Immun 74:1896–1906

Lafond RE, Lukehart SA (2006) Biological basis for syphilis. Clin Microbiol Rev 19:29–49

LaFond RE, Molini BJ, Van Voorhis WC, Lukehart SA (2006b) Antigenic variation of TprK V regions abrogates specific antibody binding in syphilis. Infect Immun 74:6244–6251

Lee YH, Deka RK, Norgard MV, Radolf JD, Hasemann CA (1999) *Treponema pallidum* TroA is a periplasmic zinc-binding protein with a helical backbone. Nature Structural Biology 6:628–633

Lee YH, Dorwart MR, Hazlett KR, Deka RK, Norgard MV, Radolf JD, Hasemann CA (2002) The crystal structure of Zn(II)-free *Treponema pallidum* TroA, a periplasmic metal-binding protein, reveals a closed conformation. J Bacteriol 184:2300–2304

Li W, Wen L, Li C, Chen R, Ye Z, Zhao J, Pan J (2016) Contribution of the outer membrane protein OmpW in *Escherichia coli* to complement resistance from binding to factor H. Microb Pathog 98:57–62

Lithgow KV, Hof R, Wetherell C, Phillips D, Houston S and Cameron CE (2017) A defined syphilis vaccine candidate inhibits dissemination of *Treponema pallidum* subps. *pallidum*. Nat Commun 8:14272

Liu J, Lin T, Botkin DJ, McCrum E, Winkler H, Norris SJ (2009) Intact flagellar motor of *Borrelia burgdorferi* revealed by cryo-electron tomography: evidence for stator ring curvature and rotor/C-ring assembly flexion. J Bacteriol 191:5026–5036

Liu J, Howell JK, Bradley SD, Zheng Y, Zhou ZH, Norris SJ (2010) Cellular architecture of *Treponema pallidum*: novel flagellum, periplasmic cone, and cell envelope as revealed by cryo electron tomography. J Mol Biol 403:546–561

Lukehart SA, Miller JN (1978) Demonstration of the *in vitro* phagocytosis of *Treponema pallidum* by rabbit peritoneal macrophages. J Immunol 121:2014–2024

Lukehart SA, Shaffer JM, Baker-Zander SA (1992) A subpopulation of *Treponema pallidum* is resistant to phagocytosis: possible mechanism of persistence. J Infect Dis 166:1449–1453

Lukehart SA, Marra CM (2007) Isolation and laboratory maintenance of *Treponema pallidum*. Curr Protoc Microbiol Chapter 12:Unit 12A 11

Luthra A, Zhu G, Desrosiers DC, Eggers CH, Mulay V, Anand A, McArthur FA, Romano FB, Caimano MJ, Heuck AP, Malkowski MG, Radolf JD (2011) The transition from closed to open conformation of *Treponema pallidum* outer membrane-associated lipoprotein TP0453 involves membrane sensing and integration by two amphipathic helices. J Biol Chem 286:41656–41668

Luthra A, Anand A, Hawley KL, LeDoyt M, La Vake CJ, Caimano MJ, Cruz AR, Salazar JC, Radolf JD (2015a) A homology model reveals novel structural features and an immunodominant surface loop/opsonic target in the *Treponema pallidum* BamA ortholog TP_0326. J Bacteriol 197:1906–1920

Luthra A, Anand A, Radolf JD (2015b) *Treponema pallidum* in gel microdroplets: A method for topological analysis of BamA (TP0326) and localization of rare outer membrane proteins. Methods Mol Biol 1329:67–75

Machius M, Brautigam CA, Tomchick DR, Ward P, Otwinowski Z, Blevins JS, Deka RK, Norgard MV (2007) Structural and biochemical basis for polyamine binding to the Tp0655 lipoprotein of *Treponema pallidum*: putative role for Tp0655 (TpPotD) as a polyamine receptor. J Mol Biol 373:681–694

Magnuson HJ, Eagle H, Fleischman R (1948) The minimal infectious inoculum of *Spirochaeta pallida* (Nichols strain) and a consideration of its rate of multiplication *in vivo*. Am J Syph Gonorrhea Vener Dis 32:1–18

Marchler-Bauer A, Derbyshire MK, Gonzales NR, Lu S, Chitsaz F, Geer LY, Geer RC, He J, Gwadz M, Hurwitz DI, Lanczycki CJ, Lu F, Marchler GH, Song JS, Thanki N, Wang Z, Yamashita RA, Zhang D, Zheng C, Bryant SH (2015) CDD: NCBI's conserved domain database. Nucleic Acids Res 43:D222–D226

Morgan CA, Lukehart SA, Van Voorhis WC (2002a) Immunization with the N-terminal portion of *Treponema pallidum* repeat protein K attenuates syphilitic lesion development in the rabbit model. Infect Immun 70:6811–6816

Morgan CA, Molini BJ, Lukehart SA, Van Voorhis WC (2002b) Segregation of B and T cell epitopes of *Treponema pallidum* repeat protein K to variable and conserved regions during experimental syphilis infection. J Immunol 169:952–957

Morgan CA, Lukehart SA, Van Voorhis WC (2003) Protection against syphilis correlates with specificity of antibodies to the variable regions of *Treponema pallidum* repeat protein K. Infect Immun 71:5605–5612

Myint M, Bashiri H, Harrington RD, Marra CM (2004) Relapse of secondary syphilis after benzathine penicillin G: molecular analysis. Sex Transm Dis 31:196–199

Nandi B, Nandy RK, Sarkar A, Ghose AC (2005) Structural features, properties and regulation of the outer-membrane protein W (OmpW) of Vibrio cholerae. Microbiology 151:2975–2986

Narita S, Tokuda H (2007) Amino acids at positions 3 and 4 determine the membrane specificity of *Pseudomonas aeruginosa lipoproteins.* J Biol Chem 282:13372–13378

Nath A, Atkins WM, Sligar SG (2007) Applications of phospholipid bilayer nanodiscs in the study of membranes and membrane proteins. Biochemistry 46:2059–2069

Nelson RA Jr, Mayer MM (1949) Immobilization of *Treponema pallidum* in vitro by antibody produced in syphilitic infection. J Exp Med 89:369–393

Nikaido H (2003) Molecular basis of bacterial outer membrane permeability revisited. Microbiol Mol Biol Rev 67:593–656

Nikaido H, Takatsuka Y (2009) Mechanisms of RND multidrug efflux pumps. Biochim Biophys Acta 1794:769–781

Noinaj N, Kuszak AJ, Gumbart JC, Lukacik P, Chang H, Easley NC, Lithgow T, Buchanan SK (2013) Structural insight into the biogenesis of β-barrel membrane proteins. Nature 501:385–390

Norgard MV, Miller JN (1983) Cloning and expression of *Treponema pallidum* (Nichols) antigen genes in *Escherichia coli.* Infect Immun 42:435–445

Norris SJ, Cox DL, Weinstock GM (2001) Biology of *Treponema pallidum*: correlation of functional activities with genome sequence data. J Mol Microbiol Biotechnol 3:37–62

Okuda S, Tokuda H (2011) Lipoprotein sorting in bacteria. Annu Rev Microbiol 65:239–259

Okuda S, Sherman DJ, Silhavy TJ, Ruiz N, Kahne D (2016) Lipopolysaccharide transport and assembly at the outer membrane: the PEZ model. Nat Rev Microbiol 14:337–345

Ou YY, Gromiha MM, Chen SA, Suwa M (2008) TMBETADISC-RBF: Discrimination of b-barrel membrane proteins using RBF networks and PSSM profiles. Comput Biol Chem 32:227–231

Parker ML, Houston S, Petrosova H, Lithgow KV, Hof R, Wetherell C, Kao WC, Lin YP, Moriarty TJ, Ebady R, Cameron CE, Boulanger MJ (2016) The structure of *Treponema pallidum* Tp0751 (Pallilysin) reveals a non-canonical lipocalin fold that mediates adhesion to extracellular matrix components and interactions with host cells. PLoS Pathog 12:e1005919

Penn CW, Cockayne A, Bailey MJ (1985) The outer membrane of *Treponema pallidum*: biological significance and biochemical properties. J Gen Microbiol 131:2349–2357

Petersen TN, Brunak S, von Heijne G, Nielsen H (2011) SignalP 4.0: discriminating signal peptides from transmembrane regions. Nat Methods 8:785–786

Petrosova H, Zobanikova M, Cejkova D, Mikalova L, Pospisilova P, Strouhal M, Chen L, Qin X, Muzny DM, Weinstock GM, Smajs D (2012) Whole genome sequence of *Treponema pallidum* ssp. *pallidum*, strain Mexico A, suggests recombination between yaws and syphilis strains. PLoS Negl Trop Dis 6:e1832

Phan G, Picard M, Broutin I (2015) Focus on the Outer Membrane Factor OprM, the Forgotten Player from Efflux Pumps Assemblies. Antibiotics (Basel) 4:544–566

Pinto M, Borges V, Antelo M, Pinheiro M, Nunes A, Azevedo J, Borrego MJ, Mendonca J, Carpinteiro D, Vieira L, Gomes JP (2016) Genome-scale analysis of the non-cultivable *Treponema pallidum* reveals extensive within-patient genetic variation. Nat Microbiol 2:16190

Porcella SF, Popova TG, Hagman KE, Penn CW, Radolf JD, Norgard MV (1996) A mgl-like operon in *Treponema pallidum*, the syphilis spirochete. Gene 177:115–121

Purcell BK, Swancutt MA, Radolf JD (1990) Lipid modification of the 15 kiloDalton major membrane immunogen of *Treponema pallidum*. Mol Microbiol 4:1371–1379

Radolf JD, Fehniger TE, Silverblatt FJ, Miller JN, Lovett MA (1986) The surface of virulent Treponema pallidum: resistance to antibody binding in the absence of complement and surface association of recombinant antigen 4D. Infect Immun 52:579–585

Radolf JD, Chamberlain NR, Clausell A, Norgard MV (1988) Identification and localization of integral membrane proteins of virulent *Treponema pallidum* subsp. *pallidum* by phase partitioning with the nonionic detergent triton X-114. Infect Immun 56:490–498

Radolf JD, Moomaw C, Slaughter CA, Norgard MV (1989a) Penicillin-binding proteins and peptidoglycan of *Treponema pallidum* subsp. pallidum. Infect Immun 57:1248–1254

Radolf JD, Norgard MV, Schulz WW (1989b) Outer membrane ultrastructure explains the limited antigenicity of virulent *Treponema pallidum*. Proc Natl Acad Sci U S A 86:2051–2055

Radolf JD, Bourell KW, Akins DR, Brusca JS, Norgard MV (1994) Analysis of *Borrelia burgdorferi* membrane architecture by freeze-fracture electron microscopy. J Bacteriol 176:21–31

Radolf JD (1995) *Treponema pallidum* and the quest for outer membrane proteins. Mol Microbiol 16:1067–1073

Radolf JD, Goldberg MS, Bourell K, Baker SI, Jones JD, Norgard MV (1995a) Characterization of outer membranes isolated from *Borrelia burgdorferi*, the Lyme disease spirochete. Infect Immun 63:2154–2163

Radolf JD, Robinson EJ, Bourell KW, Akins DR, Porcella SF, Weigel LM, Jones JD, Norgard MV (1995b) Characterization of outer membranes isolated from *Treponema pallidum*, the syphilis spirochete. Infect Immun 63:4244–4252

Radolf JD, Lukehart SA (2006) Immunology of Syphilis. In Radolf JD, Lukehart SA (ed) Pathogenic Treponemes: Cellular and Molecular Biology, Caister Academic Press, Norfolk, UK, pp 285–322

Radolf JD, Hazlett KRO, Lukehart SA (2006). Pathogenesis of Syphilis. In Radolf JD, Lukehart SA (ed) Pathogenic Treponemes: Cellular and Molecular Biology, Caister Academic Press, Norfolk, UK, pp. 197–236

Radolf JD, Caimano MJ, Stevenson B, Hu LT (2012) Of ticks, mice and men: understanding the dual-host lifestyle of Lyme disease spirochaetes. Nat Rev Microbiol 10:87–99

Radolf JD, Tramont EC, Salazar JC (2014) Syphilis (*Treponema pallidum*). In Bennett JE, Dolin R, Blaser MJ (ed) Mandell, Douglas and Bennett's principles and practice of infectious diseases, Churchill Livingtone, Elsevier, Philadelphia, pp. 2684–2709

Radolf JD, Deka RK, Anand A, Smajs D, Norgard MV, Yang XF (2016) *Treponema pallidum*, the syphilis spirochete: making a living as a stealth pathogen. Nat Rev Microbiol 14:744–759

Randall A, Cheng J, Sweredoski M, Baldi P (2008) TMBpro: secondary structure, beta-contact and tertiary structure prediction of transmembrane beta-barrel proteins. Bioinformatics 24:513–520

Reid TB, Molini BJ, Fernandez MC, Lukehart SA (2014) Antigenic variation of TprK facilitates development of secondary syphilis. Infect Immun 82:4959–4967

Rollauer SE, Sooreshjani MA, Noinaj N, Buchanan SK (2015). Outer membrane protein biogenesis in Gram-negative bacteria. Philos Trans R Soc Lond B Biol Sci 370

Salazar JC, Hazlett KR, Radolf JD (2002) The immune response to infection with *Treponema pallidum*, the stealth pathogen. Microbes Infect 4:1133–1140

Sassone-Corsi M, Raffatellu M (2015) No vacancy: how beneficial microbes cooperate with immunity to provide colonization resistance to pathogens. J Immunol 194:4081–4087

Selkrig J, Belousoff MJ, Headey SJ, Heinz E, Shiota T, Shen HH, Beckham SA, Bamert RS, Phan MD, Schembri MA, Wilce MC, Scanlon MJ, Strugnell RA, Lithgow T (2015) Conserved features in TamA enable interaction with TamB to drive the activity of the translocation and assembly module. Sci rep 5:12905

Sena AC, Pillay A, Cox DL, Radolf JD (2015). *Treponema* and *Brachyspira*, human host-associated spirochetes. In Jorgensen JH, Pfaller MA, Carroll KC, Funke G, Landry ML, Richter SS, Warnock DW (ed) Manual of Clinical Microbiology, ASM Press, Washington, D.C, pp. 1055–1081

Setubal JC, Reis M, Matsunaga J, Haake DA (2006) Lipoprotein computational prediction in spirochaetal genomes. Microbiology 152:113–121

Shao L, Kinnally KW, Mannella CA (1996) Circular dichroism studies of the mitochondrial channel, VDAC, from Neurospora crassa. Biophys J 71:778–786

Shevchenko DV, Akins DR, Robinson EJ, Li M, Shevchenko OV, Radolf JD (1997) Identification of homologs for thioredoxin, peptidyl prolyl cis-trans isomerase, and glycerophosphodiester phosphodiesterase in outer membrane fractions from *Treponema pallidum*, the syphilis spirochete. Infect Immun 65:4179–4189

Shevchenko DV, Sellati TJ, Cox DL, Shevchenko OV, Robinson EJ, Radolf JD (1999) Membrane topology and cellular location of the *Treponema pallidum* glycerophosphodiester phosphodiesterase (GlpQ) ortholog. Infect Immun 67:2266–2276

Silhavy TJ, Kahne D, Walker S (2010) The bacterial cell envelope. Cold Spring Harb Perspect Biol 2:a000414

Smajs D, Norris SJ, Weinstock GM (2012) Genetic diversity in *Treponema pallidum*: implications for pathogenesis, evolution and molecular diagnostics of syphilis and yaws. Infect Genet Evol 12:191–202

Smith SG, Mahon V, Lambert MA, Fagan RP (2007) A molecular Swiss army knife: OmpA structure, function and expression. FEMS Microbiol Lett 273:1–11

Stamm LV, Bergen HL (2000) The sequence-variable, single-copy tprK gene of *Treponema pallidum* Nichols strain UNC and Street strain 14 encodes heterogeneous TprK proteins. Infect Immun 68:6482–6486

Stamm LV, Folds JD, Bassford PJ Jr (1982) Expression of *Treponema pallidum* antigens in Escherichia coli K-12. Infect Immun 36:1238–1241

Stamm LV, Greene SR, Bergen HL, Hardham JM, Barnes NY (1998) Identification and sequence analysis of *Treponema pallidum tprJ*, a member of a polymorphic multigene family. FEMS Microbiol Lett 169:155–163

Sun ES, Molini BJ, Barrett LK, Centurion-Lara A, Lukehart SA, Van Voorhis WC (2004) Subfamily I *Treponema pallidum* repeat protein family: sequence variation and immunity. Microbes Infect 6:725–737

Swancutt MA, Radolf JD, Norgard MV (1990) The 34-kilodalton membrane immunogen of *Treponema pallidum* is a lipoprotein. Infect Immun 58:384–392

Tokunaga M, Loranger JM, Wu HC (1984) A distinct signal peptidase for prolipoprotein in Escherichia coli. J Cell Biochem 24:113–120

Tsirigos KD, Peters C, Shu N, Kall L, Elofsson A (2015) The TOPCONS web server for consensus prediction of membrane protein topology and signal peptides. Nucleic Acids Res 43: W401–W407

Turner TB, Hollander DH (1957) Biology of the treponematoses. World Health Organization, Geneva

van den Berg B (2005) The FadL family: unusual transporters for unusual substrates. Curr Opin Struct Biol 15:401–407

van den Berg B (2012) Structural basis for outer membrane sugar uptake in pseudomonads. J Biol Chem 287:41044–41052

van den Berg B, Black PN, Clemons WM Jr, Rapoport TA (2004) Crystal structure of the long-chain fatty acid transporter FadL. Science 304:1506–1509

Walfield AM, Hanff PA, Lovett MA (1982) Expression of *Treponema pallidum* antigens in *Escherichia coli*. Science 216:522–523

Walker EM, Zampighi GA, Blanco DR, Miller JN, Lovett MA (1989) Demonstration of rare protein in the outer membrane of *Treponema pallidum* subsp. *pallidum* by freeze-fracture analysis. J Bacteriol 171:5005–5011

Webb CT, Heinz E, Lithgow T (2012) Evolution of the beta-barrel assembly machinery. Trends Microbiol 20:612–620

Weigel LM, Brandt ME, Norgard MV (1992) Analysis of the N-terminal region of the 47-kilodalton integral membrane lipoprotein of *Treponema pallidum*. Infect Immun 60:1568–1576

Weigel LM, Radolf JD, Norgard MV (1994) The 47-kDa major lipoprotein immunogen of *Treponema pallidum* is a penicillin-binding protein with carboxypeptidase activity. Proc Natl Acad Sci U S A 91:11611–11615

Whitfield C, Trent MS (2014) Biosynthesis and export of bacterial lipopolysaccharides. struct Rev Biochem 83:99–128

Wimley WC (2003) The versatile β-barrel membrane protein. Curr Opin Struct Biol 13:404–411

Wu XB, Tian LH, Zou HJ, Wang CY, Yu ZQ, Tang CH, Zhao FK, Pan JY (2013) Outer membrane protein OmpW of *Escherichia coli* is required for resistance to phagocytosis. Res Microbiol 164:848–855

Yang J, Zhang Y (2015) Protein structure and function prediction using I-TASSER. Curr Protoc Bioinform 52:5 8 1–15

Yonehara R, Yamashita E, Nakagawa A (2016) Crystal structures of OprN and OprJ, outer membrane factors of multidrug tripartite efflux pumps of *Pseudomonas aeruginosa*. Proteins 84:759–769

Yu CS, Chen YC, Lu CH, Hwang JK (2006) Prediction of protein subcellular localization. Proteins 64:643–651

Yu NY, Wagner JR, Laird MR, Melli G, Rey S, Lo R, Dao P, Sahinalp SC, Ester M, Foster LJ, Brinkman FS (2010) PSORTb 3.0: improved protein subcellular localization prediction with refined localization subcategories and predictive capabilities for all prokaryotes. Bioinformatics 26:1608–1615

Zeth K, Thein M (2010) Porins in prokaryotes and eukaryotes: common themes and variations. Biochem J 431:13–22

Zuckert WR (2014) Secretion of bacterial lipoproteins: through the cytoplasmic membrane, the periplasm and beyond. Biochim Biophys Acta 1843:1509–1516

Gene Regulation, Two Component Regulatory Systems, and Adaptive Responses in *Treponema Denticola*

Richard T. Marconi

Abstract The oral microbiome consists of a remarkably diverse group of 500–700 bacterial species. The microbial etiology of periodontal disease is similarly complex. Of the ~400 bacterial species identified in subgingival plaque, at least 50 belong to the genus *Treponema*. As periodontal disease develops and progresses, *T. denticola* transitions from a low to high abundance species in the subgingival crevice. Changes in the overall composition of the bacterial population trigger significant changes in the local physical, immunological and physiochemical conditions. For *T. denticola* to thrive in periodontal pockets, it must be nimble and adapt to rapidly changing environmental conditions. The purpose of this chapter is to review the current understanding of the molecular basis of these essential adaptive responses, with a focus on the role of two component regulatory systems with global regulatory potential.

Contents

The original version of this chapter was revised: Copyright line have been corrected. The erratum to this chapter is available at https://doi.org/10.1007/82_2018_95

R.T. Marconi (✉)
Department of Microbiology and Immunology, Virginia Commonwealth University Medical Center, 1112 East Clay Street, Room 101 McGuire Hall, 980678, Richmond VA 23298-0678, USA
e-mail: Richard.marconi@vcuhealth.org

Current Topics in Microbiology and Immunology (2018) 415:39–62
DOI 10.1007/82_2017_66

Published Online: 13 October 2017

1 Introduction

The oral cavity of mammals hosts a remarkably diverse bacterial population consisting of 500–700 species (Paster et al. 2001). Signature species differences in the composition of the bacterial population at specific niches have been demonstrated and shown to correlate with infections including caries, endodontic infections, gingivitis, subgingival plaque and periodontitis (Siqueira and Rocas 2004; Zijnge et al. 2010). The composition of the oral bacterial community varies significantly among individuals and changes with age, diet, hygiene, anatomical niche, health status, and other factors (Darveau 2010; Ebersole et al. 2008). As periodontal disease progresses, significant changes occur in the physiochemical and immunological environment of the gingival sulcus or subgingival crevice (the space between the tooth and gum tissue) (reviewed in McDowell et al. 2012). The anaerobic environment of deep periodontal pockets (>3 mm) favors the proliferation of a subset of the oral bacterial community that consists primarily of Gram-negative anaerobes and anaerobic spirochetes (Dewhirst et al. 2010, 2012). All oral spirochetes belong exclusively to the genus *Treponema* (Paster et al. 2001). Here we present an overview of the genetic regulatory mechanisms and adaptive responses that may contribute to the ability of *T. denticola* to compete and thrive in the competitive environment of subgingival plaque in periodontal pockets.

2 Microbial Etiology of Periodontal Disease: It Takes a Village to Cause Disease!

2.1 "An Unbelievably Great Company of Living Animalcules"

Much of our molecular understanding of the oral microflora is derived from analyses of bacteria that have been successfully cultivated. However, this represents only a small subset of the total oral population (Paster et al. 2001). The complexity and remarkable nature of the bacterial population in the oral cavity were noted over

340 years ago by Antonie van Leeuwenhoek, a "non-scientist" merchant in Holland who documented his observations in unpublished personal notes later recounted by others (Amako 2014). Through microscopic examination of "*spittle*", van Leeuwenhoek described the presence of "*....an unbelievably great company of living animalcules a swimming more nimbly than any I had seen up to this time. The biggest sort (whereof there were a great plenty) bent their body into curves going forward....*" One can assume from his detailed physical description of these unique "animalcules" that at least some of these organisms were spirochetes. Several hundred years after these groundbreaking early observations, Paster demonstrated the presence of 415 different bacterial species in subgingival plaque of which 76 were identified as spirochetes in the genus *Treponema* (Paster et al. 2006). A recent study suggests that the total number of *Treponema* species in the oral cavity is closer to 50 (Dewhirst et al. 2010). Of the known species, only a subset has been successfully cultivated to date (*T. denticola, T. medium, T. lecithinolyticum, T. vincentii, T. socranskii, T. amylovorum, T. parvum, T. pectinovorum,* and *T. maltophilum*). *T. denticola* has received the most research attention and is the most extensively characterized of the oral treponemes. Figure 1 shows images of *T. denticola* strain 35405 growing in culture as visualized by dark-field, scanning and transmission electron microscopy (Panel A, B, and C, respectively). The characteristic spirochetal morphology of oral treponemes is evident in the images presented in panels A and B. The endoflagella (internal flagella) that function to provide mobility, facilitate tissue invasion and to maintain the spirochetal ultrastructure (Ruby et al. 1997) are evident in panel C.

2.2 *Deciphering the Complexity of the Oral Bacterial Population*

While the complexity of the bacterial population associated with periodontal disease and other oral infections has long been recognized, until recent years most studies of periopathogens have employed reductionist strategies (one species at a time).

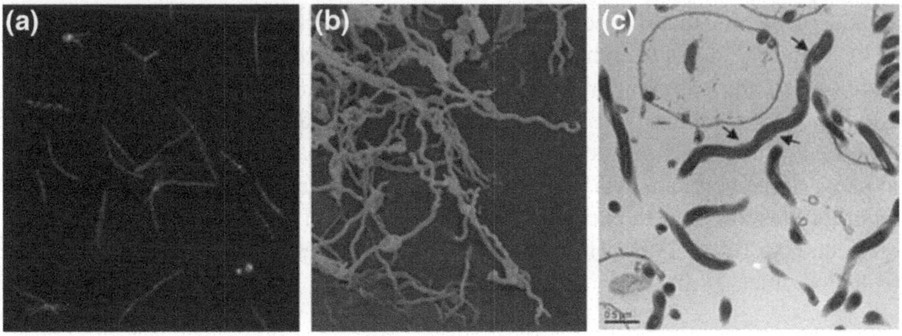

Fig. 1 Dark-field (*panel A*), scanning (*panel B*), and transmission electron micrograph (*panel C*) images of *Treponema denticola.* Endoflagella bundles are indicated by *arrows* in panel C

Socransky and colleagues changed the way that we think about the microbial component of oral diseases. Using molecular techniques, distinct microbial species groups were identified and divided into complexes differentiated by a color designation (yellow, orange, green, purple and red) (Socransky et al. 1998, 2002). Proliferation of the red microbial complex was found to correlate with the most severe forms of periodontal disease. This complex comprises *Treponema denticola, Porphyromonas gingivalis,* and *Tannerrella forsythia.*

2.3 Expansion of the Bacterial Complex Concept: The Polymicrobial Synergy and Dysbiosis Model and the Role of Keystone Pathogens

Studies over the past few decades suggest that a reassessment of the color-coded "complex" model is warranted. Hajishengallis and Lamont have proposed that the pathogenesis of oral infections "*is initiated by a synergistic and dysbiotic microbial community rather than by select 'periopathogens.....*" (Hajishengallis and Lamont 2012). At the center of this hypothesis is the concept that periodontal disease develops not simply from the co-occurrence of just a few species, but rather from a *"polymicrobial synergy"* in which "*different members or specific gene combinations within the community fulfill distinct roles that converge to shape and stabilize a disease-provoking microbiota.*" This is referred to as the polymicrobial synergy and dysbiosis model (Hajishengallis and Lamont 2012). While it is clear that it takes a "village" to cause periodontal disease, it is also clear that some species are more intimately involved in the process than others are. Bacterial species that are directly involved in influencing the environment and composition of the microbial community and disease progression as a whole have been referred to as "keystone pathogens" with other potentially less important species classified as "accessory" organisms (Darveau 2009; Hajishengallis et al. 2012). The term, keystone pathogen, in the context of periodontal disease, was first applied to *P. gingivalis* which is arguably the most intensively studied periopathogen (Mysak et al. 2014). It has been proposed that in spite of its low abundance at disease sites, it can initiate a cascade of events that is required for disease progression (Hajishengallis et al. 2011). Similarly, it has been hypothesized that *T. denticola* alters the environmental conditions of the subgingival crevice in a way that promotes the growth of the periopathogen community as a whole (Miller et al. 2016). Consistent with this *T. denticola* has also been designated as a keystone pathogen. Excellent reviews, too many to list here, are available detailing the roles that *P. gingivalis, T. denticola* and other bacteria play in periodontal disease (Darveau 2010; Dashper et al. 2010; Fenno and McBride 1998; How et al. 2016; McDowell et al. 2012; Mysak et al. 2014; Podzimek et al. 2015; Sharma 2010).

2.4 T. denticola *and Immune Dysregulation*

The subgingival crevice is bathed in complement-rich gingival crevicular fluid (Boackle et al. 1978, 1991; Stadler et al. 2016). Periopathogens in the subgingival crevice activate the alternative complement system and trigger complement amplification. Under healthy conditions, activation of the alternative complement pathway is controlled and regulated in part by factor H (FH); a highly abundant serum protein (Ruddy and Austen 1971). FH serves as a cofactor for the cleavage of C3b by the serine protease Factor I (FI), C3b is an important opsonin. FH also inhibits the formation of the C3 convertase complex and accelerates the decay of preexisting complex (reviewed in (Zipfel and Skerka 2009). Control of C3b levels is essential in order to prevent over-accumulation of C3b and resulting host tissue damage and destruction. FH also contributes to the regulation of the classical complement pathway. FH binds to a monomeric C-reactive protein (mCRP), a positive regulator of the classical pathway that is abundant at inflammatory sites (Megson et al. 2010). The FH:mCRP complex functions in part to inhibit classical pathway activation (Jarva et al. 1999). In addition, the complex promotes clearance of necrotic or apoptotic cells and inhibits the production of proinflammatory cytokines (Mihlan et al. 2009).

Many pathogens and spirochetes including *T. denticola* exploit the negative regulatory activity of FH by binding FH to the cell surface (reviewed in (McDowell et al. 2012; Zipfel et al. 2002). The sole FH receptor produced by *T. denticola* is an 11.4 kD surface-exposed lipoprotein (FhbB) (McDowell et al. 2005, 2007). FH bound to FhbB retains its ability to serve as a cofactor for FI allowing for surface localized cleavage of C3b and thus protection against complement mediated destruction. The interaction between FH and *T. denticola* is unique in that bound FH is cleaved by dentilisin (McDowell et al. 2011). Dentilisin is a multi-subunit chymotrypsin-like protease (CTLP) that is unique to some *T. denticola* isolates and not produced by other oral treponemes (Ishihara et al. 1996; Uitto et al. 1988). Its central involvement in treponemal biology and pathogenesis, including its essential role in abscess formation/tissue destruction, is well-documented (Fenno et al. 1998; Ishihara 2010; Miller et al. 2014b, 2016; Vieira and Nascimento 2016). Dentilisin cleaves a wide array of host proteins including several that regulate host immune responses and environmental homeostasis. It has been postulated that as the *T. denticola* population increases as periodontal disease progresses, local depletion of FH will ensue, triggering immune dysregulation (Miller et al. 2012). Specifically, there would be a local accumulation of C3b, enhanced proinflammatory cytokine production and impaired clearance of damaged or dead cells. This cascade of events culminates in the destruction of the periodontium and profoundly alters the physiochemical and immunological environment of the subgingival crevice. The net environmental effect would be an enlarged anaerobic zone (due to tissue degradation) that would facilitate the proliferation of *T. denticola* and other anaerobic periopathogens. The ability to adapt to this environment would most certainly correlate with significant changes in *T. denticola's* transcriptome and proteome. Below we discuss a subset of the potential genetic regulatory mechanisms that might control these critical adaptive responses.

3 Genomics and Genetic Regulatory Mechanisms of Oral Treponemes

3.1 Genetics of the Oral Treponemes

Significant insight into the pathogenic and genetic regulatory mechanisms of *T. denticola* has come from genome sequencing and bioinformatic analyses. The first *T. denticola* genome sequence determined was for the type strain, 35405 (Seshadri et al. 2004). At the time of this writing, 16 additional *T. denticola* genome sequences had been determined as well those for a limited number of other treponeme species (https://www.ncbi.nlm.nih.gov/genome/genomes). Information derived from genome analyses pertinent to this chapter is summarized in Table 1 and described briefly below. Treponemes display considerable range in G+C content indicative of early branching of spirochetes from the evolutionary tree. The G+C content of the *T. denticola* 35405 genome is 38% with minimal variation among isolates. The G+C range among oral treponemes is 38% (*T. denticola*) to 49% (*T. socranskii*). By comparison, the G+C content of *T. pallidum* (the causative agent of syphilis) is 53% (Seshadri et al. 2004). There is minimal synteny between the genomes of *T. denticola* and other spirochetal genera. *T. denticola* encodes type 1 and type II restriction modification systems. These may serve to maintain "treponemal identity" by providing a barrier against unwanted genetic advances in the complex polymicrobial community of the oral cavity. Interestingly, there is some variation in restriction modification systems among *T. denticola* isolates. Isolate 33520 harbors a type 1 system only. With regard to cell wall composition, genes that encode proteins required for lipopolysaccharide synthesis are lacking. Consistent with this absence, lipoproteins make up the dominant component of the outer leaflet of the outer membrane. Like *T. pallidum*, *T. denticola* has limited biosynthetic potential (Seshadri et al. 2004). It possesses intact pathways for gluconeogenesis and glycolysis but is incomplete in its tricarboxylic acid pathway. While it has more genes dedicated to metabolism than many other spirochete species, reductive evolution appears to have led down a path that has increased *T. denticola*'s dependence on host-association. Consistent with this hypothesis, *T. denticola* encodes a rather large cohort of ABC-type uptake and efflux transporter systems. Important to both metabolism and virulence, *T. denticola* strains encode an array of 18–25 proteases. Two proteases that have been experimentally characterized in some detail are dentilisin (Chi et al. 2003; Ellen et al. 2000; Ishihara et al. 1998; McDowell et al. 2009; Miller et al. 2016; Yamazaki et al. 2006) and cystalysin (Chu et al. 2002, 2008; Krupka et al. 2000; Zhang et al. 2010). Both play important roles in *T. denticola* pathogenesis. The influence of dentilisin on the immunological properties of the subgingival crevice is discussed in detail above.

3.2 Two Component Regulatory Systems and Adaptive Responses of T. denticola

Environmental transitions in the subgingival crevice as periodontal disease develops have been demonstrated to trigger significant alterations in both the host and *T. denticola* transcriptome (Bakthavatchalu et al. 2011; Haririan et al. 2014; Jonsson et al. 2011). However, the regulatory pathways and effectors that control adaptive responses are relatively unknown. Frederick et al. (2011) sought to gain additional insight into the potential regulatory networks involved in environmental adaptation through a bioinformatics-based assessment of the *T. denticola* 35405 genome. In bacteria, two component regulatory (TCR) systems are key players in adaptive responses. A prototypical TCR system consists of a sensor histidine kinase (HK) and a response regulator (RR). HK and RR proteins are widely distributed across bacteria and archae and are among the most abundant proteins cataloged in databases (Zschiedrich et al. 2016). The domain organization of HK and RR proteins varies considerably (Galperin 2006; Zschiedrich et al. 2016).

A general model for TCR system mediated regulation is that an HK sensor domain detects environmental signals and responds by autophosphorylating at a conserved His residue present within its sensor domain. The phosphoryl group is then transferred from the Nε of the His residue to an Asp residue present in the receiver (REC) domain of the cognate RR. Conformational changes and dimerization of the RR result in its activation. Most activated RRs regulate gene expression by binding to specific DNA target sequences and influencing transcription. Relative to most bacterial species, *T. denticola* encodes a limited number of TCR systems (Table 1) and regulators, of which only a few have been partially characterized (Frederick et al. 2008, 2011; Miller et al. 2014a; Sarkar et al. 2010). *T. denticola* 35405 encodes 6 HKs, 7RRs, and 2 HK-RR hybrids (Frederick et al. 2011). Among the oral treponemes, the AtcS-AtcR and Hpk2-Rrp2 TCR systems of *T. denticola* have received the most attention (Frederick et al. 2008; Miller et al. 2014a; Sarkar et al. 2010). The domain organizations for AtcR, AtcS, Rrp2, and Hpk2 are depicted in Fig. 2.

3.3 The AtcSR Two Component Regulatory System and the LytTR Domain

ORFs TDE0032 and TDE0033 of *T. denticola* 35405 encode a sensor HK-related ATPase (AtcS; 248 aa; 28.6 kD) and a DNA binding RR (AtcR; 239 aa; 27.9 kD), respectively. Frederick et al. (2008) demonstrated, using recombinant protein and

Table 1 Summary of the genetic properties of cultivable oral treponemes

Species and strain designation	Genome size (Mb)	Genomes available	G+C	Genes	HKs	DNA binding RR	RR (no DNA binding domain)	LytTR	PAS	HK/RR hybrids
T. denticola (ATCC35405)	2.74–2.99	17	37.8–38.1	2470–2767	6	4	2	1	1–2	3
T. medium (ATCC700293)	2.72	1	44.3	2420	2	1	2	0	2	0
T. maltophilum (ATCC51939)	2.53	1	47.9	2351	4	6	2	0	1	1
T. vincentii (F0403)	2.52–2.69	2	45.5–45.7	2306–2489	2	0	2	0	2	0
T. lecithinolyticum (ATCC700332)	2.34	1	43.8	2125	1	2	0	0	2	0
T. socranskii (ATCC35535)	2.74–2.83	3	48.3–49.2	2533–2628	5	5	3	0	1	1
T. putidum (ATCC700334T)	2.81 (3.6kB plasmid)	1	37.3	2272	3	2	3	0	0	1

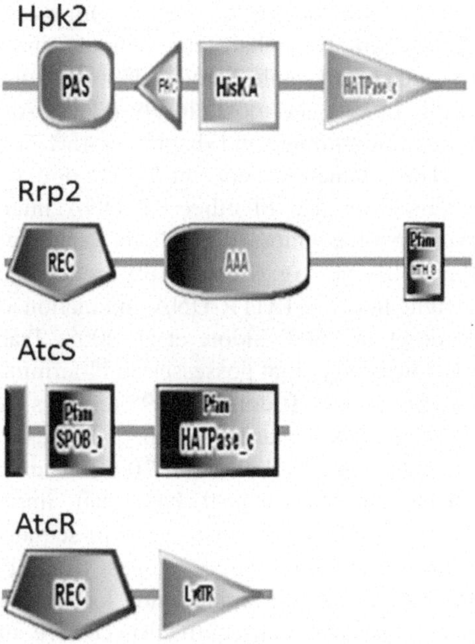

Fig. 2 Domain architecture of the AtcRS and Hpk2-Rrp2 two component regulatory system proteins. Conserved domains are indicated in the figure

in vitro assays, that AtcS undergoes autophosphorylation and transfers phosphate to AtcR. AtcS is an inner membrane anchored protein with an alpha helical type sensor kinase domain (SpoOB domain) and an HTPase_c_5 domain (ATPase domain) (Fig. 2). This domain architecture is relatively uncommon and found only in a handful of other bacterial proteins including some from the *Clostridiales* and *Lactobacillales.* AtcR differs from the majority of treponemal RRs in that it possesses a unique C-terminal LytTR domain. LytTR domains derive their name from the *Bacillus subtilis* LytT and *Staphylococcus aureus* LytR RRs (Nikolskaya and Galperin 2002). Although LytTR domains are widespread among human and plant bacterial pathogens, they are present in only a small subset of RRs (typically 1 or 2 per bacterial genome) (Galperin 2008). LytTR RRs contribute to the regulation of diverse processes and activities, including biofilm formation (Lizewski et al. 2004), toxin production (Ba-Thein et al. 1996; Lyristis et al. 1994), type IV pili synthesis (Belete et al. 2008), antimicrobial peptide production (Risoen et al. 1998), flagellar assembly, flagellar function (Martin et al. 2013), natural competence (*Streptococcus pneumoniae*) (de Saizieu et al. 2000), extracellular polysaccharide biosynthesis (*Pseudomonas aeruginosa*) (Lizewski et al. 2004; Morici et al. 2007), overall fitness and virulence gene expression (Abdelnour et al. 1993; Martin et al. 2013). It is noteworthy that among the *Spirochaetaceae*, only *T. denticola, T. bryantii* (identifier WP_022932649) and *Sphaerochaeta globosa* (WP_013608153) encode LytTR domain-containing proteins (Frederick et al. 2008, 2011). LytTR RRs are absent in other important spirochetal pathogens including *Borrelia* (tick borne relapsing fever) and *Borreliella* (Lyme disease). The LytTR domains of *S. globosa*, a free living bacterium isolated from river bed sediment (Ritalahti et al. 2012) and *T. bryantii*, a saccharolytic obligate anaerobe isolated from the rumen of cows (Stanton and Canale-Parola 1980; Stanton 1984), show only minimal similarity to AtcR. In fact, the most closely related AtcR and AtcS homologs are found in the order *Clostridiales* and display ~45% sequence identity.

The interaction between LytTR domains and their DNA recognition sequence differs from that of other RR-DNA interactions. Most RRs interact with DNA through a recognition alpha helix located within a DNA binding domain. Studies of the *S. aureus* AgrA (accessory gene regulator A), a LytTR RR, have provided insight into the LytTR-DNA interaction (Koenig et al. 2004; Nicod et al. 2014; Rode et al. 2007; Sidote et al. 2008; Traber and Novick 2006). AgrA is a cytoplasmic protein that possesses an N-terminal REC domain of 124 amino acids and a C-terminal LytTR domain of 98 residues. As in other LytTR RRs, the DNA binding domain consists largely of beta-strands (10 stranded beta fold) interspersed by variable loops (Sidote et al. 2008). The DNA recognition sequence for the AgrA dimer consists of a 9 base pair imperfect direct repeats separated by an 11-nucleotide spacer that varies in sequence. Among most bacterial LytTR regulated genes, this spacer ranges from 11 to 13 nucleotides (Galperin 2008). Positively charged amino acids located within loop domains of AgrA monomers insert into two consecutive successive major grooves of the DNA recognition sequence (Sidote et al. 2008) allowing specific side chains to insert within the intervening minor groove. Salt bridges between AgrA and the DNA phosphate backbone further stabilize the interaction. AgrA binding triggers DNA bending which regulates transcription (Sanchez et al. 2015). Studies reviewed below detail the identification of the AtcR-DNA recognition sequence and demonstrate the existence of three different promoter architectures for AtcR regulated genes.

3.4 Structural Modeling of AtcR

While the structure of AtcR has not yet been determined, threading of the AtcR sequence onto the AgrA:DNA complex structure suggests that AtcR and AgrA have similar structures (Miller et al. 2014a; Sidote et al. 2008). Ramachandran plotting provided strong support for the predicted AtcR structure as 89.36% of the residues were calculated to have favorable bond angles (Miller et al. 2014a). The predicted AtcR structural model suggests that the interaction of AtcR with DNA is also likely to be similar to that of AgrA with strong potential for residues within the loops of AtcR to insert into the major grooves of potential LytTR-DNA motifs. AgrA residues H169, N201, and R233 have been demonstrated to be in direct contact with DNA (Sidote et al. 2008). Sequence alignment revealed that AtcR residues H157, R191, and R221 are positionally analogous to the AgrA-DNA contact residues. It has been demonstrated through mutagenesis and structure determination that the AgrA Y229 residue is also a major determinant that is required for maximal activation of transcription (Leonard et al. 2012). Residue Y223 of AtcR is positionally analogous to the AgrA Y229 residue and thus may also be a key determinant for maximal transcriptional regulation potential.

3.5 *Identification of* T. denticola *Genes Potentially Regulated by AtcR*

In our laboratory, efforts to generate *atcR* gene deletion mutants for possible microarray or RNAseq analyses have proved unsuccessful. In contrast to the demonstrated nonessential role reported for AgrA (Gao et al. 2007), AtcR may have essential functions that preclude us from obtaining viable *AtcR* gene deletion mutants. The accessory role of AgrA in *S. aureus*, but the potentially essential requirement for AtcR in *T. denticola*, may simply reflect the greater number of "options" for transcriptional regulation available to *S. aureus*. As a second approach toward identifying genes potentially regulated by AtcR, Miller et al. (2014a) used a bioinformatics approach to search for potential LytTR-DNA recognition motifs. The *T. denticola* 35405 genome was queried with a consensus LytTR binding recognition motif. However, no significant hits were detected. In light of the significant evolutionary distance separating spirochetes from other bacteria, this was perhaps not unexpected. To address this further, search analyses were conducted using the fuzznuc algorithm (http://www.proteomics.leeds.ac.uk/emboss/fuzznuc.html) and a degenerate LytTR-DNA recognition motif query sequence of (TAC)(AC)(AC)GTTN(TAG)(TG) (Nikolskaya and Galperin 2002; Risoen et al. 1998; Sidote et al. 2008). Over 150 hits were detected and of these 81 had the query motif in the form of an imperfect repeat separated by spacers of 4–20 nucleotides. This set of hits was narrowed further by focusing on those in which the sequence motif was appropriately positioned upstream of an ORF or operon. Oligonucleotide duplexes representing 6 of the 81 hits were synthesized and tested for binding of phosphorylated and non-phosphorylated AtcR using electrophoretic mobility shift assays (EMSA). AtcR-DNA complexes were demonstrated and the specificity of the interaction verified through standard approaches using specific and nonspecific competitor DNA. The studies described above were an important step forward, as they paved the way for efforts to identify the AtcR regulon.

The influence of AtcR phosphorylation state and concentration (AtcR:DNA molar ratio) on complex formation was further assessed using PCR amplicons consisting of the upstream sequences of ORFs TDE1514, TDE1221, TDE0025, TDE1971, TDE1998, and TDE2000 (Table 2). Both phosphorylation-dependent and -independent AtcR-DNA complexes were detected and the banding patterns for some genes differed depending on the AtcR concentration (Miller et al. 2014a). The AtcR-DNA recognition motifs associated with ORFs TDE1514 and TDE1221, which formed AtcR phosphorylation-dependent complexes, were used to further define the AtcR recognition sequence. To accomplish this, MEME (Bailey et al. 2009), FIMO (Grant et al. 2011) and Virtual Footprinting algorithms (Munch et al.

2005) were employed as detailed in Miller et al. 2014a. Hits present on coding strands that were located within 500 base pairs of a start codon were selected for further analysis. These studies led to the identification of a *T. denticola* LytTR recognition sequence of CAGTT(CG)GGN15(TA)CAGTTTA(GT). This sequence element, designated as "motif 1", consists of a 9 base pair imperfect repeat separated by a 15 nucleotide spacer. The spacer is slightly longer than the 9–11 nucleotide spacer typically found in LytTR recognition motifs. Motif 1 was found in the upstream-untranslated region of 25 ORFs and operons that encode proteins of known or unknown function (Table 2). Additional analyses identified a second motif shared by some of these genes, consisting of an AAAAAA(CG)TTTTTT direct repeat that we designated as "motif 2".

3.6 Different LytTR Promoter Architectures for AtcR Regulated Genes

DNA motifs 1 and or 2 exist in three different configurations or promoter architectures and were designated as promoter types LytTR1, LytTR2, and LytTR3. EMSA analyses confirmed that DNA fragments individually harboring these three different promoter architectures complex with AtcR. LytTR1 promoters harbor motif 1 alone while LytTR2 promoters possess both motifs 1 and 2. The positioning between motifs 1 and 2 in genes with LytTR2 promoters is conserved. Motif 1 is 5′ to motif 2 and they are separated by an approximately 68 base pair spacer. The organization of LytTR3 promoters is similar to that of LytTR2. Motif 1 is 5′ to motif 2 with spacers of approximately 69 bp. A distinct difference in motif 2 of LytTR3 promoters is the presence of a conserved insertion sequence of 65 nucleotides that is located in the center of motif 2. Subsequent transcriptional start site (TSS) analyses using 5′ RACE of representative genes with LytTR1 and LytTR2 promoters (ORFs TDE1210 and TDE1971, respectively) revealed the existence of σ54 recognition sequences just upstream of the TSSs. The untranslated regions of mRNAs derived from representative LytTR2 and LytTR3 promoter associated genes, but not LytTR1 promoter genes, revealed the potential for regulatory secondary structures (Miller et al. 2014a). Specifically, extended and highly stable hairpin loops are possible with ΔG values ranging from −23 to −45 kcal/mol. Hence, it is possible that the post-transcriptional control of AtcR regulated genes may also be in play.

It is important to note that due to the stringent thresholds set in the promoter identification analyses described above it is highly probable that the putative AtcR regulon summarized in Table 2 is incomplete and that there are additional AtcR genes yet to be identified. The validity of the AtcR regulon defined here is supported by studies of the AgrA regulon of *C. difficile* (Martin et al. 2013). There is considerable overlap in the functional classification of genes regulated by AgrA and AtcR, most notably between those that encode proteins involved in flagellar

Table 2 The putative AtcR regulon of *T. denticola* 35405

Promoter	ORF	Protein designation and summary information
LytTR1	TDE0465	PIN domain containing nuclease (cleaves single-stranded RNA in a sequence dependent manner); potential component of a toxin-antitoxin system
LytTR1	TDE0870	2′3′-cyclic phosphodiesterase; thought to play a scavenging role during RNA hydrolysis by converting the non-transportable nucleotides produced by RNaseI to nucleosides that can enter a cell and served as a carbon source.
LytTR1	TDE1006	FlgG-1; flagellar basal body protein*
	TDE1007	FlgG-2; flagellar basal body protein*
LytTR1	TDE1210	HslV; ATP-dependent protease peptide subunit
	TDE1211	HslU; ATP-dependent protease peptide subunit
LytTR1	TDE1298	cyclic nucleotide binding protein
	TDE1297	LysM/M23/M37 peptidase
LytTR1	TDE1625	GcvP1; glycine hydrogenase subunit 1
	TDE1624	GcvP2; glycine hydrogenase subunit 2
LytTR1	TDE1655	hypothetical; not present in most *Treponemal* species
LytTR1	TDE1849	LolA; outer membrane lipoprotein carrier protein
	TDE1848	hypothetical; present in other *Treponema* and many spirochete species
LytTR1	TDE2220	prolyl-tRNA synthetase
	TDE2221	conserved hypothetical protein; present in other *Treponema* species
	TDE2222	glycerol-3-phosphate O-acyltransferase*
LytTR1	TDE2131	cobalamin biosynthesis CbiM
LytTR2	TDE0064	phosphofructokinase
LytTR2	TDE0641	MurA; peptidogylcan synthesis
LytTR2	TDE1514	ABC transporter permease
	TDE1515	ABC transporter permease
	TDE1516	ABC transporter, ATP-binding protein
	TDE1517	conserved hypothetical protein
	TDE1518	ABC transporter permease
LytTR2	TDE1693	conserved hypothetical protein (present in *T. denticola* and *T. putidum*)
LytTR2	TDE1903	ABC-2 family transporter protein
LytTR2	TDE1971	DNA polymerase III domain protein
	TDE1970	Hpk2; HK (properties are detailed in the text)
	TDE1969	Rrp2; RR (properties are discussed in the text)
	TDE1968	FtsJ; cell division protein
LytTR2	TDE2093	conserved hypothetical; unique to *T. denticola*
LytTR2	TDE2392	conserved hypothetical; possible flavin reductase
LytTR2	TDE2484	conserved hypothetical protein; unique to *T. denticola*
	TDE2483	conserved hypothetical protein; TPM domain,

(continued)

Table 2 (continued)

Promoter	ORF	Protein designation and summary information
LytTR2	TDE2496	Methyl accepting chemotaxis protein
LytTR2	TDE2770	FliK*; flagella synthesis regulator
	TDE2769	FlgD*; flagellar hook assembly scaffold
	TDE2768	FlgE*; flagellar hook protein
	TDE2767	FlgD*; flagellar protein
	TDE2766	MotA*; flagellar motor rotation protein A
	TDE2765	MotB*; flagellar motor rotation protein B
	TDE2764	FliL*; flagellar protein,
	TDE2763	FliM*; flagellar motor switch protein
	TDE2762	FliY; flagellar motor switch protein,
	TDE2761	FliO*; flagellar biosynthesis protein
	TDE2760	FliP*; flagellar biosynthesis protein
LytTR3	TDE0522	conserved hypothetical; found in plants and eucaryotes; uncharacterized
LytTR3	TDE0693	ThiD; hydroxymethylpyrimidine/phosphomethylpyrimidine kinase
LytTR3	TDE1132	AroC; chorismate synthase
LytTR3	TDE2225	acyl-CoA thioesterase

For operons, the promoter designation is shown only for the first gene in the operon. Genes that are part of the LytTR regulon of *C. difficile* (Martin et al. 2013) are indicated by an asterisk. This updated table is derived information presented in an earlier study (Miller et al. 2014a)

assembly, structure, and function. In *C. difficile*, AgrA is a transcriptional activator of flagellar synthesis (Martin et al. 2013). Overlap was also observed for genes involved in metabolism and transport. Genes that overlap between the AtcR and AgrA regulons are indicated in Table 2 by asterisks. Additional studies are required to identify the full regulon, to determine the extent to which different promoter architectures affect transcriptional regulation mediated by AtcR and to assess the potential for post-transcriptional regulation of genes that have LytTR2 and LytTR3 promoters. Lastly, the broader biological question to address is "does the carriage of a LytTR RR by *T. denticola* ascribes to this organism a unique and defining role in sensing specific environmental cues for the oral treponeme community as a whole?"

4 The Rrp2-Hpk2 Two Component Regulatory System

4.1 Domain Organization, Unique Features, and Properties of Hpk2 and Rrp2

Hpk2 (TDE1970; 415 aa; 46 kD;) is a sensory HK and Rrp2 (TDE1969; 473 aa; 53 kD) is a σ54-dependent Fis family DNA binding RR. Using recombinant proteins and *in vitro* autophosphorylation and phosphotransfer assays, it was demonstrated that Hpk2 and Rrp2 constitute a functional TCR (Sarkar et al. 2010). Orthologs of Rrp2 are widely distributed among spirochetes. They can be found in most *Treponema* species and in the distantly related *Borreliella* and *Borrelia.* In fact, the Rrp2 designation for ORF TDE1969 was originally selected based on its shared properties with the Rrp2 protein of the Lyme disease spirochetes (Yang et al. 2003). *T. denticola* Rrp2 shares >55% identity with Rrp2 of *B. burgdorferi*. The *B. burgdorferi* Rrp2 protein is also a σ54-dependent RR and is an activator of the RpoN-RpoS regulatory pathway (Ouyang et al. 2008; Samuels 2011; Yang et al. 2003). Mutational analyses of *B. burgdorferi* Rrp2 demonstrated this protein to be a key transcriptional regulator of several important virulence factors including the intensively studied OspC protein (reviewed in Samuels 2011). However, the regulatory network of Rrp2 in *T. denticola* is likely to be quite different, since *T. denticola* lacks RpoS.

4.2 The PAS Domain of Hpk2

Our initial interest in Hpk2 stemmed from the presence of a predicted heme-binding PAS domain (residues 30–66) and PAC motif (PAS associated C-terminal domain; residues 116–169). PAS is an acronym for Per (period circadian protein), Armt (Ah receptor nuclear translocator protein and Sim (single-minded protein) (Moglich et al. 2009). PAS domains serve as sensors of diverse environmental stimuli and those with heme-binding pockets represent the most common sensing strategy for detecting changes in concentrations of oxygen, carbon monoxide or nitric oxide. The binding or dissociation of oxygen (or other diatomic gases) from the heme iron complex triggers a structural rearrangement that transduces signals to the C-terminal functional domain of the protein, switching it to the on or off state. While PAS domains are widely distributed with over 99,000 annotated in the genomes of Archae, Eucaryota, and Bacteria, they are not universal. *T. denticola* has 1 while many other oral treponemes, as well as *T. pallidum*, lack discernable PAS domains. In contrast to the oral treponemes, the termite gut endosymbiont species *T. azotonutricium* and *T. primitia* encode 19 and 9 PAS domain-containing proteins, respectively. A distinct molecular feature of the *T. denticola* Hpk2 PAS domain is that it possesses a unique 15-aa insertion that has not been found in other PAS-domain-containing proteins, including those of other *Treponema* species (Sarkar et al. 2010). The presence of this unique insertion may convey structural and functional flexibility to Hpk2.

Most PAS domain proteins possess conserved PAC motifs of 40–45 aa that are located C-terminal to PAS domains (http://smart.embl-heidelberg.de/smart/do_annotation.pl?DOMAIN=PAC). PAC motifs contribute to the PAS domain fold (Hefti et al. 2004). Based on aa similarity, the Hpk2 nomenclature was originally based on its amino acid similarity with the HK2 protein of *B. burgdorferi* (BB_0764; strain B31; 382 aa; 44 kD). However, the SMART PAS domain database does not identify *B. burgdorferi* HK2 as a PAS domain-containing protein.

Both Hpk2 and HK2 belong to the BaeS protein superfamily. It has been demonstrated that many of these sensory kinases regulate adaptive transcriptional responses to stress (Konovalova et al. 2016). Hpk2, as well as HK2, appear to lack an extracytoplasmic sensing domain. Hpk2 has been demonstrated to localize to the protoplasmic cylinder of *T. denticola* in Triton X-114 phase partitioning and extraction experiments (Sarkar et al. 2010). Collectively, information available suggests that Hpk2 and its orthologs belong to a limited subset of HK proteins that are cytoplasmic.

4.3 Mutational Analyses of the Hpk2 PAS Domain

Sarkar et al. (2010) directly investigated the contribution of the PAS domain in Hpk2 autophosphorylation and phosphotransfer by generating a recombinant truncated protein lacking the N-terminal 122 aa that encompasses the PAS domain (Hpk2ΔPAS). In a creative series of experiments, autophosphorylation and phosphotransfer were assessed under anaerobic or bench top conditions. For the "anaerobic" experiments, all reagents were equilibrated in an anaerobic chamber and the assays were conducted in the chamber itself. The results obtained from assays conducted in these distinctly different environments were strikingly different. The level of autophosphorylation observed over time with the full-length Hpk2 protein was significantly higher under anaerobic conditions that under bench top conditions. Deletion of the PAS domain did not influence autophosphorylation when the assays were conducted under bench top conditions, but a significant decrease in autophosphorylation was observed under anaerobic conditions. These analyses suggest that the PAS domain is required for optimal Hpk2 activation during anaerobic growth. As discussed above, bioinformatic analyses predict a heme-binding pocket in Hpk2. Recent studies in our lab have confirmed that Hpk2 does in fact bind heme (Marconi, RT; unpublished data) and that heme binding influences the kinetics of Hpk2 autophsophorylation. While more research is required, the data obtained to date support a role for Hpk2 in the sensing of diatomic gas levels. From this, it can be extrapolated that Hpk2 likely plays an important role in sensing the dramatic environmental changes that occur in the subgingival crevice as periodontal disease progresses.

5 Transcriptional Analyses of the Rrp2-Hpk2 and AtcSR Two Component Regulatory System

The transcriptional expression patterns of *rrp2-hpk2* and *atcSR* during *in vitro* cultivation have been defined (Frederick et al. 2008; Sarkar et al. 2010). Both TCR systems are transcribed as polycistronic mRNAs that include some genes that are 5′ or 3′ to each TCR system (Frederick et al. 2008; Sarkar et al. 2010).

TSS mapping using 5′-RACE also demonstrated at least two distinct TSSs for each TCR system, indicating the potential to regulate the production of these regulators and adjacent genes through the use of different promoters. RT-PCR and northern blot analyses revealed that the dominant *atcRS* transcript is approximately 1500 nucleotides, indicative of an *atcS-atcR* bicistronic mRNA. Transcription of the operon was up-regulated with increasing cell density (Frederick et al. 2008). For *hpk2-rrp2*, a 4- gene polycistronic mRNA spanning ORFs TDE1971 through TDE1968 is the dominant transcript. The transcription of the *hpk2-rrp2* operon is also up-regulated with increasing cell density. In light of the putative functions of the other proteins cotranscribed with *hpk2-rrp2,* it is possible that upregulation with cell density could significantly influence the growth rate of *T. denticola* in a periodontal pocket. Other proteins encoded by the operon are involved in cellular functions that are important to rapid population expansion. The *hpk2-rrp2* operon also encodes proteins involved in DNA replication (DnaX) and translational efficiency (FtsJ). When expressed as a 7 gene operon, genes that encode proteins involved in cell wall synthesis (MurG) and colicin production (CvpA) would also be transcribed along with an ORF encoding a hypothetical protein. CvpA is annotated as a colicin V production factor. Colicin production could inhibit the growth of competing organisms and allow *T. denticola* to thrive. To date, colicin production, potential activity and target species specificity have not yet been assessed. There appears to be a regulatory linkage between the Rrp2 and AtcR regulons. The analyses above indicate that the transcription of the 4 gene *TDE1971-hpk2-rrp2-TDE1968* transcript is controlled by Rrp2. Upstream of TDE1971 (the first gene in this 4 gene operon) is a LytTR2 promoter (Miller et al. 2014a). The EMSA analyses discussed above directly demonstrated that a DNA fragment containing this promoter binds phosphorylated AtcR. While additional evidence is required to firmly establish the linkage between the AtcR and Rrp2 regulons, the picture that is emerging is one of a complex regulatory mesh in which separate regulatory systems interact.

6 Remaining Thoughts and Future Directions

The goal of this chapter was to provide an update to the current understanding of the molecular basis of adaptive responses of *T. denticola.* In spite of the staggering health consequences and socioeconomic costs associated with periodontal disease, we know very little about the mechanisms employed by *T. denticola* and other oral treponemes that allow these bacteria to be successful in adapting to, and thriving in, the highly competitive environment of subgingival plaque. It is evident from the literature that studies focused on regulatory pathways in oral treponemes are few and far between. With that in mind, the primary focus of this discussion was directed at TCR systems that have been at least partially characterized, specifically the Hpk-Rrp2 and AtcSR systems. The proteins that comprise these TCR systems have unique features (detailed above) that suggest their potential to regulate adaptive responses at the global level. However, the point must be stressed that there are certainly additional key regulatory proteins and regulatory networks that remain unexplored. Perhaps most notable is the potential global regulatory role of the secondary messenger, c-di-GMP, which most certainly plays an important role in the biology and pathogenesis of *T. denticola* (Bian et al. 2013).

C-di-GMP mediated regulation has been studied in depth in a long and diverse list of bacteria, including spirochete species associated with Lyme disease and relapsing fever (Caimano et al. 2015; Kostick et al. 2011; Mallory et al. 2016; Pitzer et al. 2011; Rogers et al. 2009; Ryjenkov et al. 2005; Sultan et al. 2010). C-di-GMP is produced by diguanylate cyclases (DGC) which can be identified by the presence of the signature GGDEF domain (Hengge 2009). *T. denticola* 35405 encodes several GGDEF domain-containing proteins including TDE0125, TDE1456, TDE1685, TDE2580, TDE2581, TDE2725, and TDE2726 as well as two DGC/c-di-GMP phosphodiesterases (PDE) hybrids (TDE2075 and TDE0128). The opposing activities of DGCs and PDEs control the intracellular concentration of c-di-GMP that can exert its regulatory potential through RNA binding or by interaction with effector c-di-GMP receptors (TDE1214 and TDE1318). The largest class of c-di-GMP receptors is those that possess the signature PilZ domain (Amikam and Galperin 2006). Two such c-di-GMP receptors have been identified in *T. denticola*: TDE0214 and TDE1318. Hence, bioinformatics analyses suggest that *T. denticola* (as well as other oral treponemes) possess one or more complete c-di-GMP regulatory networks. Consistent with this hypothesis, Kostick et al. (2011) demonstrated production of c-di-GMP by *T. denticola* during cultivation, but the specific DGC associated with this activity was not defined. The only component of a c-di-GMP regulatory network protein that has been experimentally characterized in an oral treponeme is the c-di-GMP receptor/PilZ domain protein, TDE0214. Insertional inactivation of TDE0214 attenuated motility, biofilm formation, and virulence and based on this it was speculated that the attenuation of these activities was related to its inability to bind c-di-GMP (Bian et al. 2013). However, it is important to note that TDE0214 is not a standalone PilZ domain protein, as it also possesses tetratricopeptide repeat (TPR). TPR proteins 23 of which are annotated in the *T. denticola* 35405 genome, can mediate specific protein-protein interactions as well as the assembly of some multi-protein complexes (D'Andrea and Regan 2003). Hence, it remains to be determined if the phenotype associated with inactivation of TDE0214 is due to loss of a c-di-GMP related function or is due to loss of an important TPR-mediated function (or both). It is clear that the investigation of c-di-GMP networks and other TCR systems and regulators is important and offers a fertile area of research.

7 Conclusions

Studies conducted to date, coupled with information derived from genome analyses now provide us with the necessary information base upon which informed decisions can be made in designing research strategies to advance our understanding of the adaptive response of oral treponemes. However, it is also clear that there is significant variation in the set of potential regulatory factors encoded by different *T. denticola* strains. Intra-species differences indicate that the *T. denticola* population is not homogenous. Similar to what we have learned about the complexity of the oral microflora at the species level, we must also consider complexity at the intra-species level. A better understanding of the inter- and intra-species population dynamics coupled with the identification of critical regulatory and communication pathways will open the door to the development of new therapeutics and preventatives for periodontal disease and other oral infections.

References

Abdelnour A, Arvidson S, Bremell T, Ryden C, Tarkowski A (1993) The accessory gene regulator (*agr*) controls *Staphylococcus aureus* virulence in a murine arthritis model. Infect Immun 61:3879–3885

Amako K (2014) Little animals observed by Antony van Leeuwenhoek. Nihon Saikingaku Zasshi 69:315–330

Amikam D, Galperin MY (2006) PilZ domain is part of the bacterial c-di-GMP binding protein. Bioinformatics 22:3–6

Bailey TL, Boden M, Buske FA, Frith M, Grant CE, Clementi L, Ren J, Li WW, Noble WS (2009) MEME SUITE: tools for motif discovery and searching. Nucleic Acids Res 37:W202–W208

Bakthavatchalu V, Meka A, Mans JJ and other authors (2011) Polymicrobial periodontal pathogen transcriptomes in calvarial bone and soft tissue. Mol Oral Microbiol 26:303–320

Ba-Thein W, Lyristis M, Ohtani K, Nisbet IT, Hayashi H, Rood JI, Shimizu T (1996) The *virR/virS* locus regulates the transcription of genes encoding extracellular toxin production in *Clostridium perfringens*. J Bacteriol 178:2514–2520

Belete B, Lu H, Wozniak DJ (2008) *Pseudomonas aeruginosa* AlgR regulates type IV pilus biosynthesis by activating transcription of the *fimU-pilVWXY1Y2E* operon. J Bacteriol 190:2023–2030

Bian J, Liu X, Cheng YQ, Li C (2013) Inactivation of cyclic Di-GMP binding protein TDE0214 affects the motility, biofilm formation, and virulence of *Treponema denticola*. J Bacteriol 195:3897–3905

Boackle RJ (1991) The interaction of salivary secretions with the human complement system–a model for the study of host defense systems on inflamed mucosal surfaces. Crit Rev Oral Biol Med 2:355–367

Boackle RJ, Pruitt KM, Silverman MS, Glymph JL Jr (1978) The effects of human saliva on the hemolytic activity of complement. J Dent Res 57:103–110

Caimano MJ, Dunham-Ems S, Allard AM, Cassera MB, Kenedy M, Radolf JD (2015) Cyclic di-GMP modulates gene expression in lyme disease spirochetes at the tick-mammal interface to promote spirochete survival during the blood meal and tick-to-mammal transmission. Infect Immun 83:3043–3060

Chi B, Qi M, Kuramitsu HK (2003) Role of dentilisin in *Treponema denticola* epithelial cell layer penetration. Res Microbiol 154:637–643

Chu L, Dong Z, Xu X, Cochran DL, Ebersole JL (2002) Role of glutathione metabolism of *Treponema denticola* in bacterial growth and virulence expression. Infect Immun 70:1113–1120
Chu L, Lai Y, Xu X, Eddy S, Yang S, Song L, Kolodrubetz D (2008) A 52-kDa leucyl aminopeptidase from *Treponema denticola* is a cysteinylglycinase that mediates the second step of glutathione metabolism. J Biol Chem 283:19351–19358
D'Andrea LD, Regan L (2003) TPR proteins: the versatile helix. Trends Biochem Sci 28:655–662
Darveau RP (2009) The oral microbial consortium's interaction with the periodontal innate defense system. DNA Cell Biol 28:389–395
Darveau RP (2010) Periodontitis: a polymicrobial disruption of host homeostasis. Nat Rev Microbiol 8:481–490
Dashper SG, Seers CA, Tan KH, Reynolds EC (2010) Virulence factors of the oral spirochete *Treponema denticola*. J Dent Res 90:691–703
de Saizieu A, Gardes C, Flint N, Wagner C, Kamber M, Mitchell TJ, Keck W, Amrein KE, Lange R (2000) Microarray-based identification of a novel *Streptococcus pneumoniae* regulon controlled by an autoinduced peptide. J Bacteriol 182:4696–4703
Dewhirst FE, Chen T, Izard J, Paster BJ, Tanner AC, Yu WH, Lakshmanan A, Wade WG (2010) The human oral microbiome. J Bacteriol 192:5002–5017
Dewhirst FE, Klein EA, Thompson EC and other authors (2012) The canine oral microbiome. PLoS One 7:e36067
Ebersole JL, Steffen MJ, Gonzalez-Martinez J, Novak MJ (2008) Effects of age and oral disease on systemic inflammatory and immune parameters in nonhuman primates. Clin Vaccine Immunol 15:1067–1075
Ellen RP, Ko KS, Lo CM, Grove DA, Ishihara K (2000) Insertional inactivation of the prtP gene of *Treponema denticola* confirms dentilisin's disruption of epithelial junctions. J Mol Microbiol Biotech 2:581–586
Fenno J, McBride B (1998) Virulence factors of oral treponemes. Anaerobe 4:1–17
Fenno JC, Hannam PM, Leung WK, Tamura M, Uitto VJ, McBride BC (1998) Cytopathic effects of the major surface protein and the chymotrypsinlike protease of *Treponema denticola*. Infect Immun 66:1869–1877
Frederick JR, Rogers EA, Marconi RT (2008) Analysis of a growth-phase-regulated two-component regulatory system in the periodontal pathogen *Treponema denticola*. J Bacteriol 190:6162–6169
Frederick JR, Sarkar J, McDowell JV, Marconi RT (2011) Molecular signaling mechanisms of the periopathogen, *Treponema denticola*. J Dent Res 90:1155–1163
Galperin MY (2006) Structural classification of bacterial response regulators: diversity of output domains and domain combinations. J Bacteriol 188:4169–4182
Galperin MY (2008) Telling bacteria: do not LytTR. Structure 16:657–659
Gao R, Mack TR, Stock AM (2007) Bacterial response regulators: versatile regulatory strategies from common domains. Trends Biochem Sci 32:225–234
Grant CE, Bailey TL, Noble WS (2011) FIMO: scanning for occurrences of a given motif. Bioinformatics 27:1017–1018
Hajishengallis G, Liang S, Payne MA and other authors (2011) Low-abundance biofilm species orchestrates inflammatory periodontal disease through the commensal microbiota and complement. Cell Host Microbe 10:497–506
Hajishengallis G, Lamont RJ (2012) Beyond the red complex and into more complexity: the polymicrobial synergy and dysbiosis (PSD) model of periodontal disease etiology. Mol Oral Microbiol 27:409–419
Hajishengallis G, Darveau RP, Curtis MA (2012) The keystone-pathogen hypothesis. Nat Rev Microbiol 10:717–725
Haririan H, Andrukhov O, Bertl K, Lettner S, Kierstein S, Moritz A, Rausch-Fan X (2014) Microbial analysis of subgingival plaque samples compared to that of whole saliva in patients with periodontitis. J Periodontol 85:819–828

Hefti MH, Francoijs KJ, de Vries SC, Dixon R, Vervoort J (2004) The PAS fold. A redefinition of the PAS domain based upon structural prediction. Eur J Biochem 271:1198–1208

Hengge R (2009) Principles of c-di-GMP signalling in bacteria. Nat Rev Microbiol 7:263–273

How KY, Song KP, Chan KG (2016) *Porphyromonas gingivalis*: An overview of periodontopathic pathogen below the Gum Line. Front Microbiol 7:53

Ishihara K (2010) Virulence factors of *Treponema denticola*. Periodontol 2000(54):117–135

Ishihara K, Miura T, Kuramitsu HK, Okuda K (1996) Characterization of the *Treponema denticola prtP* gene encoding a prolyl-phenylalanine-specific protease (dentilisin). Infect Immun 64:5178–5186

Ishihara K, Kuramitsu HK, Miura T, Okuda K (1998) Dentilisin activity affects the organization of the outer sheath of *Treponema denticola*. J Bacteriol 180:3837–3844

Jarva H, Jokirant TS, Hellwage J, Zipfel PF, Meri S (1999) Regulation of complement activation by C-reactive protein: Targeting then complement regulatory activity of factor H by an interaction with short consensus repeat domains 7 and 8–11. J Immunol 163:3957–3962

Jonsson D, Ramberg P, Demmer RT, Kebschull M, Dahlen G, Papapanou PN (2011) Gingival tissue transcriptomes in experimental gingivitis. J Clin Periodontol 38:599–611

Koenig RL, Ray JL, Maleki SJ, Smeltzer MS, Hurlburt BK (2004) *Staphylococcus aureus* AgrA binding to the RNAIII-*agr* regulatory region. J Bacteriol 186:7549–7555

Konovalova A, Schwalm JA, Silhavy TJ (2016) A Suppressor mutation that creates a faster and more robust sigmae envelope stress response. J Bacteriol 198:2345–2351

Kostick JL, Szkotnicki LT, Rogers EA, Bocci P, Raffaelli N, Marconi RT (2011) The diguanylate cyclase, Rrp1, regulates critical steps in the enzootic cycle of the Lyme disease spirochetes. Mol Microbiol 81:219–231

Krupka HI, Huber R, Holt SC, Clausen T (2000) Crystal structure of cystalysin from *Treponema denticola*: a pyridoxal 5'-phosphate-dependent protein acting as a haemolytic enzyme. EMBO J 19:3168–3178

Leonard PG, Bezar IF, Sidote DJ, Stock AM (2012) Identification of a hydrophobic cleft in the LytTR domain of AgrA as a locus for small molecule interactions that inhibit DNA binding. Biochem 51:10035–10043

Lizewski SE, Schurr JR, Jackson DW, Frisk A, Carterson AJ, Schurr MJ (2004) Identification of AlgR-regulated genes in *Pseudomonas aeruginosa* by use of microarray analysis. J Bacteriol 186:5672–5684

Lyristis M, Bryant AE, Sloan J, Awad MM, Nisbet IT, Stevens DL, Rood JI (1994) Identification and molecular analysis of a locus that regulates extracellular toxin production in *Clostridium perfringens*. Mol Microbiol 12:761–777

Mallory KL, Miller DP, Oliver LD Jr, Freedman JC, Kostick-Dunn JL, Carlyon JA, Marion JD, Bell JK, Marconi RT (2016) Cyclic-di-GMP binding induces structural rearrangements in the PlzA and PlzC proteins of the Lyme disease and relapsing fever spirochetes: a possible switch mechanism for c-di-GMP-mediated effector functions. Pathog Dis 74:1–8

Martin MJ, Clare S, Goulding D, Faulds-Pain A, Barquist L, Browne HP, Pettit L, Dougan G, Lawley TD, Wren BW (2013) The *agr* locus regulates virulence and colonization genes in *Clostridium difficile* 027. J Bacteriol 195:3672–3681

McDowell JV, Lankford J, Stamm L, Sadlon T, Gordon DL, Marconi RT (2005) Demonstration of factor H-like protein 1 binding to *Treponema denticola*, a pathogen associated with periodontal disease in humans. Infect Immun 73:7126–7132

McDowell JV, Frederick J, Stamm L, Marconi RT (2007) Identification of the gene encoding the FhbB protein of *Treponema denticola*, a highly unique factor H-like protein 1 binding protein. Infect Immun 75:1050–1054

McDowell JV, Huang B, Fenno JC, Marconi RT (2009) Analysis of a unique interaction between the complement regulatory protein factor H and the periodontal pathogen *Treponema denticola*. Infect Immun 77:1417–1425

McDowell JV, Frederick J, Miller DP, Goetting-Minesky MP, Goodman H, Fenno JC, Marconi RT (2011) Identification of the primary mechanism of complement evasion by the periodontal pathogen, *Treponema denticola*. Mol Oral Microbiol 26:140–149

McDowell JV, Miller DP, Mallory KL Marconi RT (2012) *Treponema denticola*: FhbB, dentilisin, complement evasion and the paradox of factor H cleavage. In: Embers ME (ed) The pathogenic spirochetes: strategies for evasion of host immunity and persistance. New York: Springer Science+Business Media, pp 43–62

Megson E, Fitzsimmons T, Dharmapatni K, Mark Bartold P (2010) C-reactive protein in gingival crevicular fluid may be indicative of systemic inflammation. J Clin Periodontol 37:797–804

Mihlan M, Stippa S, Jozsi M, Zipfel PF (2009) Monomeric CRP contributes to complement control in fluid phase and on cellular surfaces and increases phagocytosis by recruiting factor H. Cell Death Differ 16:1630–1640

Miller DP, Bell JK, McDowell JV, Conrad DH, Burgner JW, Heroux A, Marconi RT (2012) Structure of factor H binding protein B (FhbB) of the periopathogen, *Treponema denticola*: insights into the progression of periodontal disease. J Biol Chem 287:12715–12722

Miller DP, Frederick JR, Sarkar J, Marconi RT (2014a) The Treponema denticola AtcR LytTR domain-containing response regulator interacts with three architecturally distinct promoter elements: implications for understanding the molecular signaling mechanisms that drive the progression of periodontal disease. Mol Oral Microbiol 29:219–232

Miller DP, McDowell JV, Bell JK, Goetting-Minesky MP, Fenno JC, Marconi RT (2014b) Analysis of the complement sensitivity of oral treponemes and the potential influence of FH binding, FH cleavage and dentilisin activity on the pathogenesis of periodontal disease. Mol Oral Microbiol 29:194–207

Miller DP, Oliver LD, Jr, Tegels BK and other authors (2016) The *Treponema denticola* FhbB protein is a dominant early antigen that elicits FhbB variant-specific antibodies that block factor H binding and cleavage by dentilisin. Infect Immun 84:2051–2058

Moglich A, Ayers RA, Moffat K (2009) Structure and signaling mechanism of Per-ARNT-Sim domains. Structure 17:1282–1294

Morici LA, Carterson AJ, Wagner VE and other authors (2007) *Pseudomonas aeruginosa* AlgR represses the Rhl quorum-sensing system in a biofilm-specific manner. J Bacteriol 189: 7752–7764

Munch R, Hiller K, Grote A, Scheer M, Klein J, Schobert M, Jahn D (2005) Virtual Footprint and PRODORIC: an integrative framework for regulon prediction in prokaryotes. Bioinformatics 21:4187–4189

Mysak J, Podzimek S, Sommerova P, Lyuya-Mi Y, Bartova J, Janatova T, Prochazkova J, Duskova J (2014) *Porphyromonas gingivalis*: major periodontopathic pathogen overview. J Immunol Res 2014, article ID: 476068

Nicod SS, Weinzierl RO, Burchell L, Escalera-Maurer A, James EH, Wigneshweraraj S (2014) Systematic mutational analysis of the LytTR DNA binding domain of *Staphylococcus aureus* virulence gene transcription factor AgrA. Nucleic Acids Res 42:12523–12536

Nikolskaya AN, Galperin MY (2002) A novel type of conserved DNA-binding domain in the transcriptional regulators of the AlgR/AgrA/LytR family. Nucleic Acids Res 30:2453–2459

Ouyang Z, Blevins JS, Norgard MV (2008) Transcriptional interplay among the regulators Rrp2, RpoN and RpoS in *Borrelia burgdorferi*. Microbiol 154:2641–2658

Paster BJ, Boches SK, Galvin JL, Ericson RE, Lau CN, Levanos VA, Sahasrabudhe A, Dewhirst FE (2001) Bacterial diversity in human subgingival plaque. J Bacteriol 183: 3770–3783

Paster BJ, Olsen I, Aas JA, Dewhirst FE (2006) The breadth of bacterial diversity in the human periodontal pocket and other oral sites. Periodontol 2000(42):80–87

Pitzer JE, Sultan SZ, Hayakawa Y, Hobbs G, Miller MR, Motaleb MA (2011) Analysis of the *Borrelia burgdorferi* cyclic-di-GMP binding protein PlzA reveals a role in motility and virulence. Infect Immun 79:1815–1825

Podzimek S, Mysak J, Janatova T, Duskova J (2015) C-Reactive protein in peripheral blood of patients with chronic and aggressive periodontitis, gingivitis, and gingival recessions. Mediators Inflamm 2015:564858

Risoen PA, Havarstein LS, Diep DB, Nes IF (1998) Identification of the DNA-binding sites for two response regulators involved in control of bacteriocin synthesis in *Lactobacillus plantarum* C11. Mol Gen Genet 259:224–232

Ritalahti KM, Justicia-Leon SD, Cusick KD, Ramos-Hernandez N, Rubin M, Dornbush J, Loffler FE (2012) *Sphaerochaeta globosa* gen. nov., sp. nov. and *Sphaerochaeta pleomorpha* sp. nov., free-living, spherical spirochaetes. Int J Syst Evol Microbiol 62:210–216

Rode TM, Langsrud S, Holck A, Moretro T (2007) Different patterns of biofilm formation in *Staphylococcus aureus* under food-related stress conditions. Int J Food Microbiol 116:372–383

Rogers EA, Terekhova D, Zhang HM, Hovis KM, Schwartz I, Marconi RT (2009) Rrp1, a cyclic-di-GMP-producing response regulator, is an important regulator of *Borrelia burgdorferi* core cellular functions. Mol Microbiol 71:1551–1573

Ruby JD, Li H, Kuramitsu H, Norris SJ, Goldstein SF, Buttle KF, Charon NW (1997) Relationship of *Treponema denticola* periplasmic flagella to irregular cell morphology. J Bacteriol 179:1628–1635

Ruddy S, Austen KF (1971) C3b inactivator of man. II. Fragments produced by C3b inactivator cleavage of cell-bound or fluid phase C3b. J Immunol 107:742–750

Ryjenkov DA, Tarutina M, Moskvin OV, Gomelsky M (2005) Cyclic diguanylate is a ubiquitous signaling molecule in bacteria: insights into biochemistry of the GGDEF protein domain. J Bacteriol 187:1792–1798

Samuels DS (2011) Gene regulation in *Borrelia burgdorferi*. Annu Rev Microbiol 65:479–499

Sanchez D, Boudes M, van Tilbeurgh H, Durand D, Quevillon-Cheruel S (2015) Modeling the ComD/ComE/comcde interaction network using small angle X-ray scattering. FEBS J 282:1538–1553

Sarkar J, Frederick J, Marconi RT (2010) The Hpk2-Rrp2 two-component regulatory system of *Treponema denticola*: a potential regulator of environmental and adaptive responses. Mol Oral Microbiol 25:241–251

Seshadri R, Myers GSA, Tettelin H and other authors (2004) Comparison of the genome of the oral pathogen *Treponema denticola* with other spirochete genomes. PNAS 101:5646–5651

Sharma A (2010) Virulence mechanisms of Tannerella forsythia. Periodontol 2000(54):106–116

Sidote DJ, Barbieri CM, Wu T, Stock AM (2008) Structure of the *Staphylococcus aureus* AgrA LytTR domain bound to DNA reveals a beta fold with an unusual mode of binding. Structure 16:727–735

Siqueira JF, Rocas IN (2004) Treponema species associated with abscesses of endodontic origin. Oral Microbiol Immunol 19:336–339

Socransky SS, Haffajee AD, Cugini MA, Smith C, Kent RL Jr (1998) Microbial complexes in subgingival plaque. J Clin Periodontol 25:134–144

Socransky SS, Smith C, Haffajee AD (2002) Subgingival microbial profiles in refractory periodontal disease. J Clin Periodontol 29:260–268

Stadler AF, Angst PD, Arce RM, Gomes SC, Oppermann RV, Susin C (2016) Gingival crevicular fluid levels of cytokines/chemokines in chronic periodontitis: a meta-analysis. J Clin Periodontol 43:727–745

Stanton TB (1984) Glucose metabolism of *Treponema bryantii*, an anaerobic rumen spirochete. Can J Microbiol 30:526–531

Stanton TB, Canale-Parola E (1980) *Treponema bryantii* sp. nov., a rumen spirochete that interacts with cellulolytic bacteria. Arch Microbiol 127:145–156

Sultan SZ, Pitzer JE, Miller MR, Motaleb MA (2010) Analysis of a *Borrelia burgdorferi* phosphodiesterase demonstrates a role for cyclic-di-guanosine monophosphate in motility and virulence. Mol Microbiol 77:128–142

Traber K, Novick R (2006) A slipped-mispairing mutation in AgrA of laboratory strains and clinical isolates results in delayed activation of *agr* and failure to translate delta- and alpha-haemolysins. Mol Microbiol 59:1519–1530

Uitto VJ, Grenier D, Chan EC, McBride BC (1988) Isolation of a chymotrypsinlike enzyme from *Treponema denticola*. Infect Immun 56:2717–2722

Vieira ML, Nascimento AL (2016) Interaction of spirochetes with the host fibrinolytic system and potential roles in pathogenesis. Crit Rev Microbiol 42:573–587

Yamazaki T, Miyamoto M, Yamada S, Okuda K, Ishihara K (2006) Surface protease of *Treponema denticola* hydrolyzes C3 and influences function of polymorphonuclear leukocytes. Microbes Infect 8:1758–1763

Yang XF, Alani SM, Norgard MV (2003) The response regulator Rrp2 is essential for the expression of major membrane lipoproteins in *Borrelia burgdorferi*. Proc Natl Acad Sci USA 100:11001–11006

Zhang JH, Dong Z, Chu L (2010) Hydrogen sulfide induces apoptosis in human periodontium cells. J Periodontal Res 45:71–78

Zijnge V, van Leeuwen MB, Degener JE, Abbas F, Thurnheer T, Gmur R, Harmsen HJ (2010) Oral biofilm architecture on natural teeth. PLoS ONE 5:e9321

Zipfel PF, Skerka C (2009) Complement regulators and inhibitory proteins. Nat Rev Immunol 9:729–740

Zipfel PF, Skerka C, Hellwage J, Jokiranta ST, Meri S, Brade V, Kraiczy P, Noris M, Remuzzi G (2002) Factor H family proteins: on complement, microbes and human diseases. Biochem Soc Trans 30:971–978

Zschiedrich CP, Keidel V, Szurmant H (2016) Molecular mechanisms of two-component signal transduction. J Mol Biol 428:3752–3775

Physiologic and Genetic Factors Influencing the Zoonotic Cycle of *Borrelia burgdorferi*

Philip E. Stewart and Patricia A. Rosa

Abstract *Borrelia burgdorferi* is a symbiont of ticks of the *Ixodes ricinus* complex. These ticks serve as vectors to disseminate the spirochete to a variety of susceptible vertebrate hosts, which, in turn, act as reservoirs for naïve ticks to become infected, perpetuating the infectious life cycle of *B. burgdorferi*. The pivotal role of ticks in this life cycle and tick–spirochete interactions are the focus of this chapter. Here, we describe the challenging physiological environment that spirochetes encounter within *Ixodes* ticks, and the genetic factors that *B. burgdorferi* uses to successfully infect, persist, and be transmitted from the vector.

Contents

P.E. Stewart (✉) · P.A. Rosa (✉)
Laboratory of Zoonotic Diseases, Rocky Mountain Laboratories,
National Institute of Allergy and Infectious Diseases, National Institutes of Health,
903 South 4th St., Hamilton, MT 59840, USA
e-mail: pestewart@niaid.nih.gov

P.A. Rosa
e-mail: prosa@niaid.nih.gov

Current Topics in Microbiology and Immunology (2018) 415:63–82
DOI 10.1007/82_2017_43

Published Online: 02 September 2017

1 Introduction

Members of the genus *Borrelia* have a number of characteristics that unify them, though they are generally divided into two groups based on the diseases they cause in humans. Relapsing fever *Borrelia* spp. (RFB), as the name implies, induces a recurring fever and have historically resulted in enormous morbidity and mortality [reviewed in (Bryceson et al. 1970)]. Lyme disease *Borrelia* spp. (LDB) rarely cause fatal disease in humans, but can produce severe symptoms including carditis, facial nerve palsy, and musculoskeletal disorders such as arthritis. Some species of the LDB (such as *B. andersonii* and *B. valaisiana*) are not known to cause disease in humans, but based on genetic similarity are classified with this group.

Despite the general division of the genus, a variety of traits are shared by all *Borrelia*. All species maintain a conserved life cycle as obligate symbionts. That is to say, no free-living state for members of this genus has been observed. Although various vertebrate species act as reservoir hosts for maintaining spirochete populations, all *Borrelia* spp. require arthropods to transfer the infection from one host to another. In biological terms, arthropods act as vectors of dissemination for these spirochetes, acquiring the borreliae while feeding on infected vertebrates and later transmitting them to naïve hosts during subsequent blood feedings. The lone exception to these generalities is *Borrelia recurrentis*, which is transmitted by the human clothing louse *Pediculus humanis*. This RFB spirochete requires the crushing of the louse for successful transmission and is not delivered via a blood feeding, as are the other *Borrelia* spp.

To be successful in the infectious cycle, these spirochetes require mechanisms to cope with, evade or avoid the hazards associated with infecting and persisting in both the hosts and the vectors. *Borrelia* must be able to acquire nutrients for growth and replication while avoiding destruction by the immune systems of the vertebrate hosts and arthropod vectors. Additionally, some *Borrelia* species are capable of persisting for months in the nutrient-depleted environment of the midgut before the arthropod feeds again. For details of *Borrelia* infection and persistence in the host the reader is referred to other chapters of this volume. Here, we focus on the genetic factors that borreliae use to successfully infect the arthropod vector, persist throughout the physiological changes that occur during the vector life cycle, and mechanisms utilized for transmission to new hosts. *B. burgdorferi*, the main agent of North American Lyme disease, is the most well characterized regarding these aspects and therefore this chapter concentrates heavily on this species; but, where appropriate, we will compare it with mechanisms utilized by other *Borrelia* species.

2 Enzootic Cycle of *B. burgdorferi* in North America

The obligate symbiotic life cycle of *B. burgdorferi* confines its geographic distribution to that of its vector, namely hard-bodied ticks of the *Ixodes* complex. However, this range of *B. burgdorferi* is further restricted by the availability of competent hosts that can sustain the spirochete population long enough for naïve *Ixodes* ticks to feed upon and acquire the infection, thereby completing the infectious cycle from tick to host to a new tick. The ability of a host to maintain a *B. burgdorferi* infection may represent a subset of the animals that the tick needs to complete its life cycle. For example, adult female *I. scapularis* ticks (also referred to as *I. dammini*) are dependent on the presence of white-tailed deer (*Odocoileus virginianus*) to acquire their final blood meal, but these mammals may not be competent hosts for *B. burgdorferi* (Telford et al. 1988).

Four different species of *Ixodes* ticks can maintain *B. burgdorferi* in different geographic regions. *I. scapularis* is the primary vector of the Lyme disease spirochetes in the Eastern and mid-Western U.S., *I. pacificus* in the Western U.S., *I. ricinus* in Europe and parts of western Asia, and *I. persulcatus* in Asia (Brown et al. 2005). In contrast, RFB displays a fascinating vector specificity generally requiring a cognate species of ticks for successful maintenance and transmission (Davis 1942). *B. hermsii*, for instance, is vectored only by the soft-bodied tick *Ornithodoros hermsi*, while *B. parkeri* is maintained by *O. parkeri*. The nature of the RFB: tick specificity remains unknown, but may be similar to the mechanism of vector specificity characterized between *Leishmania* and sandfly species wherein the unique lipophosphoglycan produced by each protozoan species mediates binding only to the midgut lectin of its cognate sandfly vector (Pimenta et al. 1994; Sacks et al. 1995). However, *B. burgdorferi* appears to be more promiscuous, being capable of using multiple *Ixodes* species as vectors compared to the RFB.

All *Ixodes* species have three life stages, larval, nymphal, and adult, and each requires a blood meal in order to proceed to the next developmental state (Fig. 1). The feeding process requires several days, after which the tick detaches from its host and develops on the ground. After completing the digestion of the blood, the ticks molt to the next stage of the life cycle. The molt can last weeks or months, during which *I. scapularis* sheds its cuticle, forms new cuticle to accommodate the increase in body size, and to develop physiologically. Although *B. burgdorferi* can be acquired by any life stage of *I. scapularis*, the infection is not passed on to the progeny (Magnarelli et al. 1987; Rollend et al. 2013), which means that freshly hatched larvae are uninfected. In contrast, RFB can be passed transovarially (Burgdorfer and Varma 1967; Dutton and Todd 1905). A major reservoir host for *B. burgdorferi* in nature is the white-footed mouse, *Peromyscus leucopus*, which is also an important maintenance host of larval and nymphal *I. scapularis*. Although a variety of small, medium, and large vertebrates can be infected by *B. burgdorferi*, as well as birds and some reptiles (Anderson and Magnarelli 1984; Anderson et al. 1986; Brown and Lane 1992; Clark et al. 2005; Donahue et al. 1987; Levin et al. 1996; Levine et al. 1985; Takano et al. 2010), it is generally agreed that *P. leucopus*

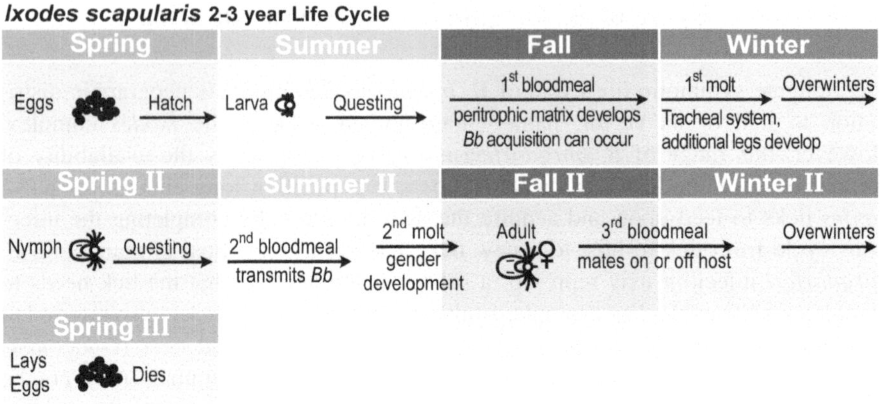

Fig. 1 The *Ixodes scapularis* life cycle. *I. scapularis* ticks have a 2–3 year life cycle, depending on climate conditions and host availability. All stages of this tick require a blood meal to proceed to the next developmental phase. The length of questing, the process by which the tick acquires its host, is determined in part by the availability of appropriate animals upon which it can feed. Each developmental stage of the tick feeds upon progressively larger hosts, but as an opportunistic parasite, *I. scapularis* will feed on a variety of animal species if given the chance. After successful feeding, the tick drops from its host and returns to the environment to molt to the next stage. The adult females feed, but the males do not, and mating can occur on or off the host

is the primary host by which larval ticks become infected. Once the tick is infected, the spirochetes persist in the midgut throughout the remaining developmental stages and can be transmitted to new hosts during subsequent feedings. However, for successful transmission to occur, *B. burgdorferi* must escape from the midgut, traverse the hemocoel (essentially the body cavity of the tick), and invade the salivary glands from which borreliae are released from the tick with the discharge of saliva. All of this occurs within the timeframe of a single tick feeding. Spirochetes that remain outside of the midgut after the tick has completed feeding appear to be destroyed, presumably by the innate immune defenses of *I. scapularis* (Johns et al. 2001). Again, RFB differs substantively in this regard, as these spirochetes chronically infect the salivary glands of their tick vectors (Raffel et al. 2014; Schwan and Hinnebusch 1998), allowing for a much more rapid transmission rate to the host that is measured in seconds rather than days (Boyle et al. 2014).

3 Physiological Conditions within *I. scapularis*

The life cycle of *I. scapularis* and the developmental changes that occur during the molt impose specific obstacles to symbiotic organisms that seek to inhabit such a difficult environment. For example, the extended times between blood meals means prolonged periods devoid of nutrients for *B. burgdorferi*, which has limited biosynthetic abilities and is reliant on obtaining the necessary compounds from its surrounding environment (Casjens et al. 2000; Fraser et al. 1997). A fresh blood meal relieves the starvation conditions for the tick and the spirochete, but produces different dangers: reactive nitrogen species are produced within the tick midgut and salivary glands during feeding, and can lead to bacterial DNA damage (Bourret et al. 2016); during the migration to the salivary glands *B. burgdorferi* is exposed to attack by the phagocytic hemocyte cells of the tick, and *Ixodes* ticks wall off the digesting blood from the midgut epithelial cells by producing a peritrophic membrane. This structure acts as a physical barrier to protect the epithelial midgut cells from abrasive particles in the blood (such as hemoglobin crystals) and to prevent dissemination of pathogens acquired with the blood (Peters 1992; Rudzinska et al. 1982; Zhu et al. 1991). After feeding, major portions of the ixodid salivary glands atrophy and are remodeled. The molting process allows for growth and development of the tick, which requires substantial cellular remodeling, such as the synthesis of new cuticle, the production of a tracheal system and an additional pair of legs during the larval molt, and sexual differentiation into either males or females during the nymphal molt.

In addition to the physiological changes associated with the tick life cycle, *B. burgdorferi* must survive the vector's mechanisms to prevent infection. To avoid being trapped by the peritrophic membrane, it is presumed that the spirochete disseminates before the membrane solidifies. *I. scapularis* (*dammini*) requires approximately 24–28 h for the peritrophic membrane to completely form (Rudzinska et al. 1982), and this may be sufficient time for enough spirochetes to escape and establish residency in the gap between the membrane and the cells that line the internal face of the midgut. *I. scapularis* innate immune defenses in the hemolymph (circulatory fluid in the hemocoel) are capable of phagocytosing and killing *B. burgdorferi*, but not in sufficient numbers or timeframes to prevent infection of the salivary glands (Johns et al. 2001). Therefore, it appears to be a race between how quickly and how many bacteria can traverse the hemocoel and reach the salivary glands and avoiding immune clearance.

Nutrients represent a changing objective for *B. burgdorferi* within the tick. When the blood meal is ingested by the tick, nutrients are plentiful for *B. burgdorferi*, but this nutrient-rich period is brief and followed by lengthy famines where *I. scapularis* has digested the blood meal and enters the molt or overwinters until it can attach and feed upon a new host. During these nutrient-limited periods, *B. burgdorferi* may switch to using glycerol as an energy source (Pappas et al. 2011), as this carbohydrate is produced by the tick to act as an antifreeze mechanism (Vandyk et al. 1996).

Within the tick midgut, do other microorganisms interact with or affect *B. burgdorferi*? Ticks coinfected with different *Borrelia* species occur both within the U.S. and within Europe (Hamsikova et al. 2016; Herrmann et al. 2013; Pritt et al. 2016; Sytykiewicz et al. 2015). The fed midgut environment allows spirochetes to reach high densities while in close proximity to each other, conditions that may allow for horizontal gene transfer between members of the genus, although this has not yet been experimentally demonstrated. Ticks, however, coinfected with multiple *B. burgdorferi* strains transfer only a random subset of these strains during feeding, essentially acting as a population bottleneck during transmission (Rego et al. 2014). Other microorganisms are well-known members of the tick microbiota, including intracellular bacterial species such as *Rickettsia* and *Wolbachia*, with the former being the most abundant organism detected in the midgut (Zolnik et al. 2016). Yet, *B. burgdorferi* is strictly an extracellular symbiont and it is not clear that there would be any significant interaction between these prokaryotes. However, Narasimhan and colleagues suggested that alteration of the tick microbiota may impact *B. burgdorferi* densities, indicating some level of interaction between bacterial populations, albeit possibly an indirect one (Narasimhan et al. 2014).

Temperature and pH are other factors that change during the tick feeding and are critical cues that *B. burgdorferi* senses, presumably in anticipation of entering the next phase of its infectious cycle. Ticks are cold-blooded and their body temperature fluctuates with environmental conditions, but once attached to a warm-blooded host and ingesting blood, the tick temperature also rises. Temperature shift experiments of cultured spirochetes demonstrated that production of a major outer surface protein (OspC) was regulated, at least in part, by temperature (Schwan et al. 1995). Subsequently, other *B. burgdorferi* gene products were demonstrated to respond to temperature changes (Ojaimi et al. 2003; Stevenson et al. 1995). Variations in pH also induce or repress a variety of proteins, as the tick presents a more basic environment than that of the mammal (Carroll et al. 1999, 2000). So, nutrients from the blood and changes in temperature and pH are sensory cues that are detected by *B. burgdorferi* and can be manipulated in the laboratory to mimic the environment of *I. scapularis*.

What other physiological conditions occurring within the tick vector does *B. burgdorferi* respond to? Oxygen and CO_2 levels, osmolarity, and bacterial cell densities may serve as environmental cues for *B. burgdorferi* to alter its protein profile for progressing through the infectious cycle (Bontemps-Gallo et al. 2016; Hyde et al. 2007; Seshu et al. 2004). A variety of *B. burgdorferi* proteins, some influenced by the environmental cues described above, aid the spirochete in surviving within *I. scapularis* and during the transition to the mammalian host.

4 *B. burgdorferi* Genetic Factors that Contribute to Successful Acquisition, Persistence, and Transmission

As naïve larvae feed on an infected host, *B. burgdorferi* within the skin or blood tissues can be acquired by the tick. Presumably, this occurs by an active process of the spirochete such as chemotaxis toward the feeding lesion, perhaps sensing compounds injected by the tick to prevent coagulation or to diminish the host immune response (Shih et al. 2002). After ingestion, *B. burgdorferi* establishes residence within the midgut, acquiring nutrients from the digestion of the blood meal, escaping the peritrophic membrane trap that walls off the blood meal during digestion, and replicating to higher densities. Spirochetes reach populations of several thousands/larvae shortly after feeding, and increase with each successive stage of the tick, reaching between 10^4 and 10^5/tick during the nymphal feeding (Piesman et al. 1990).

With the advances made in the genetic manipulation of *B. burgdorferi*, a variety of spirochete genes have now been identified that contribute to the successful colonization, survival, and transmission from the tick vector (Table 1). To date, the only non-regulatory protein that has been implicated in *I. scapularis* acquisition of the spirochete from infected vertebrates is BBA52 (Kumar et al. 2010). Genetic inactivation of *bba52* resulted in lower levels of spirochetes detected in feeding ticks at early timepoints, though the mutant eventually rose to equivalent numbers as wild-type *B. burgdorferi*. The *bba52* mutant strain was also defective in transmission to naïve mice, suggesting that this protein plays a role in transition periods to and from the tick vector. No mechanism has been proposed for the function of BBA52.

Once acquired, *B. burgdorferi* chronically infects the tick midgut, and multiple proteins contribute to this important function. Outer surface protein (Osp) A has been well characterized and shown to be required for tick midgut colonization (Neelakanta et al. 2007; Yang et al. 2004). OspA has been proposed to recognize and bind the *I. scapularis* midgut protein TROSPA, allowing the spirochete to attach to the midgut epithelial cells and preventing discharge with the tick waste products (Pal et al. 2004a). However, OspA also provides a protective role within the tick midgut by shielding *B. burgdorferi* antigens from host antibodies acquired when the tick feeds on an immune animal (Battisti et al. 2008; Tilly et al. 2016). The function and necessity of OspB, a protein related in sequence and structure to OspA (Becker et al. 2005; Bergström et al. 1989), is less clear. Although in an operon with *ospA* (Howe et al. 1986), conflicting data on the requirement for OspB during tick infection have been reported (Neelakanta et al. 2007; Tilly et al. 2016). A third plasmid-encoded gene, *bbe16*, was reported necessary for *B. burgdorferi* persistence in ticks (Revel et al. 2005), though again, no function or mechanism was proposed. In addition to these plasmid-encoded proteins, three *B. burgdorferi* chromosomal gene products (BB0243, BB0690, and BB0365) were reported to be important for persistence within the tick. Loss of BB0243, a GlpD homolog, results in lowered spirochete burden in the midgut of the tick. Presumably, glycerol is an alternative carbohydrate source within the tick that can be exploited by *B. burgdorferi* to generate energy in the form of ATP during the nutrient-depleted stages during and after the molt (Pappas et al. 2011). BB0690 is a Dps homolog (DNA-binding protein from starved bacteria) proposed to protect *B. burgdorferi* from oxidative damage during the long dormant stages of the unfed tick (Li et al. 2007b). Inactivation of the gene encoding BB0365, a protein with unknown function, decreased the spirochete burden throughout the tick portion of the life cycle (Pal et al. 2008; Yang et al. 2013).

After the establishment of a chronic infection of the midgut, successful transmission of *B. burgdorferi* to a new host requires that the spirochetes dissociate from the midgut, traverse the hemocoel, and penetrate the salivary glands. The

Table 1 *Borrelia burgdorferi* proteins with potential roles in the tick vector

Locus Designation[a]	Phenotype	Cellular location	Genome location	Reference(s)
BBA03	Facilitates transmission from tick vector	Outer membrane	lp54	Bestor et al. (2012)
BBA07	Transmission from tick vector	Outer membrane	lp54	Xu et al. (2010)
BBA15 (OspA)	Antibody shielding during tick feeding; attachment to the tick midgut	Outer membrane	lp54	Battisti et al. (2008); Yang et al. (2004)
BBA16 (OspB)	Attachment to the tick midgut	Outer membrane	lp54	Neelakanta et al. (2007)
BBA52	Facilitates tick acquisition and transmission	Predicted outer membrane	lp54	Kumar et al. (2010)
BBA62 (lp6.6)	Facilitates transmission from tick vector	Outer membrane (Lahdenne et al. 1997)	lp54	Promnares et al. (2009)
BBA64 (P35)	Efficient transmission from tick vector	Outer membrane (Brooks et al. 2006)	lp54	Gilmore et al. (2010)
BBA66	Facilitates transmission from tick vector	Outer membrane (Brooks et al. 2006)	lp54	Patton et al. (2013)
BBE16 (BptA)	Persistence in tick	Outer membrane	lp25	Revel et al. (2005)
BBE31	Enables migration from midgut to hemolymph	Outer membrane	lp25	Zhang et al. (2011)
BB0243 (GlpD)	Persistence in tick	Cytoplasmic	chromosome	Pappas et al. (2011)
BB0365 (La7/P22)	Persistence in tick	Primarily inner membrane (von Lackum et al. 2007)	chromosome	Pal et al. (2008); Yang et al. (2013)
BB0690 (Dps/NapA)	Persistence in unfed tick	Cytoplasmic	chromosome	Li et al. (2007b)

[a]Locus designations follow the nomenclature established for the *B. burgdorferi* type strain B31 (Casjens et al. 2000; Fraser et al. 1997). Regulatory proteins are discussed in the text

mechanism of escape from the midgut appears to be triggered by the incoming blood, but the exact cues that are detected have not been identified. Several theories have been proposed, though, for how *B. burgdorferi* escapes the midgut. The plasmid-encoded BBE31 spirochete protein may aid in migration by binding the midgut receptor protein TRE31 and facilitating the escape from the gut, though no mechanism was demonstrated (Zhang et al. 2011). Fluorescent microscopic studies of dissected midguts infected with *B. burgdorferi* suggested that spirochete motility is arrested, but the nonmotile bacteria are associated with tick epithelial cells that migrate toward the basement membrane of the midgut and away from the blood, then transition to a motile state and breach the midgut barrier (Dunham-Ems et al. 2009). This highly interesting observational theory stands in contrast to the long-held view that motility would be required throughout the migration process (Charon and Goldstein 2002). Additional studies are required to clarify the processes employed by *B. burgdorferi* to escape the midgut confines.

The final stage in the *B. burgdorferi–Ixodes* cycle is transmission to the vertebrate host. Multiple gene products have been implicated in the successful transmission of *B. burgdorferi* (see Table 1) and a variety of interesting techniques was employed in these studies, but none of the five proteins have known functions or proposed mechanisms for how they might aid in transmission.

Several proteins initially thought to be important for *B. burgdorferi* survival in the tick vector either have roles in other portions of the infectious cycle or their functions remain enigmatic. Of these, OspC is the most well characterized and perhaps the most heavily studied protein produced by *B. burgdorferi*. OspC was initially described as "pC" by Wilske and colleagues in reference to a major and immunodominant 22 kDa protein made by many European *B. burgdorferi* isolates in culture (Wilske et al. 1986). Following the nomenclature of OspA and OspB, pC was renamed "OspC" after its surface localization was demonstrated by immunoelectron microscopy (Wilske et al. 1993). The *ospC* gene was also present in *B. burgdorferi* strains that did not typically make OspC in culture, but from which OspC-producing variants could be derived, consistent with a mechanism for repression of *ospC* expression in wild-type organisms (Sadziene et al. 1993; Wilske et al. 1993). In general, reciprocal synthesis of OspA/B versus OspC was observed in different strains or between spontaneously switching variants of a particular strain. Unlike OspA/B, OspC represented an important (but polymorphic) diagnostic antigen of the early immune response of Lyme borreliosis patients in both Europe and North America (Fung et al. 1994; Padula et al. 1993; Wilske et al. 1986, 1988), indicating that OspC was made during *B. burgdorferi* infection of the mammalian host.

The biological basis for the reciprocal synthesis of OspA/B and OspC in vitro, along with their disparate immunogenic natures during natural infection, was elucidated by the seminal observation of Schwan and colleagues in 1995 (Schwan et al. 1995). Using immunofluorescence assays and specific antibodies to characterize the surface proteins of *B. burgdorferi* in vivo, Schwan et al. demonstrated that spirochetes in the midguts of unfed ticks stained only for OspA/B. However, OspA/B disappeared and OspC was detected on a proportion of spirochetes in the midgut after ticks had attached to a murine host and ingested blood (Schwan et al. 1995). This pattern is reversed after ticks were fully engorged and had detached from the host, when OspA/B replaced OspC and again predominated on the surface of the spirochete (Schwan and Piesman 2000).

The first indication that OspC played a role at the initial stage of infection of the mammalian host was provided in a careful study by Ohnishi and colleagues (Ohnishi et al. 2001), in which they analyzed the Osp phenotype of the population of spirochetes transmitted from an infected tick to a murine host. These investigators quantitatively followed the production of OspA and OspC by spirochetes in the tick midgut and salivary glands, and deposited in the dermis of the host throughout tick feeding. They also assessed the outcome of infection in the mice on which ticks had fed for various lengths of time. These researchers confirmed the earlier observation that spirochetes in the midgut of an infected tick begin to produce OspC during tick feeding. However, they also noted that this change in surface phenotype was not required by the spirochete to be deposited in the skin by the feeding tick, counter to an existing model, in which the switch from OspA to OspC preceded their exit from the midgut and dissemination to the salivary glands (Pal et al. 2000). Ohnishi and colleagues found that spirochetes making OspA were deposited in the mouse dermis at early time points, but mouse infection ensued only after OspC-producing spirochetes were transmitted (Ohnishi et al. 2001).

Conflicting reports about the role of OspC in the tick vector followed in 2004. Pal and colleagues reported that genetically modified spirochetes lacking OspC multiplied in the tick midgut, but were unable to disseminate to the salivary glands, thereby preventing transmission (Pal et al. 2004b). In contrast, Grimm and colleagues noted no defect in tick colonization or salivary gland migration by *ospC* mutant spirochetes, but found that these organisms were unable to infect mice by either tick transmission or needle inoculation, a defect that was fully reversed when a wild-type copy of the *ospC* gene was reintroduced at the endogenous plasmid locus (Grimm et al. 2004). The basis for these discrepant findings was not resolved, but subsequent investigations (Dunham-Ems et al. 2012; Fisher et al. 2005) found that although changes to *B. burgdorferi* induced during tick feeding were indeed necessary for spirochete migration from the midgut to the salivary glands; this process did not require OspC, supporting the previous findings of Grimm et al. and Ohnishi et al.

The absolute requirement for OspC production by *B. burgdorferi* at the initial stage of host infection has now been well documented, but the essential function provided by OspC is unclear (Liang et al. 2004; Stewart et al. 2006; Tilly et al. 2006, 2007, 2013). Various components of the innate and acquired immune response have been interrogated, but a clear picture of the mechanism by which OspC permits spirochete infection has not emerged. It is assumed that OspC provides protection from some aspect of the innate immune response, as spirochetes lacking OspC are cleared shortly after entering a naïve host, irrespective of the route of inoculation (Grimm et al. 2004). A recent report by Carrasco and colleagues found that *ospC* mutant *B. burgdorferi* remained non-infectious in mice lacking B cells, T cells, NK cells, and lytic complement (Carrasco et al. 2015). However, these investigators reported that depletion of macrophages in the skin at the site of inoculation permitted survival and disseminated infection by *ospC* mutant spirochetes, suggesting an anti-phagocytic function for OspC (Carrasco et al. 2015). Whatever critical role this protein plays at the initial stage of infection, it is clear that the requirement for OspC is transient. Consistent with its original description as a sero-diagnostic marker of early Lyme disease, OspC is recognized and targeted by neutralizing antibodies of the host acquired immune response, and the *ospC* gene must be down-regulated and can be lost, during persistent infection (Liang et al. 2004; Tilly et al. 2006, 2007).

Several other proteins were presumed to be important for spirochete survival in the tick vector but were subsequently shown to be dispensable for all phases of the infectious cycle. The gene encoding OspD, a major outer surface protein, was the most highly expressed ORF at 23 °C in a microarray study using a condition mimicking the unfed tick (Ojaimi et al. 2003). However, inactivation of *ospD* did not significantly affect *B. burgdorferi* at any stage of the infectious cycle (Li et al. 2007a; Stewart et al. 2008). Likewise, ChbC, a chitobiose transporter subunit, was predicted to be critical to *B. burgdorferi* persistence within the tick. Chitobiose is a tick-produced carbohydrate that can be metabolized by *B. burgdorferi* as an energy source, but a *chbC* mutant survived through all phases of the infectious cycle (Tilly et al. 2004). BBA74 is produced during larval and nymphal blood meals but, similar to the other proteins described, appears to be dispensable to *B. burgdorferi* during the mouse–tick infectious cycle (Mulay et al. 2009). BB0249, a homolog of the *E. coli* hibernation-promoting factor (HPF) that dimerizes and inactivates ribosomes, was predicted to be important for *B. burgdorferi* survival during the long periods between blood meals in the tick (Fazzino et al. 2015). However, this protein, too, was shown to be unnecessary during the infectious cycle. Together, these findings suggest that either the laboratory model of the infectious cycle does not accurately depict all aspects of the natural enzootic cycle of *B. burgdorferi* or that other spirochete proteins can compensate for the loss of these functions.

5 *B. burgdorferi* Regulatory Systems Important for Survival within *I. scapularis*

For *B. burgdorferi* to successfully persist and maneuver through the phases of the tick life cycle, the spirochete must sense particular environmental signals and regulate gene expression appropriately. This implies that *B. burgdorferi* has mechanisms for detecting specific cues (such as temperature, pH, specific nutrients, etc.) and conducting that signal to produce the desired response. These signaling cascades have begun to be deciphered in *B. burgdorferi* and here we will provide an overview of tick-specific gene regulation, but we refer the reader to the chapter by Stevenson and Seshu in this volume for greater detail.

Rel_{Bbu} controls production of the nucleotide alarmones collectively known as (p)ppGpp and these signaling molecules modulate gene expression under nutrient-limited conditions in vitro and within the tick (Drecktrah et al. 2015).

Inactivation of the gene encoding Rel_{Bbu} reduced *B. burgdorferi* survival compared to wild-type under nutrient-limited conditions in vitro and after nymphal feeding, suggesting a role in regulating gene expression in response to nutrient stress. Transcriptome changes under these in vitro conditions identified a large number of genes controlled by Rel_{Bbu}, some involved in the transition from the fed tick to the host, such as OspC, and several involved in binding components of the host extracellular matrix.

Another key regulatory cascade is initiated by the two-component histidine kinase and response regulatory proteins Hk1 and Rrp1, respectively, during tick feeding. This sensory–regulatory complex directs the synthesis of cyclic diguanylate to control gene expression (Caimano et al. 2015; He et al. 2011; Rogers et al. 2009). Although various studies reported different results, perhaps because of the differences in strains, culture conditions, and techniques used, glycerol utilization genes were uniformly upregulated by the Hk1/Rrp1 pathway. This two-component system appears to be active during both the larval and nymphal feedings, and mutants lacking either Hk1 or Rrp1 were killed during the blood meal (Caimano et al. 2015; He et al. 2011). How the Hk1/Rrp1 regulatory cascade interfaces with the Rel_{Bbu} system is still being deciphered.

The characterization of a previously designated hypothetical protein, BBD18, identified it as a novel regulator of *ospC* expression (Sarkar et al. 2011). Further investigation indicated that BBD18 acts in a global regulatory capacity and this function is critical for *B. burgdorferi* acquisition by feeding ticks (Hayes et al. 2014). The exact mechanism of control remains unknown, but BBD18 appears to post-transcriptionally down-regulate the alternative sigma factor RpoS, a crucial event necessary in the transition from the host to the tick vector (Dulebohn et al. 2014).

Small RNAs (sRNAs) are known to regulate gene expression in both eukaryotes and prokaryotes, but have only begun to be studied in *B. burgdorferi*. Recent studies have identified sRNAs from in vitro cultivated *B. burgdorferi* under conditions that represent the unfed tick temperature, 23 °C, or from different growth phases (Arnold et al. 2016; Popitsch et al. 2017). Notably, the genes encoding the Hk1/Rrp1 two-component system have several sRNAs associated with them, as do OspB and GlpF. The function of these and other sRNAs and their role in the *B. burgdorferi* infectious cycle have not been characterized yet, but clearly represent a new area of research in spirochete gene regulation.

6 Laboratory Techniques for Studying *B. burgdorferi*: Tick Interactions

Techniques in the manipulation of the *I. scapularis* vector as a model organism have been slowly advancing. The first such technique was an artificial feeding method whereby *B. burgdorferi* was acquired by *I. scapularis* via glass capillary tubes (Hu et al. 1992). Subsequently, *B. burgdorferi* was introduced internally into the tick by injection (Johns et al. 2001), and later modified (Pal et al. 2004b). These approaches were used successfully in multiple studies but, while functional, were labor-intensive and limited to the study of the nymphal and adult stages of the tick. The microinjection technique also required specialized instruments and a certain level of technical proficiency to avoid killing the tick. A significant improvement to artificial infection came with the development of infection by immersion, where ticks were submerged in medium containing the desired spirochete strain (Policastro and Schwan 2003). This method did not require any special instrumentation or technical expertise, could be applied to any developmental stage of the tick, and did not physically damage the arthropod. Because the microinjection technique was labor-intensive, it usually resulted in relatively few infected ticks produced, but immersion is a simple procedure that can process hundreds of ticks in a short duration of time. More recently, artificial membrane feeding chambers have been successfully used to infect *I. scapularis* (Oliver et al. 2016). This method requires special instruments and can be technically demanding to learn. However, spirochetes are acquired via natural ingestion and without physical damage to the tick, and the numbers of spirochetes transmitted during subsequent feedings can be quantitatively assessed.

Advances in RNA interference (RNAi) technology have made it feasible to manipulate gene expression in the tick to study protein–protein interactions that occur between the bacterium and the vector. RNAi constructs can target and down-regulate specific *I. scapularis* genes and allowed researchers to study the interactions of OspA with TROSPA, and other tick proteins (Narasimhan et al. 2007; Pal et al. 2004a; Ramamoorthi et al. 2005; Zhang et al. 2011).

The overwhelming success of CRISPR/Cas9-mediated genetic studies in other eukaryotic systems suggests that this may be another method to genetically manipulate the tick chromosomes. Further aiding in these techniques has been the release of the *I. scapularis* genome, providing the raw sequence data to identify genetic targets (Gulia-Nuss et al. 2016). Together, these technologies now make it possible to manipulate the vector and assess how *B. burgdorferi* responds to the altered conditions of the tick.

7 Conclusions and Future Directions

B. burgdorferi was discovered in 1981 (Burgdorfer et al. 1982) and has been intensively studied because of its impact on human and veterinary health. In that time, great strides have been made in elucidating its enzootic life cycle, its ability to invade animal hosts and the immune responses elicited, and in the ability to genetically manipulate the spirochete. However, the interactions between *B. burgdorferi* and its vector, *I. scapularis*, have lagged behind the studies that have focused on spirochete–mammalian interactions. This is not surprising given the human illness caused by infection with *B. burgdorferi*, but studies on the spirochete–vector association have much to offer to both human health (the only human vaccine marketed for Lyme disease was based on OspA, an antigen produced in the tick) and to our broader understanding of vector-borne diseases. Research has identified spirochete genes that facilitate the successful infection and persistence of *B. burgdorferi* in the tick. Unfortunately, only a few of these gene products have been studied beyond the initial description, and fewer still have a known function. Besides identifying spirochetal genes involved in the tick portion of the infectious cycle, more follow-up studies are also needed to define the mechanisms by which these *B. burgdorferi* proteins enable acquisition of the bacterium by the tick, long-term survival in the nutrient-depleted environment of the unfed midgut, and the environmental signals that dictate replication and migration out of the tick. With the development of RNAi techniques for manipulating *I. scapularis* gene expression and with evermore sensitive microscopy-based techniques and transcriptomic and proteomic technologies, these questions can now be addressed in vivo rather than in artificial laboratory conditions.

Acknowledgements We are grateful to Tom Schwan for insights and discussions of tick biology and critical reading of the manuscript, and to Austin Athman for graphical expertise. The authors were supported by the Intramural Research Program of the NIAID, NIH.

References

Anderson JF, Magnarelli LA (1984) Avian and mammalian hosts for spirochete-infected ticks and insects in a Lyme disease focus in Connecticut. Yale J Biol Med 57:627–641

Anderson JF, Johnson RC, Magnarelli LA, Hyde FW (1986) Involvement of birds in the epidemiology of the Lyme disease agent *Borrelia burgdorferi*. Infect Immun 51:394–396

Arnold WK, Savage CR, Brissette CA, Seshu J, Livny J, Stevenson B (2016) RNA-Seq of *Borrelia burgdorferi* in multiple phases of growth reveals insights into the dynamics of gene expression, transcriptome architecture, and noncoding RNAs. PLoS ONE 11:e0164165

Battisti JM, Bono JL, Rosa PA, Schrumpf ME, Schwan TG, Policastro PF (2008) Outer surface protein A protects Lyme disease spirochetes from acquired host immunity in the tick vector. Infect Immun 76:5228–5237

Becker M, Bunikis J, Lade BD, Dunn JJ, Barbour AG, Lawson CL (2005) Structural investigation of *Borrelia burgdorferi* OspB, a bactericidal Fab target. J Biol Chem 280:17363–17370

Bergström S, Bundoc VG, Barbour AG (1989) Molecular analysis of linear plasmid-encoded major surface proteins, OspA and OspB, of the Lyme disease spirochete *Borrelia burgdorferi*. Mol Microbiol 3:479–486

Bestor A, Rego RO, Tilly K, Rosa PA (2012) Competitive advantage of *Borrelia burgdorferi* with outer surface protein BBA03 during tick-mediated infection of the mammalian host. Infect Immun 80:3501–3511

Bontemps-Gallo S, Lawrence K, Gherardini FC (2016) Two different virulence-related regulatory pathways in *Borrelia burgdorferi* are directly affected by osmotic fluxes in the blood meal of feeding *Ixodes* ticks. PLoS Pathog 12:e1005791

Bourret TJ, Lawrence KA, Shaw JA, Lin T, Norris SJ, Gherardini FC (2016) The nucleotide excision repair pathway protects *Borrelia burgdorferi* from nitrosative stress in *Ixodes scapularis* ticks. Front Microbiol 7:1397

Boyle WK, Wilder HK, Lawrence AM, Lopez JE (2014) Transmission dynamics of *Borrelia turicatae* from the arthropod vector. PLoS Negl Trop Dis 8:e2767

Brooks CS, Vuppala SR, Jett AM, Akins DR (2006) Identification of *Borrelia burgdorferi* outer surface proteins. Infect Immun 74:296–304

Brown RN, Lane RS (1992) Lyme disease in California: a novel enzootic transmission cycle of *Borrelia burgdorferi*. Science 256:1439–1442

Brown RN, Lane RS, Dennis DT (2005) Geographic distributions of tick-borne diseases and their vectors. In: Goodman JL, Dennis DT, Sonenshine DE (eds) Tick-borne diseases of humans. Washington, D.C. ASM Press. pp 363–392

Bryceson AD, Parry EH, Perine PL, Warrell DA, Vukotich D, Leithead CS (1970) Louse-borne relapsing fever. Q J Med 39:129–170

Burgdorfer W, Varma MGR (1967) Trans-stadial and transovarial development of disease agents in arthropods. Annu Rev Entomol 12:347–376

Burgdorfer W, Barbour AG, Hayes SF, Benach JL, Grunwaldt E, Davis JP (1982) Lyme disease—a tick-borne spirochetosis? Science 216:1317–1319

Caimano MJ, Dunham-Ems S, Allard AM, Cassera MB, Kenedy M, Radolf JD (2015) Cyclic di-GMP modulates gene expression in Lyme disease spirochetes at the tick-mammal interface to promote spirochete survival during the blood meal and tick-to-mammal transmission. Infect Immun 83:3043–3060

Carrasco SE, Troxell B, Yang Y, Brandt SL, Li H, Sandusky GE, Condon KW, Serezani CH, Yang XF (2015) Outer surface protein OspC is an anti-phagocytic factor that protects *Borrelia burgdorferi* from phagocytosis by macrophages. Infect Immun 83:4848–4860

Carroll JA, Garon CF, Schwan TG (1999) Effects of environmental pH on membrane proteins in *Borrelia burgdorferi*. Infect Immun 67:3181–3187

Carroll JA, Cordova RM, Garon CF (2000) Identification of 11 pH-regulated genes in *Borrelia burgdorferi* localizing to linear plasmids. Infect Immun 68:6677–6684

Casjens S, Palmer N, van Vugt R, Huang WM, Stevenson B, Rosa P, Lathigra R, Sutton G, Peterson J, Dodson RJ, Haft D, Hickey E, Gwinn M, White O, Fraser C (2000) A bacterial genome in flux: the twelve linear and nine circular extrachromosomal DNAs in an infectious isolate of the Lyme disease spirochete *Borrelia burgdorferi*. Mol Microbiol 35:490–516

Charon NW, Goldstein SF (2002) Genetics of motility and chemotaxis of a fascinating group of bacteria: the spirochetes. Annu Rev Genet 36:47–73

Clark K, Hendricks A, Burge D (2005) Molecular identification and analysis of *Borrelia burgdorferi* sensu lato in lizards in the southeastern United States. Appl Environ Microbiol 71:2616–2625

Davis GE (1942) Species unity or plurality of the relapsing fever spirochetes. In: Moulton FR (ed) Relapsing fever in the Americas. Washington, D.C. American Association of the Advancement of Science. pp 41–47

Donahue JG, Piesman J, Spielman A (1987) Reservoir competence of white-footed mice for Lyme disease spirochetes. Am J Trop Med Hyg 36:92–96

Drecktrah D, Lybecker M, Popitsch N, Rescheneder P, Hall LS, Samuels DS (2015) The *Borrelia burgdorferi* RelA/SpoT homolog and stringent response regulate survival in the tick vector and global gene expression during starvation. PLoS Pathog 11:e1005160

Dulebohn DP, Hayes BM, Rosa PA (2014) Global repression of host-associated genes of the Lyme disease spirochete through post-transcriptional modulation of the alternative sigma factor RpoS. PLoS ONE 9:e93141

Dunham-Ems SM, Caimano MJ, Pal U, Wolgemuth CW, Eggers CH, Balic A, Radolf JD (2009) Live imaging reveals a biphasic mode of dissemination of *Borrelia burgdorferi* within ticks. J Clin Invest 119:3652–3665

Dunham-Ems SM, Caimano MJ, Eggers CH, Radolf JD (2012) *Borrelia burgdorferi* requires the alternative sigma factor RpoS for dissemination within the vector during tick-to-mammal transmission. PLoS Pathog 8:e1002532

Dutton JE, Todd JL (1905) The nature of human tick-fever in the eastern part of the Congo Free State. Mem Liverpool Sch Trop Med 17:1–18

Fazzino L, Tilly K, Dulebohn DP, Rosa PA (2015) Long-term survival of *Borrelia burgdorferi* lacking the hibernation promotion factor homolog in the unfed tick vector. Infect Immun 83:4800–4810

Fisher MA, Grimm D, Henion AK, Elias AF, Stewart PE, Rosa PA, Gherardini FC (2005) *Borrelia burgdorferi* sigma54 is required for mammalian infection and vector transmission but not for tick colonization. Proc Natl Acad Sci U S A 102:5162–5167

Fraser CM, Casjens S, Huang WM, Sutton GG, Clayton R, Lathigra R, White O, Ketchum KA, Dodson R, Hickey EK, Gwinn M, Dougherty B, Tomb J-F, Fleischmann RD, Richardson D, Peterson J, Kerlavage AR, Quackenbush J, Salzberg S, Hanson M, van Vugt R, Palmer N, Adams MD, Gocayne J, Weidmann J, Utterback T, Watthey L, McDonald L, Artiach P, Bowman C, Garland S, Fujii C, Cotton MD, Horst K, Roberts K, Hatch B, Smith HO, Venter JC (1997) Genomic sequence of a Lyme disease spirochaete, *Borrelia burgdorferi*. Nature 390:580–586

Fung BP, McHugh GL, Leong JM, Steere AC (1994) Humoral immune response to outer surface protein C of *Borrelia burgdorferi* in Lyme disease: role of the immunoglobulin M response in the serodiagnosis of early infection. Infect Immun 62:3213–3221

Gilmore RD Jr, Howison RR, Dietrich G, Patton TG, Clifton DR, Carroll JA (2010) The *bba64* gene of *Borrelia burgdorferi*, the Lyme disease agent, is critical for mammalian infection via tick bite transmission. Proc Natl Acad Sci U S A 107:7515–7520

Grimm D, Tilly K, Byram R, Stewart PE, Krum JG, Bueschel DM, Schwan TG, Policastro PF, Elias AF, Rosa PA (2004) Outer-surface protein C of the Lyme disease spirochete: A protein induced in ticks for infection of mammals. Proc Natl Acad Sci U S A 101:3142–3147

Gulia-Nuss M, Nuss AB, Meyer JM, Sonenshine DE, Roe RM, Waterhouse RM, Sattelle DB, de la Fuente J, Ribeiro JM, Megy K, Thimmapuram J, Miller JR, Walenz BP, Koren S, Hostetler JB, Thiagarajan M, Joardar VS, Hannick LI, Bidwell S, Hammond MP, Young S, Zeng Q, Abrudan JL, Almeida FC, Ayllon N, Bhide K, Bissinger BW, Bonzon-Kulichenko E, Buckingham SD, Caffrey DR, Caimano MJ, Croset V, Driscoll T, Gilbert D, Gillespie JJ, Giraldo-Calderon GI, Grabowski JM, Jiang D, Khalil SM, Kim D, Kocan KM, Koci J, Kuhn RJ, Kurtti TJ, Lees K, Lang EG, Kennedy RC, Kwon H, Perera R, Qi Y, Radolf JD, Sakamoto JM, Sanchez-Gracia A, Severo MS, Silverman N, Simo L, Tojo M, Tornador C, Van Zee JP, Vazquez J, Vieira FG, Villar M, Wespiser AR, Yang Y, Zhu J, Arensburger P, Pietrantonio PV, Barker SC, Shao R, Zdobnov EM, Hauser F, Grimmelikhuijzen CJ, Park Y, Rozas J, Benton R, Pedra JH, Nelson DR, Unger MF, Tubio JM, Tu Z, Robertson HM, Shumway M, Sutton G, Wortman JR, Lawson D, Wikel SK, Nene VM, Fraser CM, Collins FH, Birren B, Nelson KE, Caler E, Hill CA (2016) Genomic insights into the *Ixodes scapularis* tick vector of Lyme disease. Nat Commun 7:10507

Hamsikova Z, Coipan C, Mahrikova L, Minichova L, Sprong H, Kazimirova M (2016) *Borrelia miyamotoi* and co-infection with *Borrelia afzelii* in *Ixodes ricinus* ticks and rodents from Slovakia. Microb Ecol 73:1000–1008

Hayes BM, Dulebohn DP, Sarkar A, Tilly K, Bestor A, Ambroggio X, Rosa PA (2014) Regulatory protein BBD18 of the lyme disease spirochete: essential role during tick acquisition? MBio 5: e01017–e01014

He M, Ouyang Z, Troxell B, Xu H, Piesman J, Norgard MV, Gomelsky M, Yang XF (2011) Cyclic di-GMP is essential for the survival of the Lyme disease spirochete in ticks. PLoS Pathog 7:e1002133

Herrmann C, Gern L, Voordouw MJ (2013) Species co-occurrence patterns among Lyme borreliosis pathogens in the tick vector *Ixodes ricinus*. Appl Environ Microbiol 79:7273–7280

Howe TR, LaQuier FW, Barbour AG (1986) Organization of genes encoding two outer membrane proteins of the Lyme disease agent *Borrelia burgdorferi* within a single transcriptional unit. Infect Immun 54:207–212

Hu CM, Gern L, Aeschlimann A (1992) Changes in the protein profile and antigenicity of different *Borrelia burgdorferi* strains after reintroduction to *Ixodes ricinus* ticks. Parasite Immunol 14:415–427

Hyde JA, Trzeciakowski JP, Skare JT (2007) *Borrelia burgdorferi* alters its gene expression and antigenic profile in response to CO_2 levels. J Bacteriol 189:437–445

Johns R, Ohnishi J, Broadwater A, Sonenshine DE, De Silva AM, Hynes WL (2001) Contrasts in tick innate immune responses to *Borrelia burgdorferi* challenge: immunotolerance in *Ixodes scapularis* versus immunocompetence in *Dermacentor variabilis* (Acari: Ixodidae). J Med Entomol 38:99–107

Kumar M, Yang X, Coleman AS, Pal U (2010) BBA52 facilitates *Borrelia burgdorferi* transmission from feeding ticks to murine hosts. J Infect Dis 201:1084–1095

Lahdenne P, Porcella SF, Hagman KE, Akins DR, Popova TG, Cox DL, Katona LI, Radolf JD, Norgard MV (1997) Molecular characterization of a 6.6-kilodalton *Borrelia burgdorferi* outer membrane-associated lipoprotein (lp6.6) which appears to be downregulated during mammalian infection. Infect Immun 65:412–421

Levin M, Levine JF, Yang S, Howard P, Apperson CS (1996) Reservoir competence of the southeastern five-lined skink (*Eumeces inexpectatus*) and the green anole (*Anolis carolinensis*) for *Borrelia burgdorferi*. Am J Trop Med Hyg 54:92–97

Levine JF, Wilson ML, Spielman A (1985) Mice as reservoirs of the Lyme disease spirochete. Am J Trop Med Hyg 34:355–360

Li X, Neelakanta G, Liu X, Beck DS, Kantor FS, Fish D, Anderson JF, Fikrig E (2007a) Role of outer surface protein D in the *Borrelia burgdorferi* life cycle. Infect Immun 75:4237–4244

Li X, Pal U, Ramamoorthi N, Liu X, Desrosiers DC, Eggers CH, Anderson JF, Radolf JD, Fikrig E (2007b) The Lyme disease agent *Borrelia burgdorferi* requires BB0690, a Dps homologue, to persist within ticks. Mol Microbiol 63:694–710

Liang FT, Yan J, Mbow ML, Sviat SL, Gilmore RD, Mamula M, Fikrig E (2004) *Borrelia burgdorferi* changes its surface antigenic expression in response to host immune responses. Infect Immun 72:5759–5767

Magnarelli LA, Anderson JF, Fish D (1987) Transovarial transmission of *Borrelia burgdorferi* in *Ixodes dammini* (Acari:Ixodidae). J Infect Dis 156:234–236

Mulay VB, Caimano MJ, Iyer R, Dunham-Ems S, Liveris D, Petzke MM, Schwartz I, Radolf JD (2009) *Borrelia burgdorferi bba74* is expressed exclusively during tick feeding and is regulated by both arthropod- and mammalian host-specific signals. J Bacteriol 191:2783–2794

Narasimhan S, Sukumaran B, Bozdogan J, Thomas V, Liang X, DePonte K, Marcantonio N, Koski RA, Anderson JF, Kantor FS, Fikrig E (2007) A tick antioxidant facilitates the Lyme disease agent's successful migration from the mammalian host to the arthropod vector. Cell Host Microbe 2:7–18

Narasimhan S, Rajeevan N, Liu L, Zhao YO, Heisig J, Pan J, Eppler-Epstein R, Deponte K, Fish D, Fikrig E (2014) Gut microbiota of the tick vector *Ixodes scapularis* modulate colonization of the Lyme disease spirochete. Cell Host Microbe 15:58–71

Neelakanta G, Li X, Pal U, Liu X, Beck DS, Deponte K, Fish D, Kantor FS, Fikrig E (2007) Outer surface protein B is critical for *Borrelia burgdorferi* adherence and survival within *Ixodes* ticks. PLoS Pathog 3:e33

Ohnishi J, Piesman J, de Silva AM (2001) Antigenic and genetic heterogeneity of *Borrelia burgdorferi* populations transmitted by ticks. Proc Natl Acad Sci U S A 98:670–675

Ojaimi C, Brooks C, Casjens S, Rosa P, Elias A, Barbour AG, Jasinskas A, Benach J, Katona L, Radolf J, Caimano M, Skare J, Swingle K, Akins D, Schwartz I (2003) Profiling temperature-induced changes in *Borrelia burgdorferi* gene expression using whole genome arrays. Infect Immun 71:1689–1705

Oliver JD, Lynn GE, Burkhardt NY, Price LD, Nelson CM, Kurtti TJ, Munderloh UG (2016) Infection of immature *Ixodes scapularis* (Acari: Ixodidae) by membrane feeding. J Med Entomol 53:409–415

Padula SJ, Sampieri A, Dias F, Szczepanski A, Ryan RW (1993) Molecular characterization and expression of p23 (OspC) from a North American strain of *Borrelia burgdorferi*. Infect Immun 61:5097–5105

Pal U, de Silva AM, Montgomery RR, Fish D, Anguita J, Anderson JF, Lobet Y, Fikrig E (2000) Attachment of *Borrelia burgdorferi* within *Ixodes scapularis* mediated by outer surface protein A. J Clin Invest 106:561–569

Pal U, Li X, Wang T, Montgomery RR, Ramamoorthi N, de Silva AM, Bao F, Yang X, Pypaert M, Pradhan D, Kantor FS, Telford S, Anderson JF, Fikrig E (2004a) TROSPA, an *Ixodes scapularis* receptor for *Borrelia burgdorferi*. Cell 119:457–468

Pal U, Yang X, Chen M, Bockenstedt LK, Anderson JF, Flavell RA, Norgard MV, Fikrig E (2004b) OspC facilitates *Borrelia burgdorferi* invasion of *Ixodes scapularis* salivary glands. J Clin Invest 113:220–230

Pal U, Dai J, Li X, Neelakanta G, Luo P, Kumar M, Wang P, Yang X, Anderson JF, Fikrig E (2008) A differential role for BB0365 in the persistence of *Borrelia burgdorferi* in mice and ticks. J Infect Dis 197:148–155

Pappas CJ, Iyer R, Petzke MM, Caimano MJ, Radolf JD, Schwartz I (2011) *Borrelia burgdorferi* requires glycerol for maximum fitness during the tick phase of the enzootic cycle. PLoS Pathog 7:e1002102

Patton TG, Brandt KS, Nolder C, Clifton DR, Carroll JA, Gilmore RD (2013) *Borrelia burgdorferi bba66* gene inactivation results in attenuated mouse infection by tick transmission. Infect Immun 81:2488–2498

Peters W (1992) Peritrophic membranes. In: Bradshaw SD, Burggren W, Heller HC, Ishii S, Langer H, Neuweiler G, Randall DJ (eds) Zoophysiology. Berlin, Springer. p 12

Piesman J, Oliver JR, Sinsky RJ (1990) Growth kinetics of the Lyme disease spirochete (*Borrelia burgdorferi*) in vector ticks (*Ixodes dammini*). Am J Trop Med Hyg 42:352–357

Pimenta PF, Saraiva EM, Rowton E, Modi GB, Garraway LA, Beverley SM, Turco SJ, Sacks DL (1994) Evidence that the vectorial competence of phlebotomine sand flies for different species of *Leishmania* is controlled by structural polymorphisms in the surface lipophosphoglycan. Proc Natl Acad Sci U S A 91:9155–9159

Policastro PF, Schwan TG (2003) Experimental infection of *Ixodes scapularis* larvae (Acari: Ixodidae) by immersion in low passage cultures of *Borrelia burgdorferi*. J Med Entomol 40:364–370

Popitsch N, Bilusic I, Rescheneder P, Schroeder R, Lybecker M (2017) Temperature-dependent sRNA transcriptome of the Lyme disease spirochete. BMC Genom 18:28

Pritt BS, Mead PS, Johnson DK, Neitzel DF, Respicio-Kingry LB, Davis JP, Schiffman E, Sloan LM, Schriefer ME, Replogle AJ, Paskewitz SM, Ray JA, Bjork J, Steward CR, Deedon A, Lee X, Kingry LC, Miller TK, Feist MA, Theel ES, Patel R, Irish CL, Petersen JM (2016) Identification of a novel pathogenic *Borrelia* species causing Lyme borreliosis with unusually high spirochaetaemia: a descriptive study. Lancet Infect Dis 16:556–564

Promnares K, Kumar M, Shroder DY, Zhang X, Anderson JF, Pal U (2009) *Borrelia burgdorferi* small lipoprotein Lp6.6 is a member of multiple protein complexes in the outer membrane and facilitates pathogen transmission from ticks to mice. Mol Microbiol 74:112–125

Raffel SJ, Battisti JM, Fischer RJ, Schwan TG (2014) Inactivation of genes for antigenic variation in the relapsing fever spirochete *Borrelia hermsii* reduces infectivity in mice and transmission by ticks. PLoS Pathog 10:e1004056

Ramamoorthi N, Narasimhan S, Pal U, Bao F, Yang XF, Fish D, Anguita J, Norgard MV, Kantor FS, Anderson JF, Koski RA, Fikrig E (2005) The Lyme disease agent exploits a tick protein to infect the mammalian host. Nature 436:573–577

Rego RO, Bestor A, Stefka J, Rosa PA (2014) Population bottlenecks during the infectious cycle of the Lyme disease spirochete *Borrelia burgdorferi*. PLoS ONE 9:e101009

Revel AT, Blevins JS, Almazan C, Neil L, Kocan KM, de la Fuente J, Hagman KE, Norgard MV (2005) *bptA* (*bbe16*) is essential for the persistence of the Lyme disease spirochete, *Borrelia burgdorferi*, in its natural tick vector. Proc Natl Acad Sci U S A 102:6972–6977

Rogers EA, Terekhova D, Zhang HM, Hovis KM, Schwartz I, Marconi RT (2009) Rrp1, a cyclic-di-GMP-producing response regulator, is an important regulator of *Borrelia burgdorferi* core cellular functions. Mol Microbiol 71:1551–1573

Rollend L, Fish D, Childs JE (2013) Transovarial transmission of *Borrelia* spirochetes by *Ixodes scapularis*: a summary of the literature and recent observations. Ticks Tick Borne Dis 4:46–51
Rudzinska MA, Spielman A, Lewengrub S, Piesman J, Karakashian S (1982) Penetration of the peritrophic membrane of the tick by *Babesia microti*. Cell Tissue Res 221:471–481
Sacks DL, Pimenta PF, McConville MJ, Schneider P, Turco SJ (1995) Stage-specific binding of *Leishmania donovani* to the sand fly vector midgut is regulated by conformational changes in the abundant surface lipophosphoglycan. J Exp Med 181:685–697
Sadziene A, Wilske B, Ferdows MS, Barbour AG (1993) The cryptic *ospC* gene of *Borrelia burgdorferi* B31 is located on a circular plasmid. Infect Immun 61:2192–2195
Sarkar A, Hayes BM, Dulebohn DP, Rosa PA (2011) Regulation of the virulence determinant OspC by *bbd18* on linear plasmid lp17 of *Borrelia burgdorferi*. J Bacteriol 193:5365–5373
Schwan TG, Piesman J, Golde WT, Dolan MC, Rosa PA (1995) Induction of an outer surface protein on *Borrelia burgdorferi* during tick feeding. Proc Natl Acad Sci U S A 92:2909–2913
Schwan TG, Hinnebusch BJ (1998) Bloodstream- versus tick-associated variants of a relapsing fever bacterium. Science 280:1938–1940
Schwan TG, Piesman J (2000) Temporal changes in outer surface proteins A and C of the Lyme disease-associated spirochete, *Borrelia burgdorferi*, during the chain of infection in ticks and mice. J Clin Microbiol 39:382–388
Seshu J, Boylan JA, Gherardini FC, Skare JT (2004) Dissolved oxygen levels alter gene expression and antigen profiles in *Borrelia burgdorferi*. Infect Immun 72:1580–1586
Shih CM, Chao LL, Yu CP (2002) Chemotactic migration of the Lyme disease spirochete (*Borrelia burgdorferi*) to salivary gland extracts of vector ticks. Am J Trop Med Hyg 66:616–621
Stevenson B, Schwan TG, Rosa PA (1995) Temperature-related differential expression of antigens in the Lyme disease spirochete, *Borrelia burgdorferi*. Infect Immun 63:4535–4539
Stewart PE, Wang X, Bueschel DM, Clifton DR, Grimm D, Tilly K, Carroll JA, Weis JJ, Rosa PA (2006) Delineating the requirement for the *Borrelia burgdorferi* virulence factor OspC in the mammalian host. Infect Immun 74:3547–3553
Stewart PE, Bestor A, Cullen JN, Rosa PA (2008) Tightly regulated surface protein of *Borrelia burgdorferi* is not essential to the mouse-tick infectious cycle. Infect Immun 76:1970–1978
Sytykiewicz H, Karbowiak G, Chorostowska-Wynimko J, Szpechcinski A, Supergan-Marwicz M, Horbowicz M, Szwed M, Czerniewicz P, Sprawka I (2015) Coexistence of *Borrelia burgdorferi s.l.* genospecies within *Ixodes ricinus* ticks from central and eastern Poland. Acta Parasitol 60:654–661
Takano A, Goka K, Une Y, Shimada Y, Fujita H, Shiino T, Watanabe H, Kawabata H (2010) Isolation and characterization of a novel *Borrelia* group of tick-borne borreliae from imported reptiles and their associated ticks. Environ Microbiol 12:134–146
Telford SR 3rd, Mather TN, Moore SI, Wilson ML, Spielman A (1988) Incompetence of deer as reservoirs of the Lyme disease spirochete. Am J Trop Med Hyg 39:105–109
Tilly K, Grimm D, Bueschel DM, Krum JG, Rosa P (2004) Infectious cycle analysis of a *Borrelia burgdorferi* mutant defective in transport of chitobiose, a tick cuticle component. Vector Borne Zoonotic Dis 4:159–168
Tilly K, Krum JG, Bestor A, Jewett MW, Grimm D, Bueschel D, Byram R, Dorward D, Stewart P, Rosa P (2006) *Borrelia burgdorferi* OspC protein required exclusively in a crucial early stage of mammalian infection. Infect Immun 74:3554–3564
Tilly K, Bestor A, Jewett MW, Rosa P (2007) Rapid clearance of Lyme disease spirochetes lacking OspC from skin. Infect Immun 75:1517–1519
Tilly K, Bestor A, Rosa PA (2013) Lipoprotein succession in *Borrelia burgdorferi*: similar but distinct roles for OspC and VlsE at different stages of mammalian infection. Mol Microbiol 89:216–227
Tilly K, Bestor A, Rosa PA (2016) Functional equivalence of OspA and OspB, but not OspC, in tick colonization by *Borrelia burgdorferi*. Infect Immun 84:1565–1573
Vandyk JK, Bartholomew DM, Rowley WA, Platt KB (1996) Survival of *Ixodes scapularis* (Acari: Ixodidae) exposed to cold. J Med Entomol 33:6–10

von Lackum K, Ollison KM, Bykowski T, Nowalk AJ, Hughes JL, Carroll JA, Zuckert WR, Stevenson B (2007) Regulated synthesis of the *Borrelia burgdorferi* inner-membrane lipoprotein IpLA7 (P22, P22-A) during the Lyme disease spirochaete's mammal-tick infectious cycle. Microbiology 153:1361–1371

Wilske B, Preac-Mursic V, Schierz G, Busch KV (1986) Immunochemical and immunological analysis of European *Borrelia burgdorferi* strains. Zentralbl Bakteriol Hyg A 263:92–102

Wilske B, Preac-Mursic V, Schierz G, Kuhbeck R, Barbour AG, Kramer M (1988) Antigenic variability of *Borrelia burgdorferi*. In: Benach JL, Bosler EM (eds) Lyme Disease and Related Disorders. New York, Academy of Sciences. pp 126–143

Wilske B, Preac-Mursic V, Jauris S, Hofmann A, Pradel I, Soutschek E, Schwab E, Will G, Wanner G (1993) Immunological and molecular polymorphisms of OspC, an immunodominant major outer surface protein of *Borrelia burgdorferi*. Infect Immun 61:2182–2191

Xu H, He M, He JJ, Yang XF (2010) Role of the surface lipoprotein BBA07 in the enzootic cycle of *Borrelia burgdorferi*. Infect Immun 78:2910–2918

Yang X, Hegde S, Shroder DY, Smith AA, Promnares K, Neelakanta G, Anderson JF, Fikrig E, Pal U (2013) The lipoprotein La7 contributes to *Borrelia burgdorferi* persistence in ticks and their transmission to naïve hosts. Microbes Infect 15:729–737

Yang XF, Pal U, Alani SM, Fikrig E, Norgard MV (2004) Essential role for OspA/B in the life cycle of the Lyme disease spirochete. J Exp Med 199:641–648

Zhang L, Zhang Y, Adusumilli S, Liu L, Narasimhan S, Dai J, Zhao YO, Fikrig E (2011) Molecular interactions that enable movement of the Lyme disease agent from the tick gut into the hemolymph. PLoS Pathog 7:e1002079

Zhu Z, Gern L, Aeschlimann A (1991) The peritrophic membrane of *Ixodes ricinus*. Parasitol Res 77:635–641

Zolnik CP, Prill RJ, Falco RC, Daniels TJ, Kolokotronis SO (2016) Microbiome changes through ontogeny of a tick pathogen vector. Mol Ecol 25:4963–4977

Regulation of Gene and Protein Expression in the Lyme Disease Spirochete

Brian Stevenson and Janakiram Seshu

Abstract The infectious cycle of *Borrelia burgdorferi* necessitates persistent infection of both vertebrates and ticks, and efficient means of transmission between those two very different types of hosts. The Lyme disease spirochete has evolved mechanisms to sense its location in the infectious cycle, and use that information to control production of the proteins and other factors required for each step. Numerous components of borrelial regulatory pathways have been characterized to date. Their effects are being pieced together, thereby providing glimpses into a complex web of cooperative and antagonistic interactions. In this chapter, we present a broad overview of *B. burgdorferi* gene and protein regulation during the natural infectious cycle, discussions of culture-based methods for elucidating regulatory mechanisms, and summaries of many of the known regulatory proteins and small molecules. We also highlight areas that are in need of substantially more research.

Contents

B. Stevenson (✉)
Department of Microbiology, Immunology, and Molecular Genetics,
MS 421 Chandler Medical Center, University of Kentucky College of Medicine,
40536-0298 Lexington, KY, USA
e-mail: brian.stevenson@uky.edu

J. Seshu (✉)
Department of Biology, University of Texas at San Antonio, 78249 San Antonio, TX, USA
e-mail: j.seshu@utsa.edu

Current Topics in Microbiology and Immunology (2018) 415:83–112
DOI 10.1007/82_2017_49

Published Online: 25 October 2017

1 Introduction

In order to survive in nature, *Borrelia burgdorferi* must colonize two very distinct types of hosts. Each poses a unique set of challenges to the bacterium. For examples, different sets of bacterial adhesins are required for colonizing each niche, and carbon source availability varies from glucose and other high energy sugars in the vertebrate to energy-poor glycerol in the unfed tick. Moreover, the Lyme disease bacterium must efficiently transmit between its vertebrate and tick hosts. The spirochete needs to sense when a vector tick is feeding, then cease production of tick-specific adhesins, migrate to the tick's salivary glands, and begin producing vertebrate-specific adhesins and other host-interactive factors. Even within the vertebrate host, precise regulation is necessary, requiring the downregulation of early antigens such as OspC and CspA. Deciphering the sensory and regulatory networks of *B. burgdorferi* is providing insights into infection and pathogenesis mechanisms, as well as identifying potential therapeutic targets that will broadly inhibit critical bacterial functions. In addition, studies of borrelial gene control have identified regulatory proteins that are widely conserved among Eubacteria but had not previously been functionally defined, such as EbfC and SpoVG, thereby expanding our understanding across the domain Fig. 1.

To overview briefly, *B. burgdorferi* within the midgut of an unfed tick produces surface proteins such as OspA and OspB, which facilitate adhesion to midgut tissues, and transporters/enzymes for metabolism of glycerol and chitobiose

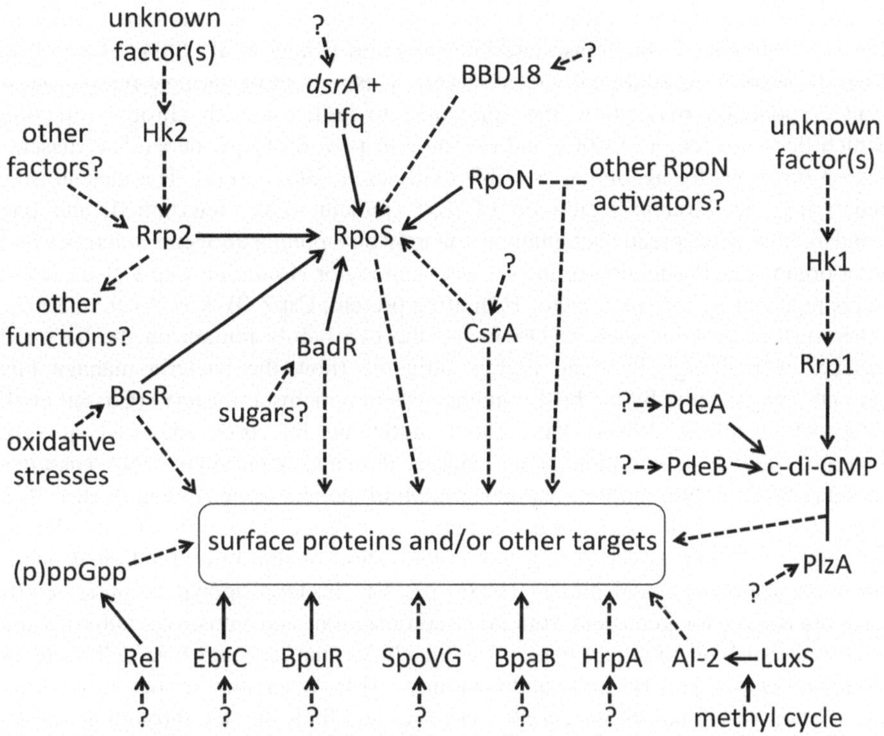

Fig. 1 A schematic overview of known and hypothesized interactions among *B. burgdorferi* regulatory factors. Continued studies are revealing a network of interactions that is far more complex than previously thought (e.g. Radolf et al. 2012). For simplicity, protein nomenclature is used in the figure, even though interactions may involve regulation of transcription, translation, or protein stability. Known direct interactions are indicated by *solid lines. Dashed lines* show interactions that have not been confirmed to be direct, occur through unknown mechanisms, or are hypothetical. In order for a factor to be a regulator, it must itself be regulated, either at the level of synthesis or through factor–factor interactions. For the vast majority of identified borrelial regulatory proteins, the mechanisms by which they are controlled have not been defined (indicated by *question marks*). Without a doubt, additional factors are also involved in coordination of *B. burgdorferi* protein expression patterns

(Neelakanta et al. 2007; Pal et al. 2004; Pappas et al. 2011; Tilly et al. 2004). When the tick attaches to a vertebrate host and begins ingesting blood, the spirochetes repress OspA and OspB while turning on production of surface proteins such as OspC, Erps, DbpB, VlsE, and CspA (Bykowski et al. 2006; Cassatt et al. 1998; Grimm et al. 2004; Iyer et al. 2015; Miller et al. 2003; Schwan et al. 1995; Schwan and Piesman 2000; von Lackum et al. 2005). Within the first few weeks of infection, OspC production is terminated (Tilly et al. 2006, 2013). It appears that only a subset of infecting *B. burgdorferi* downregulates OspC, and that spirochetes that do not repress OspC production are killed by host antibodies: infection of immunodeficient mice does not result in OspC repression, and transcriptional fusions between an *ospC* promoter and the luciferase gene showed continual synthesis of cytoplasmic luciferase late into infection (Liang et al. 2004; Skare et al. 2016; Stewart et al. 2006; Tilly et al. 2009). This system of variable regulation of OspC production may allow the spirochete to both establish chronic infection (which does not require OspC) and continue to perform OspC-dependent dissemination functions for as long as possible (Tilly et al. 2007, 2009). The mechanisms underlying the variable regulation of *ospC* remain to be determined, and one wonders how widespread such phenomena may be among pathogenic microbes and other organisms. Production of the complement factor H-binding CspA also ceases, and is replaced by a distinct factor H-binding protein, CspZ (Bykowski et al. 2007). Other surface proteins such as DbpB and the Erp family remain on the bacterial surface, even though they are highly antigenic (how the bacteria manage this without being cleared by the host immune system remains a mystery) (Cassatt et al. 1998; Miller et al. 2003). Throughout vertebrate infection, the VlsE protein undergoes antigenic variation of its external domains, with silent DNA cassettes continually recombining into a variable region of the *vlsE* gene (Zhang et al. 1997). Recombination in the *vlsE* locus is tightly controlled, and only occurs during vertebrate infection, never during tick colonization or during cultivation in laboratory media (Zhang and Norris 1998a, b). The bacteria appear to prefer cartilaginous tissues for persistent vertebrate colonization, and numerous adhesins are produced that bind to components of extracellular matrices and host cell surfaces (Antonara et al. 2011; Brissette and Gaultney 2014). When another tick feeds upon the infected vertebrate, *B. burgdorferi* emerges and finds the tick through unknown mechanisms. CspZ is repressed, and CspA is once again produced (Bykowski et al. 2007). Presumably, other vertebrate-to-tick transmission proteins are produced during that time, although this aspect of the borrelial life cycle has not been very well studied. After entering the tick midgut, production of CspA and other vertebrate-specific proteins is repressed, and tick-specific proteins such as OspA and OspB are turned on (Iyer et al. 2015; Schwan and Piesma 2000; von Lackum et al. 2005).

2 Models for Studying Borrelial Gene and Protein Regulation

A further complicating point is the fact that cultured *B. burgdorferi* does not accurately represent any aspect of its natural infectious cycle (Stevenson et al. 2006). OspA and OspB are the most highly expressed borrelial proteins during cultivation, indicating that tick colonization signals are activated. Yet some mammal-specific proteins, such as OspC and VlsE, may also be produced in culture, indicating activation of those regulatory networks. Nonetheless, differential expression of a number of genes and proteins has been observed by varying culture conditions, permitting insights into control mechanisms. Recently developed technologies are also providing direct analyses of bacterial expression profiles during actual infection processes. Following, we describe commonly applied culture methods, and overviews of identified players in borrelial gene and protein regulation.

2.1 Culture Temperature Change

This was the first model to be used, and is based upon previous studies of other vector-borne bacteria such as *Yersinia pestis*. The underlying concept is that a temperature near ambient represents a condition encountered in an unfed tick, a temperature near that of blood mimics conditions encountered in the feeding tick and within a vertebrate host, and a shift from ambient to blood temperature represents conditions as the tick initiates feeding. Commonly used temperatures are 23 °C as ambient, and 34–37 °C as vertebrate. The initial studies to use this model observed that OspC was poorly expressed at 23 °C, but greatly induced when bacteria were shifted to 35 °C (Schwan et al. 1995). A subsequent study demonstrated that numerous other antigenic proteins are similarly induced by temperature increase, including members of the Erp surface lipoprotein family (Stevenson et al. 1995).

This model provided evidence that several distinct pathways serve to control borrelial gene and protein expression. Transcription of *ospC*, and production of OspC protein, is greatly enhanced when a 23 °C culture is diluted into fresh medium then grown at 35 °C. However, subsequent passage of the 35 °C culture to fresh medium at the same temperature results in repression of *ospC*, until it is essentially undetectable after 3–4 passages (Schwan et al. 1995; Schwan and Piesman 2000). Thus, it is the change in culture temperature that induces *ospC* transcription. Reflecting on the expression pattern of OspC in nature, these data lead to a hypothesis that changing conditions as a tick begins to feed (e.g., a surfeit of nutrients and warmer temperature) trigger induction of *ospC*, whereas the relatively constant conditions inside the vertebrate host are interpreted as a signal to repress *ospC* (Schwan et al. 1995). In contrast, *erp* genes are repressed during cultivation at 23 °C, are induced by a shift to 35 °C, and remain high throughout repeated passages at 35 °C (Stevenson et al. 1995, 1998). This pattern resembles that of *erp* transcription during the tick-vertebrate infectious cycle: repressed in the tick, then produced during tick -feeding and throughout the entirety of vertebrate infection (Hefty et al. 2001; Miller et al. 2003). Other studies of OspC and Erp expression found that the effects of temperature were sensitive to culture medium composition, and further supported the conclusion that those two protein families are regulated through distinct pathways (Babb et al. 2001; Yang et al. 2001).

To date, there is no firm evidence for a borrelial mechanism(s) that directly senses temperature. One hypothesis suggests that RNA secondary structure could be altered by temperature changes, although that proposal has not been experimentally tested (Archambault et al. 2013; Lybecker and Samuels 2007). Another hypothesis suggested that *B. burgdorferi* DNAs may supercoil to different extents at different temperatures. Supporting that idea, changing culture temperature can have immediate impacts on circular replicon supercoiling (Alverson et al. 2003). The duration of those supercoiling effects has not been examined. It can be expected that *B. burgdorferi*, like other organisms, adapts to temperature-induced supercoiling changes by adjusting topoisomerase function to quickly restore optimal supercoil structures. Whether supercoiling is involved with the long-term, multigenerational effects of culture temperature on transcription of *ospC*, *erp*, or other genes has yet to be investigated. A third hypothesis proposes that effects of culture temperature may be indirect, with warmer temperatures allowing faster bacterial growth and replication, and that those metabolic changes are responsible for triggering some transcriptional changes (Jutras et al. 2013c). This hypothesis is discussed below in greater detail.

There are significant caveats to the culture temperature model, the most notable being that it is not predictive of expression patterns of individual genes and proteins during the tick-mammal cycle (Stevenson et al. 2006). For example, neither OspA nor OspB are affected by culture temperature changes, even though their expression is repressed during transmission from tick to vertebrate (Schwan et al. 1995). An even more significant observation is that VlsE is induced during transmission from tick to vertebrate, yet it is expressed at substantially greater levels during culture at 23 °C than it is at 35 °C (Bykowski et al. 2006). Thus, one cannot accurately predict a gene's expression pattern based solely upon the effects of culture temperature.

2.2 Growth Rate in Culture

Another caveat to the culture temperature change model is that *B. burgdorferi* grows and divides 3–4 times faster when cultured at 35 °C than it does at 23 °C (Jutras et al. 2013c; Stevenson et al. 1995). In order to separate temperature per se from growth rate, a series of studies was performed in which *B. burgdorferi* was first cultured in incomplete media, then transferred to complete medium, all at a constant temperature of 35 °C (Jutras et al. 2013c). Two different formulations of incomplete media were used, each of which yielded a growth rate at 35 °C that was three- to fourfold slower than was obtained in complete medium. In both cases, transfer of bacteria from incomplete to complete medium resulted in induction of OspC and Erp proteins. Moreover, the media changes also had the same effects as did changes in temperature on expression of the regulatory proteins EbfC, BpuR, and BpaB, and transcription of the alternate sigma *rpoS* (Jutras et al. 2013c).

Further indication that changes in growth rate, rather than temperature per se, induce changes in borrelial protein composition was provided by studies that examined *B. burgdorferi* transferred from −80 to 23 °C. Although this is not a biologically relevant condition change, it did critically examine the role of absolute temperature on borrelial gene and protein expression: if *B. burgdorferi* has a mechanism(s) to directly sense temperature, then such a mechanism would be equally responsive to 23 °C regardless of whether the bacteria had previously experienced 35 °C or −80 °C. The results were that a shift from −80 to 23 °C induced the same transcriptional and protein expression changes as does a shift from 23 to 35 °C (Jutras et al. 2013c).

A hypothesis to explain these phenomena states that a positive change in growth rate mimics that experienced in a tick midgut when it begins to ingest blood. *B. burgdorferi* within an unfed tick has a very slow metabolism, does not move, and probably does not grow to any substantial extent (Dunham-Ems et al. 2009). This condition may last for several months in nature. However, when the tick begins to feed, the incoming blood provides substantial nutrition to the bacteria, and they divide at rates of approximately one doubling per two hours (Burkot et al. 1994; de Silva and Fikrig 1995; Dunham-Ems et al. 2009; Piesman et al. 1990, 2001; Piesman and Schneider 2002; Stevenson et al. 2006). Consistent with this hypothesis, transcription of the genes encoding regulatory proteins EbfC and BpuR appears to be controlled by DnaA, the master regulator of bacterial DNA replication (Arnold et al., unpublished results).

As with other culture models, the growth rate change model is not necessarily predictive of gene and protein expression levels in the infectious cycle. VlsE is, again, a notable exception: shifting culture temperature from 23 to 35 °C represses VlsE production, while shifting from either formulation of incomplete medium into complete medium had no discernible effect on VlsE expression levels (Jutras et al., unpublished results).

2.3 Culture pH

The initial reports of this culture phenomenon suggested that *B. burgdorferi* may encounter varying pH conditions during its vertebrate-tick infectious cycle (Carroll et al. 1999, 2000). A subsequent study noted that tick midguts acidify upon feeding, and that acidification of culture medium results in enhanced production of OspC and other transmission-associated proteins (Ramamoorthy and Scholl-Meeker 2001; Yang et al. 2000). This led to a hypothesis that the spirochetes possess a pH-sensing mechanism(s) that is responsive to changes in the tick midgut. To date, such a specific mechanism has yet to be identified. An alternate possibility is that changes in culture pH affect the charge potential across the borrelial inner membrane: a decrease in pH is an increase in proton-motive force (PMF), whereas alkalization reduces the PMF. Decreasing pH will enhance ATP production by the PMF-dependent ATP synthase. Such effects of bacterial ATP concentrations may be interpreted as differences in nutritional status, reflective of conditions in the unfed and feeding tick midgut, and during vertebrate infection. The mechanism(s) underlying culture pH effects remains a field that is ripe for productive exploration.

Although alkaline culture pH is often referred to as an “unfed tick-like” condition, and acidic pH as “feeding tick-like,” this model is not predictive of in vivo conditions (Stevenson et al. 2006). VlsE is induced during tick feeding, yet is more highly expressed in alkaline culture conditions than it is during acidic conditions (Bykowski et al. 2006).

2.4 Culture Density

Observations that borrelial transcript and protein profiles change as cultures age led to a hypothesis that the spirochetes sense bacterial density (Indest et al. 1997; Ramamoorthy and Philipp 1998). That hypothesis was demonstrated to be incorrect, and that the observed phenomena are actually due to acidification of the medium as cultures age. *B. burgdorferi* cultured in medium that is heavily buffered to remain slightly alkaline does not change its protein profile as culture density increases, and low-density cultures in acidified medium express the same proteins as do dense cultures (Carroll et al. 2000).

2.5 Polyamines

Polyamines are organic bases that are present in mammalian tissues at millimolar levels. They are known to influence resistance to oxidative stress, transcription and translation among many physiological functions by interacting with nucleic acids, proteins, and phospholipids. While *B. burgdorferi* has a polyamine transport system, PotABCD, there are no apparent homologs for enzymes involved in the biosynthesis of polyamines. Regulation of the PotABCD system is complex: while PotA, the membrane-associated ATPase, is differentially expressed in response to changes in culture temperature and pH, those conditions had no effects on PotD, the periplasmic polyamine-binding protein. Propagation of *B. burgdorferi* in the presence of 4 mM spermidine or spermine increased the levels of several proteins, including RpoS, OspC and DbpA. Similar levels of other polyamines did not have significant effects on those key borrelial virulence proteins (Lin et al. 2017). It appears that *B. burgdorferi* has evolved systems to acquire host-derived substrates, which in turn have the potential to regulate gene expression consistent with the stages of its infectious cycle. The molecular mechanisms underlying the effects of spermine and spermidine on borrelial gene expression are currently under investigation.

2.6 Other Changes to Culture Conditions

Additional methods have been applied that may also result in changes to borrelial transcription. Changes in levels of oxygen or carbon dioxide, or redox potential, may affect DNA-binding by the BosR protein, which is described in greater detail below (Hyde et al. 2007; Seshu et al. 2004a). Culture osmolarity can impact protein expression through an unknown mechanism (Bontemps-Gallo et al. 2016). Addition of blood to culture medium has been reported to affect levels of numerous *B. burgdorferi* proteins (Tokarz et al. 2004), although, since BSK-II culture medium normally contains 6% serum, it is not clear how additional blood/serum has an effect. The main purpose for inclusion of serum in BSK-II is to provide fatty acids, so additional blood may boost levels of those essential nutrients (*B. burgdorferi* cannot produce fatty acids, so production of membranes and lipoproteins is completely dependent upon scavenging lipids from its hosts) (Barbour 1984; Fraser et al. 1997; Kelly 1971; Livermore et al. 1978).

2.7 Dialysis Membrane Chambers Implanted in Rat Peritoneal Cavities

Upon transmission into a vertebrate host, *B. burgdorferi* disseminates throughout the body, and colonizes various tissues at relatively low bacterial densities. At the present time, transcriptomic methods are not sensitive enough to measure global *B. burgdorferi* transcript levels during mammalian infection (although continued refinement of RNA-Seq methodologies may soon permit such analyses). Immunological methods such as immunoblot and ELISA can provide evidence of protein production during infection, with the caveats that less-antigenic proteins may not be detected and that only proteins produced during cultivation can be detected when probing cultured bacteria.

As a compromise, the dialysis membrane chamber (DMC) method involves confining a volume of cultured bacteria within a dialysis membrane bag, then implanting it inside the peritoneum of a rat. Small, soluble components of the rat's fluids enter the dialysis bag, and interact with the bacteria. Several days later, the bag is removed, and proteomic and/or transcriptomic analyses can be performed. This is the only method by which one can reliably observe repression of the tick-specific OspA and OspB proteins, along with induction of vertebrate-specific proteins (Akins et al. 1998; Caimano 2005, 2007; Hefty et al. 2001).

There are caveats to the DMC method, however (Caimano 2005). The volume of bacteria that can be analyzed per experiment is limited by the size of the rat peritoneal cavity (one can generally examine about 8 ml), which restricts the number of analyses that can be performed on each batch of bacteria. In addition, *B. burgdorferi* cannot survive in an implanted DMC for more than a couple of weeks. A consequence of this is that bacterial stress responses as they approach death may obscure true infection-associated responses.

3 Some Known *B. Burgdorferi* Regulatory Proteins and Networks

Numerous regulatory proteins and enzymatically produced small molecules have been identified in the Lyme disease spirochete. Following, we present an overview of several such regulatory factors and known networks that connect some of them. These include at least three RNA-binding proteins that act directly on mRNAs to post-transcriptionally regulate protein production. The study of borrelial gene and protein regulation is still in its infancy, and much work remains to be done. In addition to the following list, we note that the *B. burgdorferi* genome encodes several as-yet uncharacterized proteins that are predicted to contain DNA-binding domains (Fraser et al. 1997). *B. burgdorferi* also transcribes numerous, abundant noncoding RNAs (ncRNAs) (Arnold et al. 2016; Popitsch et al. 2017). Many of these are antisense to mRNAs, and could impact protein production. The current picture is, without a doubt, incomplete and overly simplified.

3.1 The Cyclic-di-GMP Network

B. burgdorferi encodes two classical two-component regulatory systems. A transmembrane sensory histidine kinase, named Hk1, and response regulator Rrp1 are encoded in a bicistronic operon, and are therefore predicted to function together (Fraser et al. 1997). Hk1 is predicted to possess three ligand-binding domains, one of which has been crystallized and its structure solved (Bauer et al. 2015). The molecules sensed by Hk1 have yet to be identified. Since both Hk1 and Rrp1 are more highly expressed during tick colonization, and are essential for that aspect of the borrelial life cycle, it is hypothesized that Hk1 is triggered by components of the tick and/or ingested blood (Caimano et al. 2011, 2015; He et al. 2011; Kostick et al. 2011). However, it is not known at which point(s) in the infectious cycle Hk1 becomes activated.

Rrp1 is an enzyme, synthesizing the modified nucleotide cyclic-di-GMP (c-di-GMP) (Rogers et al. 2009; Ryjenkov et al. 2005). That small molecule then binds to the PilZ-domain protein, PlzA, although it is not yet known how c-di-GMP effects PlzA function (Freedman et al. 2010; He et al. 2013). Whether c-di-GMP has additional partners remains to be determined. While many other species of bacteria use c-di-GMP as a component of riboswitches, no such regulatory elements have been identified in *B. burgdorferi* (McCown et al. 2017). *B. burgdorferi* also encodes two c-di-GMP phosphodiesterases, PdeA and PdeB.

As with all other organisms that encode enzymes that both synthesize and degrade c-di-GMP, one wonders how this does not create a futile cycle of synthesis and destruction. One possible explanation is that the enzymatic functions are tightly regulated, with Rrp1 becoming functional only upon Hk1 sensing an appropriate ligand, and PdeA and PdeB subject to some sort of control. Each component may be produced at specific concentrations throughout the infectious cycle. It is also possible that each enzyme may be restricted to a specific place in the borrelial cell, such that c-di-GMP accumulates in some locations, but is degraded when it diffuses to other sites.

3.2 Guanosine Tetraphosphate and Guanosine Pentaphosphate, (p)ppGpp

B. burgdorferi encodes a single enzyme, Rel, that both synthesizes and degrades (p) ppGpp (Bugrysheva et al. 2005). Rel appears to function in manners similar to the SpoT and RelA proteins of other bacteria. Nutritional depravation of *B burgdorferi* induces production of (p)ppGpp (Bugrysheva et al. 2002, 2003; Concepcion and Nelson 2003). As with the other modified nucleotides described above, it is not currently possible to determine exactly where (p)ppGpp is produced during the borrelial infectious cycle. A Rel-deficient mutant was defective in ability to survive in ticks, and in ability to recover from stationary phase or other starvation conditions (Drecktrah et al. 2015). Mutants of *rel* exhibit global transcriptional defects, which include significant impacts on the *glpFKD* operon that encodes a transporter and enzymes necessary for catabolism of glycerol (Bugrysheva et al. 2015; Drecktrah et al. 2015). This may contribute to the inability of the *rel* mutant to survive in ticks, as glycerol is *B. burgdorferi's* major carbon source during tick colonization (Pappas et al. 2011).

3.3 RpoS

This alternate sigma factor is essential for mammalian infection. Deletion of *rpoS* has widespread effects on transcription, including that of virulence factors such as OspC and DbpB (Caimano et al. 2004, 2005, 2007; Dunham-Ems et al. 2012; Eggers et al. 2004; Elias et al. 2000; Hübner et al. 2001; Ouyang et al. 2008; Xu et al. 2012; Yang et al. 2003b). Transcription of *rpoS* is shut off during tick colonization and induced during tick feeding. Consistent with its role in virulence gene regulation, an *rpoS* mutant is unable to infect mice. It has yet to be determined which transcripts of the RpoS regulon are directly transcribed by RpoS-containing RNA polymerase, and which are regulated indirectly via RpoS-controlled factors.

The *rpoS* gene has at least two promoters, one dependent upon another alternate sigma, RpoN, and the other upon the housekeeping sigma, RpoD (Boardman et al. 2008; Burtnick et al. 2007; Smith et al. 2007). The RpoD-dependent promoter is located 5′ of the RpoN-dependent promoter. Only the RpoN-dependent promoter requires Rrp2 (see below).

In vitro binding assays revealed the presence of binding-sites for both BosR and BadR in the *rpoS* promoter region (Miller et al. 2013; Ouyang et al. 2011, 2014a; Ouyang and Zhou 2015). The BosR and BadR proteins are described in greater detail in following sections. Deletion of *bosR* inhibited RpoS production, suggesting that it is a direct activator of *rpoS* (Hyde et al. 2009, 2010; Ouyang et al. 2009). Addition of phospho-sugars inhibited the interactions of BadR with the upstream regions of *rpoS* (Miller et al. 2013), suggesting that *B. burgdorferi* might regulate gene expression in response to the presence/absence of certain carbohydrates. Much remains to be learned about the impacts of BosR and BadR on *rpoS*, which appear to have a complex interrelationship. We also note that the majority of publications on transcriptional regulation of *rpoS* have focused their discussions on the RpoN/Rrp2-dependent promoter, and largely ignored the 5′ RpoD-dependent promoter.

In culture, production of RpoS can be induced by increased temperature, bacterial division rate, and pH (Jutras et al. 2013c; Yang et al. 2000). The mechanisms by which these environmental changes are used to trigger production of RpoS protein remain unknown.

3.4 Hk2 and Rrp2

The second classical two-component regulatory system is currently hypothesized to consist of Hk2 and Rrp2 (Burtnick et al. 2007; Fraser et al. 1997; Richards et al. 2015; Yang et al. 2003a; Yin et al. 2016). However, deletion of Rrp2 is lethal, whereas deletion of Hk2 is not fatal to *B. burgdorferi* (Burtnick et al. 2007; Groshong et al. 2012; Yang et al. 2003a; Yin et al. 2016). This suggests that the two proteins do not completely overlap in function, and that Rrp2 may be controlled by an additional mechanism(s). Rrp2 is a DNA-binding protein, which appears to function as an activator for transcription by RpoN-containing RNA polymerase holoenzyme (Boardman et al. 2008; Ouyang et al. 2014b). However, deletion of either *rpoN* or its known target, *rpoS*, has no significant effect upon *B burgdorferi* survival in culture. These observations indicate that Rrp2 serves a critical function (s) in addition to transcribing *rpoS*. Insights on Rrp2 functions have been gathered from studies of site-specific mutants of the protein (Ouyang and Zhou 2017; Yang et al. 2003a; Yin et al. 2016). Fully functional Hk2 and Rrp2 are both essential for the borrelial tick-vertebrate infection cycle.

The signal(s) that activate Hk2 and Rrp2 have not been identified. Exogenously applied acetate can stimulate the system in culture, leading to a postulation that acetate is the trigger (Xu et al. 2010). Subsequent studies found that the effects of adding acetate are actually due to acid-stress responses (Richards et al. 2015; Van Laar et al. 2012).

3.5 *RpoN*

The borrelial RpoN is related to the non-classical RpoN sigma factor of *E. coli*. The name is misleading, however, as *B. burgdorferi* is not capable of fixing nitrogen, nor does it appear to specifically respond to nitrogenous compounds. The major, known function of RpoN is to activate a promoter of *rpoS*, in conjunction with Rrp2 (Burtnick et al. 2007; Hübner et al. 2001; Smith et al. 2007). Transcriptomic studies revealed overlaps between genes affected by deletion of *rpoN* and of *rpoS* (Fisher et al. 2005; Hübner et al. 2001). There was also a number of genes that were affected in *rpoN* mutants but not in *rpoS* mutants, suggesting that RpoN may transcribe additional genes. Detailed investigations of the entire RpoN regulon remain to be undertaken.

3.6 *dsrA*

Translation of *rpoS* mRNA arising from the 5′, RpoD-dependent promoter is regulated by *dsrA*, a small ncRNA, which hybridizes with the mRNA and inhibits ribosome access (Archambault et al. 2013; Lybecker and Samuels 2007). It has been speculated that *dsrA* may act as a thermosensor, pairing with *rpoS* at cool temperatures but not at warm temperatures, although that hypothesis has not been tested. The borrelial Hfq protein is involved with *dsrA-rpoS* RNA interactions, and an *hfq* mutant is unable to infect mice (Lybecker et al. 2010).

3.7 *BBD18*

This protein is encoded on the small linear replicon lp17 of type strain B31 and exerts a post-transcriptional effect on RpoS. A *bbd18* mutant expressed wild-type levels of *rpoS* mRNA, but no detectable RpoS protein, and had global transcriptional impacts that were comparable to an *rpoS* mutant (Dulebohn et al. 2014; Hayes et al. 2014; Sarkar et al. 2011). BBD18 does not appear to have an impact on translation of *rpoS*, suggesting that it is involved with RpoS stability/degradation (Dulebohn et al. 2014).

3.8 *BosR*

This DNA-binding protein shares homology with the metalloregulatory Fur and PerR proteins of other bacterial species (Boylan et al. 2003; Katona et al. 2004). Arguing against a Fur-like function, *B. burgdorferi* does not produce any enzymes or cofactors that coordinate iron (Aitken and Rouviere-Yaniv 1979; Fraser et al. 1997; Posey and Gherardini 2000; Troxell et al. 2012). On the other hand, it is possible that *B. burgdorferi* might sense presence of iron as an indication that a vector tick is feeding on blood, even if it does not otherwise use the iron. BosR shares the PerR functional characteristic of being involved with protection against oxidative stress, although BosR differs in being an apparent activator of transcription (Boylan et al. 2003; Imlay 2015). BosR can bind either zinc or copper, and thus could have different effects depending upon relative levels of those two metals (Boylan et al. 2003; Wang et al. 2017). In addition, manganese levels impact production of BosR protein (Troxell et al. 2013) A *bosR* deletion mutant exhibited altered resistance to oxidative stressors such as tertbutyl hydroperoxide and methyl viologen, with a concomitant increase in the levels of SodA, the manganese-dependent superoxide dismutase (Boylan et al. 2003; Seshu et al. 2004b). BosR functions as a homodimer. It has a DNA-binding domain, and two redox-sensor CXXC motifs at its C-terminus. BosR affects transcript levels of numerous genes, including those encoding resistance against oxidative/nitrosative stress and the infection-associated RpoS sigma factor (Esteve-Gassent et al. 2015; Hyde et al. 2006, 2009; Medrano et al. 2007; Ouyang et al. 2009; Shi et al. 2014; Wang et al. 2013).

A single conservative amino acid change from arginine to lysine at position 39 of BosR significantly altered its regulatory activities compared to the wild-type *bosR* allele, including transcription of *rpoS* (Katona 2015; Seshu et al. 2004b). The extent of these defects is highlighted by the inability of *B. burgdorferi* carrying the R39 K allele to infect mice. The molecular basis for the role of Arg39 in regulating the functions of BosR remains to be ascertained.

3.9 BadR/XylR2

The genome of *B. burgdorferi* type strain contains two genes that were originally annotated as encoding "xylose operon regulatory proteins," ORFs BB0693 and BB0831 (Fraser et al. 1997). *B. burgdorferi* is unlikely to ever encounter xylose during its tick-vertebrate life cycle, and, as would be expected, xylose cannot support growth of *B. burgdorferi* (von Lackum and Stevenson 2005).

Mutants lacking ORF BB0693 are unable to infect either immunocompetent or immunodeficient mice, leading to the naming of the ORF BB0693/"XylR1" protein as BadR (Borrelia host-adaptation Regulator) (Miller et al. 2013). BadR has a DNA-binding N-terminal domain linked to a putative carbohydrate-responsive C-terminal domain, while the ORF BB0831-endoded protein ("XylR2") is likely to be a sugar kinase with no apparent DNA-binding capabilities (Miller et al. 2013; Ouyang and Zhou 2015). Both these proteins are similar to a family of proteins classified as ROKs (repressor, ORF, kinase). BadR has features in common with ROKs of *E. coli* and *B. subtilis* that are involved in glucose (NagC) and N-acetyl glucosamine (Mlc) responsive gene expression.

A mutant in *badR* was defective for growth at low temperature/high pH conditions, indicating pleiotropic effects of the protein (Miller et al. 2013). Two targets of BadR-binding have been identified, namely the promoter regions of *rpoS* and *bosR* (Miller et al. 2013; Ouyang and Zhou 2015). Addition of phospho-sugars inhibited the interactions of BadR with those regions of *rpoS*, suggesting a role for PTS sugars in regulating gene expression (Miller et al. 2013). In a microarray study, a *badR* deletion mutant exhibited elevated RpoS levels, with significant increases in transcripts of many members of the *rpoS* regulon (Miller et al. 2013).

Transcriptomic studies found that *badR* mutants exhibit defective expression of genes associated with a wide range of metabolite uptake and utilization functions, possibly occurring independently of RpoS (Arnold et al. 2017; Miller et al. 2013). Significant impacts of BadR include transcripts for proteins involved with chitobiose and N-acetyl glucosamine import and utilization, other sugar transporters, and polyamine metabolism. Other studies have revealed links between *B. burgdorferi* carbohydrate importers/phosphotransferase systems and gene regulation (Khajanchi et al. 2015; Troy et al. 2016). The roles of BadR in such signaling remains to be discovered.

The role of putative sugar kinase "XylR2" has yet to be addressed. It might be involved with phosphorylation of nutrient and/or structural carbohydrates, such as N-acetyl glucosamine. The intercellular signal molecule autoinducer 2 (AI-2, see below) is a modified ribose, and other bacteria phosphorylate internalized AI-2 via a specific kinase, raising the possibility that "XylR2" may be involved with that regulatory system (Pereira et al. 2012; Taga et al. 2003; Xavier et al. 2007).

3.10 CsrA

This is a homolog of the carbon-storage regulatory proteins of other bacterial species. It was demonstrated to functionally overlap with similar proteins of other species, by inducing glycogen biosynthesis when overexpressed in *E. coli* (Sanjuan et al. 2009). The borrelial CsrA is an RNA-binding protein that binds to 5′-untranslated regions of specific borrelial mRNAs. Analyses of *B. burgdorferi csrA* mutants indicated that the protein is a global regulatory factor (Arnold et al. 2017; Karna et al. 2011; Sze and Li 2011). Among CsrA-affected proteins are the FlaB flagellin and virulence-associated surface proteins OspC, DbpB, and BBK32 (Karna et al. 2011; Sze et al. 2011). Recent RNA-Seq analyses indicated that CsrA affects transcript levels of *ospC*, *dbpB*, and other targets in an RpoS-independent manner, consistent with other results (Arnold et al. 2017; Ouyang et al. 2014c). Substantial overlap exists between the CsrA and BadR regulons, including proteins necessary for N-acetyl glucosamine metabolism, suggesting that both proteins control functions of one or more other regulatory factors (Arnold et al. 2017). Studies on the impact of *csrA*-deletion on *B. burgdorferi* infectivity have yielded conflicting results, with two groups finding *csrA* mutants to be unable to infect mice (Karna et al. 2013; Sze and Li 2011), while another group did not observe such an effect (Ouyang et al. 2014c). Since CsrA is an RNA-binding protein, its effects are dependent on the relative levels of the specific mRNAs that it regulates, as well as the mechanisms controlling it own expression in response to different signals. Variations in the levels of signaling or levels of target mRNAs could lead to a wide spectrum of phenotypes. Alternatively, single nucleotide polymorphisms within the CsrA-binding sites of target transcripts could affect the binding affinity of CsrA. Much remains to be done toward understanding the borrelial CsrA protein and its regulon.

CsrA/RsmA homologs in Gram-negative organisms regulate gene expression via specific and high affinity binding to 5′-untranslated regions of certain mRNAs, thereby altering their translation or stability (Vakulskas et al. 2015). CsrA levels are constitutive in these prokaryotes, and its effects are regulated/titrated via binding of cognate noncoding RNAs such as CsrB and CsrC, which sequester CsrA away from its targets. Since no homologs of small RNAs such as CsrB or CsrC have been identified in *B. burgdorferi*, it appears that its CsrA has evolved other regulatory features. Notably, expression levels of *B. burgdorferi* CsrA are not constitutive but can be influenced by changes in culture temperature and pH (Karna et al. 2013). There is evidence that CsrA participates in protein-protein interactions (Karna et al., unpublished data), similar to what has been observed with the homologs of *B. subtilis* (Mukherjee et al. 2011, 2016).

Another factor that influences *B. burgdorferi* CsrA function is a stretch of seven amino acids at the C-terminus that is absent from homologs of *Treponema pallidum, E. coli* and *B. subtilis* (Karna et al. 2013). Deletion of the seven amino acids from the borrelial CsrA resulted in phenotypes similar to those of *csrA*-deletion mutants. Moreover, replacement of eight conserved amino acids with alanines resulted in a highly stabilized version of CsrA that had a significantly higher binding affinity to 5′UTR of target *pta* mRNA. That mutant also exhibited increased levels of RpoS and members of the *rpoS* regulon (Karna et al. 2013).

A hint of the involvement of CsrA with flagellar synthesis and function is the location of its gene within a motility-associated *flbF-flgK-flgL-fliW-csrA* operon (Fraser et al. 1997; Sanjuan et al. 2009). Transformation of *B. burgdorferi* with a plasmid carrying *csrA* under the control of a constitutive promoter resulted in a "rod"-like morphology with reduced motility, rather than the "wave-like" morphology of the parental strain (Sanjuan et al. 2009). There was a concomitant reduction in the levels of FlaB protein. Consistent with those results, the *flaB* mRNA contains two CsrA-binding sites, one of which overlaps the Shine-Delgarno sequence, and CsrA inhibits *flaB* translation (Sze and Li 2011). Another CsrA-binding site within the *flgK* operon affects translation of FliW (Karna et al. 2013).

3.11 SpoVG

Homologs of this protein are produced by many spirochetes and firmicutes. Its name derives from its effect on stage V of sporulation in *Bacillus subtilis*

(Rosenbluh et al. 1981). Although SpoVG was long known in the firmicutes, it is rather surprising that its function as a DNA-binding protein was first identified in *B. burgdorferi* (Jutras et al. 2013b). The *Listeria monocytogenes* SpoVG was later demonstrated to also bind RNA (Burke and Portnoy 2016), a trait that has since been shown to also apply to the *B. burgdorferi* homolog (Savage et al., unpublished results). Mutation of *B. burgdorferi spoVG* has pleotropic effects, evidently due to both transcriptional and post-transcriptional regulatory activities (Savage et al., unpublished results). SpoVG binds to a DNA sequence adjacent to the recombination site within the *vlsE* ORF, suggesting an additional role in borrelial antigenic variation (Jutras et al. 2013b). The *spoVG* mRNA is produced at high levels during tick colonization, and at very low levels during vertebrate infection (Iyer et al. 2015).

3.12 BpuR

This small homodimeric protein folds into a "PUR" domain, a motif that is also found within eukaryotic proteins (Graebsch et al. 2010; Jutras et al. 2013a). The domain is so-named because it preferentially binds to purine-rich nucleic acids. BpuR and its eukaryotic homologs bind to double-stranded and single-stranded DNAs and to RNA. Of the borrelial targets studied to date, BpuR exhibits a stronger affinity for RNA than for any type of DNA (Jutras et al. 2013a). BpuR is highly expressed during tick colonization, and undetectable during mammalian infection (Arnold et al., unpublished results). It was originally identified as a co-repressor of *B. burgdorferi erp* operons, where it enhanced the repressive activity of BpaB (see below) (Jutras et al. 2013a). BpuR also binds to its own mRNA and represses translation, an autoregulation that prevents the bacteria from producing too much BpuR protein (Jutras et al. 2013d). RNA-Seq analyses of a *bpuR* mutant indicated little to no detectable effects on global transcript levels (Arnold et al., unpublished results). The mutant does, however, exhibit substantial differences in protein content, suggesting widespread post-transcriptional regulation through BpuR binding to mRNAs. Consistent with this hypothesis, RNA immunoprecipitation (RIP) of live *B. burgdorferi* identified a variety of in vivo RNA-binding partners for BpuR (Jutras et al., unpublished results). BpuR is an evolutionary curiosity, since homologs are found in only some spirochetes and *Bacteroides*, but in all multicellular eukaryotes.

3.13 EbfC and Other Nucleoid-Associated Proteins

EbfC is another small protein that was identified due to its effects on *erp* transcription (Riley et al. 2009). It regulates *erp* expression as an anti-repressor, competing for DNA-binding with the BpaB repressor (see below) (Jutras et al. 2012b). EbfC specifically recognizes a small palindrome, 5′-GTnAC-3′, which is present in two–three copies in all *erp* operator sites, and a single copy is found every kilobase or so throughout the *B. burgdorferi* genome (Riley et al. 2009). In vivo analyses with GFP-tagged EbfC revealed that it associates with the borrelial nucleoid (Jutras et al. 2012a). The current model holds that EbfC is primarily a structural nucleoid-associated protein, helping to appropriately shape the main chromosome and smaller DNAs. Almost all other species of Eubacteria produce a homolog of EbfC (Cooley et al. 2009; Lim et al. 2003; Wang et al. 2012a). In *B. burgdorferi* and many other species, *ebfC* is cotranscribed in a bicistronic operon with *dnaX*, which encodes the gamma subunit of DNA polymerase II (Jutras et al. 2012a). The conservation of that gene order across many bacterial phyla indicates that co-expression of EbfC and DnaX serves an important function(s). In *B. burgdorferi*, both of those genes and their proteins are highly expressed during rapid growth conditions, such as within a feeding tick or during vertebrate infection, and poorly expressed during conditions of slow growth, such as within an unfed tick (Jutras et al. 2012a). This expression pattern of EbfC is a major contributor to the growth-dependent regulation of the spirochete's *erp* operons (Jutras et al. 2013c).

B. burgdorferi encodes at least three additional nucleoid-associated proteins that are known to, or could, influence gene expression. Hbb shares some similarities with the *E. coli* IHF and HU proteins, and is involved with transcriptional regulation of the *p66* gene (Kobryn et al. 2000; Medrano et al. 2010; Mouw and Rice 2007; Tilly et al. 1996). Gac is the product of an internal translational start site within the *gyrA* gene, and bears some similarities with *E. coli* HU (Knight and Samuels 1999; Knight et al. 2000). A third protein, variously named BicA, Dps, or NapA, binds iron and copper, and is likely to be involved with DNA compaction during stationary phase, similarly to *E. coli* Dps (Codolo et al. 2010; Hyde et al. 2010; Li et al. 2007; Wang et al. 2012b).

3.14 BpaB

Whereas all of the above-described regulatory factors are encoded on the *B. burgdorferi* main chromosome, *bpaB* genes are located on the small DNAs ("plasmids"). They are adjacent to a gene that encodes a ParA-like protein, and each replicon contains a unique *parA-bpaB* pair, suggesting that they function together for plasmid replication and compatibility (Brisson et al. 2013; Casjens et al. 1997, 2000; Eggers et al. 2002; Zückert and Meyer 1996). In addition to that hypothesized function, the BpaB proteins of the *B. burgdorferi* cp32 prophages bind to specific sites on the prophage genome (Burns et al. 2010; Chenail et al. 2012). Although each replicon's BpaB is overall unique, the cp32 homologs possess a conserved sequence that is involved with DNA-binding (Burns et al. 2010). None of the BpaB proteins from non-cp32 replicons contains that motif. As a result, cp32 BpaB proteins are capable of coordinating protein expression via regulatory cross-talk between different prophages, without being affected by unrelated replicons in the bacterial cell. BpaB functions as the repressor of *erp* transcription, and as an activator of the cp32 *nucP* and *ssbP* operons (which encode the NucP nuclease and SsbP single-stranded DNA-binding protein, respectively) (Chenail et al. 2012; Jutras et al. 2012b). It is possible that BpaB homologs encoded by other borrelial replicons may likewise perform functions in addition to plasmid maintenance.

3.15 HrpA

Deletion of the gene encoding this RNA helicase resulted in significant changes to the expression of nearly 200 *B. burgdorferi* proteins (Salman-Dilgimen et al. 2011). The mechanism by which HrpA affects protein levels has yet to be defined.

3.16 LuxS/Autoinducer 2

Each of the factors described above exert profound effects upon their targets. The LuxS/Autoinducer 2 (AI-2) system is representative of a more modest regulatory system.

As do other organisms, *B. burgdorferi* synthesizes S-adenosyl methionine as the methyl donor for its methylation reactions. The byproduct of those reactions, S-adenosyl homocysteine is toxic, and so is detoxified to S-ribosyl homocysteine (SRH) by the Pfs enzyme. Some bacteria, such as the syphilis spirochete, *Treponema pallidum*, stop at that point (von Lackum et al. 2006). The *B. burgdorferi* LuxS converts SRH into homocysteine and 4,5-dihydroxy-2,3-pentanedione (AI-2) (Chen et al. 2002; Miller et al. 2004; Schauder et al. 2001). While most bacteria that produce a LuxS also have the capability to convert homocysteine into methionine, *B. burgdorferi* lacks that capacity (Babb et al. 2005; von Lackum et al. 2006). Thus, homocysteine and AI-2 accumulate as *B. burgdorferi* grows and divides. Bacteria as diverse as *Vibrio harveyi* and *Salmonella enterica* have developed mechanisms to detect accumulation of AI-2 (Pereira et al. 2012; Surette et al. 1999; Taga et al. 2003; Xavier et al. 2007). Although such sensing has been described as quorum sensing, it is more accurate to view it as a "biofeedback" mechanism, informing individuals that they are surrounded by other bacteria that are actively growing and dividing (Babb et al. 2005; Stevenson et al. 2003; Xavier and Bassler 2003). Unlike classical quorum sensing mechanisms, AI-2 is produced at maximum levels during periods of rapid growth, and is not produced during the growth-inhibiting conditions of late exponential or stationary phases (Stevenson and Babb 2002). It is not yet known how *B. burgdorferi* senses or responds to AI-2.

B. burgdorferi luxS mutants produce different levels of certain proteins than do their wild-type parents, while addition of exogenous AI-2 to *luxS* mutant cultures restores wild-type expression levels (Babb et al. 2005; Stevenson and Babb 2002). A *luxS* mutant was capable of infecting both mice and ticks, and was transmitted each way between vector and host (Blevins et al. 2004; Hübner et al. 2003). However, competition studies, in which mice were simultaneously infected with both wild-type and *luxS* mutant, revealed that the mutant was significantly impaired in ability to colonize mice (Arnold et al. 2015). This underscores the point that studies of factors may not give black/white, yes/no types of answers, and that a function need not be absolutely essential in order to have a critical role. That observation applies to all aspects of *B. burgdorferi*, from regulatory factors to outer surface adhesins.

3.17 Cyclic-di-AMP

This second cyclic dinucleotide is synthesized by CdaA, and degraded by phosphodiesterase DhhP (Savage et al. 2015; Ye et al. 2014). Both are produced simultaneously by cultured *B. burgdorferi*. Cytoplasmic extracts do not contain detectable levels of c-di-AMP, begging the same questions as asked above for c-di-GMP (Savage et al. 2015). Deletion of DhhP is lethal to *B. burgdorferi*, yet over-production of CdaA has no detectable effects on cultured spirochetes (Savage et al. 2015; Ye et al. 2014). A role(s) for c-di-AMP has yet to be identified.

4 Conclusions

The first report of *B. burgdorferi* gene regulation was published in 1995 (Schwan et al. 1995). The intervening 20-plus years have revealed that the Lyme disease spirochete differentially expresses many surface proteins, metabolic enzymes, and other factors throughout its vertebrate-tick infectious cycle. A number of regulatory proteins and small molecules have been identified that control borrelial physiology. Yet, as this review highlighted, much remains to be learned about the regulatory biology of *B. burgdorferi*. Interactions between regulons are not well understood. Little is known about how any regulatory protein is itself regulated. Even among well-studied regulons, many phenomena remain to be explored; for example, despite numerous transcriptomic and proteomic analyses of the RpoS regulon, we still do not know the consensus promoter sequence for RpoS-directed transcription, and thus do not know which components of the regulon are directly transcribed using RpoS and which are affected indirectly.

The global impacts of regulatory factors make them compelling targets for development of novel antibacterials. Continued investigation of borrelial regulatory networks therefore holds promise of enhanced therapies, in addition to expanding knowledge of the pathogenic properties of the Lyme disease agent.

Acknowledgments We are grateful to Catherine Brissette, Will Arnold, and Christina Savage for their comments.

References

Aitken A, Rouviere-Yaniv J (1979) Amino and carboxy terminal sequences of the DNA-binding protein HU from the cyanobacterium *Synechocystis* PCC 6701 (ATCC 27170). Biochem Biophys Res Commun 91:461–467

Akins DR, Bourell KW, Caimano MJ, Norgard MV, Radolf JD (1998) A new animal model for studying Lyme disease spirochetes in a mammalian host-adapted state. J Clin Invest 101:2240–2250

Alverson J, Bundle SF, Sohasky CD, Lybecker MC, Samuels DS (2003) Transcriptional regulation of the *ospAB* and *ospC* promoters of *Borrelia burgdorferi*. Mol Microbiol 48:1665–1677

Antonara S, Ristow L, Coburn J (2011) Adhesion mechanisms of *Borrelia burgdorferi*. Adv Exp Med Biol 715:35–49

Archambault L, Linscott J, Swerdlow N, Boyland K, Riley E, Schlax P (2013) Translational efficiency of *rpoS* mRNA from *Borrelia burgdorferi*: effects of the length and sequence of the mRNA leader region. Biochem Biophys Res Commun 433:73–78

Arnold WK, Savage CR, Antonicello AD, Stevenson B (2015) Apparent role for *Borrelia burgdorferi* LuxS during mammalian infection. Infect Immun 83:1347–1353

Arnold WK, Savage CR, Brissette CA, Janakiram S, Livny J, Stevenson B (2016) RNA-Seq of *Borrelia burgdorferi* in multiple phases of growth reveals insights into the dynamics of gene expression, transcriptome architecture, and noncoding RNAs. PLoS ONE 11:e0164165

Arnold WK, Lethbridge KG, Savage CR, Smith TC, Brissette CA, Livny J, Seshu J, Stevenson B (2017) Transcriptomic impacts of the virulence-controlling CsrA, BadR, RpoN, and RpoS regulatory networks in the Lyme disease spirochete. Submitted

Babb K, El-Hage N, Miller JC, Carroll JA, Stevenson B (2001) Distinct regulatory pathways control the synthesis of *Borrelia burgdorferi* infection-associated OspC and Erp surface proteins. Infect Immun 69:4146–4153

Babb K, von Lackum K, Wattier RL, Riley SP, Stevenson B (2005) Synthesis of autoinducer 2 by the Lyme disease spirochete, *Borrelia burgdorferi*. J Bacteriol 187:3079–3087

Barbour AG (1984) Isolation and cultivation of Lyme disease spirochetes. Yale J Biol Med 57:521–525

Bauer WJ, Luthra A, Zhu G, Radolf JD, Malkowski MG, Caimano MJ (2015) Structural characterization and modeling of the *Borrelia burgdorferi* hybrid histidine kinase Hk1 periplasmic sensor: a system for sensing small molecules associated with tick feeding. J Struct Biol 192:48–58

Blevins JS, Revel AT, Caimano MJ, Yang XF, Richardson JA, Hagman KE, Norgard MV (2004) The *luxS* gene is not required for *Borrelia burgdorferi* tick colonization, transmission to a mammalian host, or induction of disease. Infect Immun 72:4862–4867

Boardman BK, He M, Ouyang Z, Xu H, Pang X, Yang XF (2008) Essential role of the response regulator Rrp2 in the infectious cycle of *Borrelia burgdorferi*. Infect Immun 76:3844–3853

Bontemps-Gallo S, Lawrence K, Gherardini FC (2016) Two different virulence-related regulatory pathways in *Borrelia burgdorferi* are directly affected by osmotic fluxes in the blood meal of feeding *Ixodes* ticks. PLoS Pathog 15:e1005791

Boylan JA, Posey JE, Gherardini FC (2003) *Borrelia* oxidative stress response regulator, BosR: a distinct Zn-dependent transcriptional activator. Proc Natl Acad Sci USA 100:11684–11689

Brissette CA, Gaultney RA (2014) That's my story, and I'm sticking to it—an update on *Borrelia burgdorferi* adhesins. Front Cell Infect Microbiol 4:41, doi:10.3389/fcimb.2014.00041

Brisson D, Zhou W, Jutras BL, Casjens S, Stevenson B (2013) Distribution of cp32 prophages among Lyme disease-causing spirochetes and natural diversity of their lipoprotein-encoding *erp* loci. Appl Env Microbiol 79:4115–4128

Bugrysheva J, Dobrikova EY, Godfrey HP, Sartakova ML, Cabello FC (2002) Modulation of *Borrelia burgdorferi* stringent response and gene expression during extracellular growth with tick cells. Infect Immun 70:3061–3067

Bugrysheva J, Dobrikova EY, Sartakova ML, Caimano MJ, Daniels TJ, Radolf JD, Godfrey HP, Cabello FC (2003) Characterization of the stringent response and *rel*$_{Bbu}$ expression in *Borrelia burgdorferi*. J Bacteriol 185:957–965

Bugrysheva JV, Bruyskin AV, Godfrey HP, Cabello FC (2005) *Borrelia burgdorferi rel* is responsible for generation of guanosine-3′-diphosphate-5′-triphosphate and growth control. Infect Immun 73:4972–4981

Bugrysheva JV, Pappas CJ, Terekhova DA, Iyer R, Godfrey HP, Schwartz I, Cabello FC (2015) Characterization of the RelBbu regulon in *Borrelia burgdorferi* reveals modulation of glycerol metabolism by (p)ppGpp. PLoS ONE 10:e0118063

Burke TP, Portnoy DA (2016) SpoVG is a conserved RNA-binding protein that regulates *Listeria monocytogenes* lysozyme resistance, virulence, and swarming motility. mBio 7:e00240

Burkot TR, Piesman J, Wirtz RA (1994) Quantitation of the *Borrelia burgdorferi* outer surface protein A in *Ixodes scapularis*: fluctuations during the tick life cycle, doubling times and loss while feeding. J Infect Dis 170:883–889

Burns LH, Adams CA, Riley SP, Jutras BL, Bowman A, Chenail AM, Cooley AE, Haselhorst LA, Moore AM, Babb K, Fried MG, Stevenson B (2010) BpaB, a novel protein encoded by the Lyme disease spirochete's cp32 prophages, binds to *erp* Operator 2 DNA. Nucleic Acids Res 38:5443–5455

Burtnick MN, Downey JS, Brett PJ, Boylan JA, Frye JG, Hoover TR, Gherardini FC (2007) Insights into the complex regulation of *rpoS* in *Borrelia burgdorferi*. Mol Microbiol 65:277–293

Bykowski T, Babb K, von Lackum K, Riley SP, Norris SJ, Stevenson B (2006) Transcriptional regulation of the *Borrelia burgdorferi* antigenically variable VlsE surface protein. J Bacteriol 188:4879–4889

Bykowski T, Woodman ME, Cooley AE, Brissette CA, Brade V, Wallich R, Kraiczy P, Stevenson B (2007) Coordinated expression of *Borrelia burgdorferi* complement regulator-acquiring surface proteins during the Lyme disease spirochete's mammal-tick infection cycle. Infect Immun 75:4227–4236

Caimano MJ (2005) Cultivation of *Borrelia burgdorferi* in dialysis membrane chambers in rat peritonea. In: Coico RT, Kowalik TF, Quarles J, Stevenson B, Taylor R (eds) Current Protocols In Microbiology. Wiley, Hoboken, N.J.

Caimano MJ, Eggers CH, Hazlett KRO, Radolf JD (2004) RpoS is not central to the general stress response in *Borrelia burgdorferi* but does control expression of one or more essential virulence determinants. Infect Immun 72:6433–6445

Caimano MJ, Eggers CH, Gonzalez CA, Radolf JD (2005) Alternate sigma factor RpoS is required for the in vivo-specific repression of *Borrelia burgdorferi* plasmid lp45-borne *ospA* and *lp6.6* genes. J Bacteriol 187:7845–7852

Caimano MJ, Iyer R, Eggers CH, Gonzalez C, Morton EA, Gilbert MA, Schwartz I, Radolf JD (2007) Analysis of the RpoS regulon in *Borrelia burgdorferi* in response to mammalian host signals provides insight into RpoS function during the enzootic cycle. Mol Microbiol 65:1193–1217

Caimano MJ, Kenedy MR, Kairu T, Desrosiers DC, Harman M, Dunham-Ems S, Akins DR, Pal U, Radolf JD (2011) The hybrid histidine kinase Hk1 is part of a two-component system that is essential for survival of *Borrelia burgdorferi* in feeding *Ixodes scapularis* ticks. Infect Immun 79:3117–3130

Caimano MJ, Dunham-Ems S, Allard AM, Cassera MB, Kenedy M, Radolf JD (2015) Cyclic di-GMP modulates gene expression in Lyme disease spirochetes at the tick-mammal interface to promote spirochete survival during the blood meal and tick-to-mammal transmission. Infect Immun 83:3043–3060

Carroll JA, Garon CF, Schwan TG (1999) Effects of environmental pH on membrane proteins in *Borrelia burgdorferi*. Infect Immun 67:3181–3187

Carroll JA, Cordova RM, Garon CF (2000) Identification of eleven pH-regulated genes in *Borrelia burgdorferi* localized to linear plasmids. Infect Immun 68:6677–6684

Casjens S, van Vugt R, Tilly K, Rosa PA, Stevenson B (1997) Homology throughout the multiple 32-kilobase circular plasmids present in Lyme disease spirochetes. J Bacteriol 179:217–227

Casjens S, Palmer N, van Vugt R, Huang WM, Stevenson B, Rosa P, Lathigra R, Sutton G, Peterson J, Dodson RJ, Haft D, Hickey E, Gwinn M, White O, Fraser C (2000) A bacterial genome in flux: the twelve linear and nine circular extrachromosomal DNAs of an infectious isolate of the Lyme disease spirochete *Borrelia burgdorferi*. Mol Microbiol 35:490–516

Cassatt DR, Patel NK, Ulbrandt ND, Hanson MS (1998) DbpA, but not OspA, is expressed by *Borrelia burgdorferi* during spirochetemia and is a target for protective antibodies. Infect Immun 66:5379–5387

Chen X, Schauder S, Potier N, Van Dorsselaer A, Pelczer I, Bassler BL, Hughson FM (2002) Structural identification of a bacterial quorum-sensing signal containing boron. Nature 415:545–549

Chenail AM, Jutras BL, Adams CA, Burns LH, Bowman A, Verma A, Stevenson B (2012) *Borrelia burgdorferi* cp32 BpaB modulates expression of the prophage NucP nuclease and SsbP single-stranded DNA-binding protein. J Bacteriol 194:4570–4578

Codolo G, Papinutto E, Polenghi A, D'Elios MM, Zanotti G, de Bernard M (2010) Structure and immunomodulatory property relationship in NapA of *Borrelia burgdorferi*. Biochim Biophys Acta 1804:2191–2197

Concepcion MB, Nelson DR (2003) Expression of *spoT* in *Borrelia burgdorferi* during serum starvation. J Bacteriol 185:444–452

Cooley AE, Riley SP, Kral K, Miller MC, DeMoll E, Fried MG, Stevenson B (2009) DNA-binding by *Haemophilus influenzae* and *Escherichia coli* YbaB, members of a widely-distributed bacterial protein family. BMC Microbiol 9:137

de Silva AM, Fikrig E (1995) Growth and migration of *Borrelia burgdorferi* in *Ixodes* ticks during blood feeding. Am J Trop Med Hyg 53:397–404

Drecktrah D, Lybecker M, Popitsch N, Rescheneder P, Hall LS, Samuels DS (2015) The *Borrelia burgdorferi* RelA/SpoT homolog and stringent response regulate survival in the tick vector and global gene expression during starvation. PLoS Pathog 11:e1005160

Dulebohn DP, Hayes BM, Rosa PA (2014) Global repression of host-associated genes of the Lyme disease spirochete through post-transcriptional modulation of the alternative sigma factor RpoS. PLoS ONE 9:e93141

Dunham-Ems SM, Caimano MJ, Pal U, Wolgemuth CW, Eggers CH, Balic A, Radolf JD (2009) Live imaging reveals a biphasic mode of dissemination of *Borrelia burgdorferi* within ticks. J Clin Inv 119:3652–3665

Dunham-Ems SM, Caimano MJ, Eggers CH, Radolf JD (2012) *Borrelia burgdorferi* requires the alternative sigma factor RpoS for dissemination within the vector during tick-to-mammal transmission. PLoS Pathog 8:e1002532

Eggers CH, Caimano MJ, Clawson ML, Miller WG, Samuels DS, Radolf JD (2002) Identification of loci critical for replication and compatibility of a *Borrelia burgdorferi* cp32 plasmid and use of a cp32-based shuttle vector for the expression of fluorescent reporters in the Lyme disease spirochaete. Mol Microbiol 43:281–295

Eggers CH, Caimano MJ, Radolf JD (2004) Analysis of promoter elements involved in the transcription initiation of RpoS-dependent *Borrelia burgdorferi* genes. J Bacteriol 186:7390–7402

Elias AF, Bono JL, Carroll JA, Stewart P, Tilly K, Rosa P (2000) Altered stationary-phase response in a *Borrelia burgdorferi rpoS* mutant. J Bacteriol 182:2909–2918

Esteve-Gassent MD, Smith TC, Small CM, Thomas DP, Seshu J (2015) Absence of *sodA* increases the levels of oxidation of key metabolic determinants of *Borrelia burgdorferi*. PLoS ONE 10:e0136707

Fisher MA, Grimm D, Henion AK, Elias AF, Stewart PE, Rosa PA, Gherardini FC (2005) *Borrelia burgdorferi* σ^{54} is required for mammalian infection and vector transmission but not for tick colonization. Proc Natl Acad Sci USA 102:5162–5167

Fraser CM, Casjens S, Huang WM, Sutton GG, Clayton R, Lathigra R, White O, Ketchum KA, Dodson R, Hickey EK, Gwinn M, Dougherty B, Tomb J-F, Fleischmann RD, Richardson D,

Peterson J, Kerlavage AR, Quackenbush J, Salzberg S, Hanson M, van Vugt R, Palmer N, Adams MD, Gocayne J, Weidmann J, Utterback T, Watthey L, McDonald L, Artiach P, Bowman C, Garland S, Fujii C, Cotton MD, Horst K, Roberts K, Hatch B, Smith HO, Venter JC (1997) Genomic sequence of a Lyme disease spirochaete, *Borrelia burgdorferi*. Nature 390:580–586

Freedman JC, Rogers EA, Kostick JL, Zhang H, Iyer R, Schwartz I, Marconi RT (2010) Identification and molecular characterization of a cyclic-di-GMP effector protein, PlzA (BB0733): additional evidence for the existence of a functional cyclic-di-GMP regulatory network in the Lyme disease spirochete, *Borrelia burgdorferi*. FEMS Immunol Med Microbiol 58:285–294

Graebsch A, Roche S, Kostrewa D, Söding J, Niessing D (2010) Of bits and bugs—on the use of bioinformatics and a bacterial crystal structure to solve a eukaryotic repeat-protein structure. PLoS ONE 5:e13402

Grimm D, Tilly K, Byram R, Stewart PE, Krum JG, Bueschel DM, Schwan TG, Policastro PF, Elias AF, Rosa PA (2004) Outer-surface protein C of the Lyme disease spirochete: a protein induced in ticks for infection of mammals. Proc Natl Acad Sci USA 101:3142–3147

Groshong AM, Gibbons NE, Yang XF, Blevins JS (2012) Rrp2, a prokaryotic enhancer-like binding protein, is essential for viability of *Borrelia burgdorferi*. J Bacteriol 194:3336–3342

Hayes BM, Dulebohn DP, Sarkar A, Tilly K, Bestor A, Ambroggio X, Rosa PA (2014) Regulatory protein BBD18 of the Lyme disease spirochete: essential role during tick acquisition? MBio 5: e01017–01014

He M, Ouyang Z, Troxell B, Xu H, Moh A, Piesman J, Norgard MV, Gomelsky M, Yang XF (2011) Cyclic di-GMP is essential for the survival of the Lyme disease spirochete in ticks. PLoS Pathog 7:e1002133

He M, Zhang J, Ye M, Lou Y, Yang XF (2013) The cyclic dimeric-GMP receptor PlzA controls virulence gene expression through RpoS in *Borrelia burgdorferi*. Infect Immun 82:445–452

Hefty PS, Jolliff SE, Caimano MJ, Wikel SK, Radolf JD, Akins DR (2001) Regulation of OspE-related, OspF-related, and Elp lipoproteins of *Borrelia burgdorferi* strain 297 by mammalian host-specific signals. Infect Immun 69:3618–3627

Hübner A, Yang X, Nolen DM, Popova TG, Cabello PC, Norgard MV (2001) Expression of *Borrelia burgdorferi* OspC and DbpA is controlled by a RpoN-RpoS regulatory pathway. Proc Natl Acad Sci USA 98:12724–12729

Hübner A, Revel AT, Nolen DM, Hagman KE, Norgard MV (2003) Expression of a *luxS* gene is not required for *Borrelia burgdorferi* infection of mice via needle inoculation. Infect Immun 71:2892–2896

Hyde JA, Seshu J, Skare JT (2006) Transcriptional profiling of *Borrelia burgdorferi* containing a unique *bosR* allele identifies a putative stress regulon. Microbiology 152:2599–2609

Hyde JA, Trzeciakowski JP, Skare JT (2007) *Borrelia burgdorferi* alters its gene expression and antigenic profile in response to CO2 levels. J Bacteriol 189:4370445

Hyde JA, Shaw DK, Smith R, Trzeciakowski JP, Skare JT (2009) The BosR regulatory protein of *Borrelia burgdorferi* interfaces with the RpoS regulatory pathway and modulates both the oxidative stress response and pathogenic properties of the Lyme disease spirochete. Mol Microbiol 74:1344–1355

Hyde JA, Shaw DK, Smith R, Trzeciakowski JP, Skare JT (2010) Characterization of a conditional *bosR* mutant in *Borrelia burgdorferi*. Infect Immun 78:265–274

Imlay JA (2015) Transcription factors that defend bacteria against reactive oxygen species. Annu Rev Microbiol 69:93–108

Indest KJ, Ramamoorthy R, Sole M, Gilmore RD, Johnson BJB, Philipp MT (1997) Cell-density-dependent expression of *Borrelia burgdorferi* lipoproteins in vitro. Infect Immun 65:1165–1171

Iyer R, Caimano MJ, Luthra A, Axline D, Corona A, Iacobas DA, Radolf JD, Schwartz I (2015) Stage-specific global alterations in the transcriptomes of Lyme disease spirochetes during tick feeding and following mammalian host-adaptation. Mol Microbiol 95:509–538

Jutras BL, Bowman A, Brissette CA, Adams CA, Verma A, Chenail AM, Stevenson B (2012a) EbfC (YbaB) is a new type of bacterial nucleoid-associated protein, and a global regulator of gene expression in the Lyme disease spirochete. J Bacteriol 194:3395–3406

Jutras BL, Verma A, Adams CA, Brissette CA, Burns LH, Whetstine CR, Bowman A, Chenail AM, Zückert WR, Stevenson B (2012b) BpaB and EbfC DNA-binding proteins regulate production of the Lyme disease spirochete's infection-associated Erp surface proteins. J Bacteriol 194:778–786

Jutras BL, Chenail AM, Carroll DW, Miller MC, Zhu H, Bowman A, Stevenson B (2013a) Bpur, the Lyme disease spirochete's PUR-domain protein: identification as a transcriptional modulator and characterization of nucleic acid interactions. J Biol Chem 288:26220–26234

Jutras BL, Chenail AM, Rowland CL, Carroll D, Miller MC, Bykowski T, Stevenson B (2013b) Eubacterial SpoVG homologs constitute a new family of site-specific DNA-binding proteins. PLoS ONE 8:e66683

Jutras BL, Chenail AM, Stevenson B (2013c) Changes in bacterial growth rate govern expression of the *Borrelia burgdorferi* OspC and Erp infection-associated surface proteins. J Bacteriol 195:757–764

Jutras BL, Jones G, Verma A, Brown NA, Antonicello AD, Chenail AM, Stevenson B (2013d) Post-transcriptional autoregulation of the Lyme disease bacterium's BpuR DNA/RNA-binding protein. J Bacteriol 195:4915–4923

Karna SL, Prabhu RG, Lin YH, Miller CL, Seshu J (2013) Contributions of environmental signals and conserved residues to the functions of carbon storage regulator A of *Borrelia burgdorferi*. Infect Immun 81:2972–2985

Karna SLR, Sanjuan E, Esteve-Gassent MD, Miller CL, Maruskova M, Seshu J (2011) CsrA modulates levels of lipoproteins and key regulators of gene expression critical for pathogenic mechanisms of *Borrelia burgdorferi*. Infect Immun 79:732–744

Katona LI, Tokarz R, Kuhlow CJ, Benach J, Benach JL (2004) The Fur homologue in *Borrelia burgdorferi*. J Bacteriol 186:6443–6456

Katona LI (2015) The Fur homologue BosR requires Arg39 to activate *rpoS* transcription in *Borrelia burgdorferi* and thereby direct spirochaete infection in mice. Microbiology 161:2243–2255

Kelly R (1971) Cultivation of *Borrelia hermsi*. Science 173:443–444

Khajanchi BK, Odeh E, Gao L, Jacobs MB, Philipp MT, Lin T, Norris SJ (2015) Phosphoenolpyruvate phosphotransferase system components modulate gene transcription and virulence of Borrelia burgdorferi. Infect Immun 84:754–764

Knight SW, Samuels DS (1999) Natural synthesis of a DNA-binding protein from the C-terminal domain of DNA gyrase A in *Borrelia burgdorferi*. EMBO J 18:4875–4881

Knight SW, Kimmel BJ, Eggers CH, Samuels DS (2000) Disruption of the *Borrelia burgdorferi gac* gene, encoding the naturally synthesized GyrA C-terminal domain. J Bacteriol 182: 2048–2051

Kobryn K, Naigamwalla DZ, Chaconas G (2000) Site-specific DNA binding and bending by the *Borrelia burgdorferi* Hbb protein. Mol Microbiol 37:145–155

Kostick JL, Szkotnicki LT, Rogers EA, Bocci P, Raffaelli N, Marconi RT (2011) The diguanylate cyclase, Rrp1, regulates critical steps in the enzootic cycle of the Lyme disease spirochetes. Mol Microbiol 81:219–231

Li X, Pal U, Ramamoorthi N, Liu X, Desrosiers DC, Eggers CH, Anderson JF, Radolf JD, Fikrig E (2007) The Lyme disease agent *Borrelia burgdorferi* requires BB0690, a Dps homologue, to persist within ticks. Mol Microbiol 63:694–710

Liang FT, Yan J, Mbow ML, Sviat SL, Gilmore RD, Mamula M, Fikrig E (2004) *Borrelia burgdorferi* changes its surface antigenic expression in response to host immune reponses. Infect Immun 72:5759–5767

Lim K, Tempczyk A, Parsons JF, Bonander N, Toedt J, Kelman Z, Howard A, Eisenstein E, Herzberg O (2003) Crystal structure of YbaB from *Haemophilus influenzae* (HI0442), a protein of unknown function coexpressed with the recombinational DNA repair protein RecR. Proteins 50:375–379

Lin YH, Romo JA, Smith TC, Reyes AN, Karna SL, Miller CL, Van Laar TA, Yendapally R, Chambers JP, Seshu J (2017) Spermine and spermidine alter gene expression and antigenic profile of *Borrelia burgdorferi*. Infect Immun 85: in press

Livermore BP, Bey RF, Johnson RC (1978) Lipid metabolism of *Borrelia hermsi*. Infect Immun 20:215–220

Lybecker MC, Samuels DS (2007) Temperature-induced regulation of RpoS by a small RNA in *Borrelia burgdorferi*. Mol Microbiol 64:1075–1089

Lybecker MC, Abel CA, Feig AL, Samuels DS (2010) Identification and function of the RNA chaperone Hfq in the Lyme disease spirochete *Borrelia burgdorferi*. Mol Microbiol 78:622–635

McCown PJ, Corbino KA, Stav S, Sherlock ME, Breaker RR (2017) Riboswitch diversity and distribution. RNA 23:995–1011

Medrano MS, Ding Y, Wang XG, Lu P, Coburn J, Hu LT (2007) Regulators of expression of the oligopeptide permease A proteins of *Borrelia burgdorferi*. J Bacteriol 189:2653–2659

Medrano MS, Policastro PF, Schwan TG, Coburn J (2010) Interaction of *Borrelia burgdorferi* Hbb with the p66 promoter. Nucleic Acids Res 38:414–427

Miller CL, Karna SL, Seshu J (2013) Borrelia host adaptation regulator (BadR) regulates *rpoS* to modulate host adaptation and virulence factors in *Borrelia burgdorferi*. Mol Microbiol 88:105–124

Miller JC, von Lackum K, Babb K, McAlister JD, Stevenson B (2003) Temporal analysis of *Borrelia burgdorferi* Erp protein expression throughout the mammal-tick infectious cycle. Infect Immun 71:6943–6952

Miller ST, Xavier KB, Campagna SR, Taga ME, Semmelhack MF, Bassler BL, Hughson FM (2004) *Salmonella typhimurium* recognizes a chemically distinct form of the bacterial quorum-sensing signal AI-2. Mol Cell 15:677–687

Mouw KW, Rice PA (2007) Shaping the *Borrelia burgdorferi* genome: crystal structure and binding properties of the DNA-bending protein Hbb. Mol Microbiol 63:1319–1330

Mukherjee S, Yakhnin H, Kysela D, Sokoloski J, Babitzke P, Kearns DB (2011) CsrA-FliW interaction governs flagellin homeostasis and a checkpoint on flagellar morphogenesis in *Bacillus subtilis*. Mol Microbiol 82:447–461

Mukherjee S, Oshiro RT, Yakhnin H, Babitzke P, Kearns DB (2016) FliW antagonizes CsrA RNA binding by a noncompetitive allosteric mechanism. Proc Natl Acad Sci 113:9870–9875

Neelakanta G, Li X, Pal U, Liu X, Beck DS, DePonte K, Fish D, Kantor FS, Fikrig E (2007) Outer surface protein B is critical for *Borrelia burgdorferi* adherence and survival within *Ixodes* ticks. PLoS Pathog 3:e33

Ouyang Z, Blevins JS, Norgard MV (2008) Transcriptional interplay among the regulators Rrp2, RpoN and RpoS in *Borrelia burgdorferi*. Microbiology 154:2641–2658

Ouyang Z, Kumar M, Kariu T, Haq S, Goldberg M, Pal U, Norgard MV (2009) BosR (BB0647) governs virulence expression in *Borrelia burgdorferi*. Mol Microbiol 74:1331–1343

Ouyang Z, Deka RK, Norgard MV (2011) BosR (BB0647) controls the RpoN-RpoS regulatory pathway and virulence expression in *Borrelia burgdorferi* by a novel DNA-binding mechanism. PLoS Pathog 7:e1001272

Ouyang Z, Zhou J (2015) BadR (BB0693) controls growth phase-dependent induction of *rpoS* and *bosR* in *Borrelia burgdorferi* via recognizing TAAAATAT motifs. Mol Microbiol 98:1147–1167

Ouyang Z, Zhou J (2017) The putative Walker A and Walker B motifs of Rrp2 are required for the growth of *Borrelia burgdorferi*. Mol Microbiol 103:86–98

Ouyang Z, Zhou J, Brautigam CA, Deka RK, Norgard MV (2014a) Identification of a core sequence for the binding of BosR to the *rpoS* promoter region in *Borrelia burgdorferi*. Microbiology 160:851–862

Ouyang Z, Zhou J, Norgard MV (2014b) Synthesis of RpoS is dependent on a putative enhancer binding protein Rrp2 in Borrelia burgdorferi. PLoS ONE 9:e96917

Ouyang Z, Zhou J, Norgard MV (2014c) CsrA (BB0184) Is not involved in activation of the RpoN-RpoS regulatory pathway in *Borrelia burgdorferi*. Microbiology 82:1511–1522

Pal U, Li X, Wang T, Montgomery RR, Ramamoorthi N, deSilva AM, Bao F, Yang X, Pypaert M, Pradhan D, Kantor FS, Telford S, Anderson JF, Fikrig E (2004) TROSPA, an *Ixodes scapularis* receptor for *Borrelia burgdorferi*. Cell 119:457–468

Pappas CJ, Iyer R, Petzke MM, Caimano MJ, Radolf JD, Schwartz I (2011) *Borrelia burgdorferi* requires glycerol for maximum fitness during the tick phase of the enzootic cycle. PLoS Pathog 7:e1002102

Pereira CS, Santos AJ, Bejerano-Sagie M, Correia PB, Marques JC, Xavier KB (2012) Phosphoenolpyruvate phosphotransferase system regulates detection and processing of the quorum sensing signal autoinducer-2. Mol Microbiol 84:93–104

Piesman J, Schneider BS (2002) Dynamic changes in Lyme disease spirochetes during transmission by nymphal ticks. Exp Appl Acarol 28:141–145

Piesman J, Oliver JR, Sinsky RJ (1990) Growth kinetics of the Lyme disease spirochete (*Borrelia burgdorferi*) in vector ticks (*Ixodes dammini*). Am J Trop Med Hyg 42:352–357

Piesman J, Schneider BS, Zeidner NS (2001) Use of quantitative PCR to measure density of *Borrelia burgdorferi* in the midgut and salivary glands of feeding tick vectors. J Clin Microbiol 39:4145–4148

Popitsch N, Bilusic I, Rescheneder P, Schroeder R, Lybecker M (2017) Temperature-dependent sRNA transcriptome of the Lyme disease spirochete. BMC Genom 18:28

Posey JE, Gherardini FC (2000) Lack of a role for iron in the Lyme disease pathogen. Science 288:1651–1653

Radolf JD, Caimano MJ, Stevenson B, Hu LT (2012) Of ticks, mice, and men: understanding the dual-host lifestyle of Lyme disease spirochaetes. Nature Rev Microbiol 10:87–98

Ramamoorthy R, Philipp MT (1998) Differential expression of *Borrelia burgdorferi* proteins during growth in vitro. Infect Immun 66:5119–5124

Ramamoorthy R, Scholl-Meeker D (2001) *Borrelia burgdorferi* proteins whose expression is similarly affected by culture temperature and pH. Infect Immun 69:2739–2742

Richards CL, Lawrence KA, Su H, Yang Y, Yang XF, Dulebohn DP, Gherardini FC (2015) Acetyl-phosphate is not a global regulatory bridge between virulence and central metabolism in *Borrelia burgdorferi*. PLoS ONE 10:e0144472

Riley SP, Bykowski T, Cooley AE, Burns LH, Babb K, Brissette CA, Bowman A, Rotondi M, Miller MC, DeMoll E, Lim K, Fried MG, Stevenson B (2009) *Borrelia burgdorferi* EbfC defines a newly-identified, widespread family of bacterial DNA-binding proteins. Nucleic Acids Res 37:1973–1983

Rogers EA, Terekhova D, Zhang HM, Hovis KM, Schwartz I, Marconi RT (2009) Rrp1, a cyclic-di-GMP-producing response regulator, is an important regulator of *Borrelia burgdorferi* core cellular functions. Mol Microbiol 71:1551–1573

Rosenbluh A, Banner CD, Losick R, Fitz-James PC (1981) Identification of a new developmental locus in *Bacillus subtilis* by construction of a deletion mutation in a cloned gene under sporulation control. J Bacteriol 148:341–351

Ryjenkov DA, Tarutina M, Moskvin OV, Gomelsky M (2005) Cyclic diguanylate is a ubiquitous signaling molecule in bacteria: insights into biochemistry of the GGDEF protein domain. J Bacteriol 187:1792–1798

Salman-Dilgimen A, Hardy PO, Dresser AR, Chaconas G (2011) HrpA, a DEAH-box RNA helicase, is involved in global gene regulation in the Lyme disease spirochete. PLoS ONE 6: e22168

Sanjuan E, Esteve-Gassent MD, Maruskova M, Seshu J (2009) Overexpression of CsrA (BB0184) alters the morphology and antigen profiles of *Borrelia burgdorferi*. Infect Immun 77:5149–5162

Sarkar A, Hayes BM, Dulebohn DP, Rosa PA (2011) Regulation of the virulence determinant OspC by bbd18 on linear plasmid lp17 of *Borrelia burgdorferi*. J Bacteriol 193:5365–5373

Savage CR, Arnold WK, Gjevre-Nail A, Koestler BJ, Bruger EL, Barker JR, Waters CM, Stevenson B (2015) Intracellular concentrations of *Borrelia burgdorferi* cyclic di-AMP are not changed by altered expression of the CdaA synthase. PLoS ONE 10:e0125440

Schauder S, Shokat S, Surette MG, Bassler BL (2001) The LuxS family of bacterial autoinducers: biosynthesis of a novel quorum-sensing signal molecule. Mol Microbiol 41:463–476

Schwan TG, Piesman J, Golde WT, Dolan MC, Rosa PA (1995) Induction of an outer surface protein on *Borrelia burgdorferi* during tick feeding. Proc Natl Acad Sci USA 92:2909–2913

Schwan TG, Piesman J (2000) Temporal changes in outer surface proteins A and C of the Lyme disease-associated spirochete, *Borrelia burgdorferi*, during the chain of infection in ticks and mice. J Clin Microbiol 38:382–388

Seshu J, Boylan JA, Gherardini FC, Skare JT (2004a) Dissolved oxygen levels alter gene expression and antigen profiles in *Borrelia burgdorferi*. Infect Immun 72:1580–1586

Seshu J, Boylan JA, Hyde JA, Swingle KL, Gherardini FC, Skare JT (2004b) A conservative amino acid change alters the function of BosR, the redox regulator of *Borrelia burgdorferi*. Mol Microbiol 54:1352–1363

Shi Y, Dadhwal P, Li X, Liang FT (2014) BosR functions as a repressor of the *ospAB* operon in *Borrelia burgdorferi*. PLoS ONE 9:e109307

Skare JT, Shaw DK, Trzeciakowski JP, Hyde JA (2016) In vivo imaging demonstrates that *Borrelia burgdorferi ospC* is uniquely expressed temporally and spatially throughout experimental infection. PLoS ONE 11:e0162501

Smith AH, Blevins JS, Bachlani GN, Yang XF, Norgard MV (2007) Evidence that RpoS (σ^S) in *Borrelia burgdorferi* is controlled directly by RpoN (σ^{54}/σ^N). J Bacteriol 189:2139–2144

Stevenson B, Babb K (2002) LuxS-mediated quorum sensing in *Borrelia burgdorferi*, the Lyme disease spirochete. Infect Immun 70:4099–4105

Stevenson B, Schwan TG, Rosa PA (1995) Temperature-related differential expression of antigens in the Lyme disease spirochete, *Borrelia burgdorferi*. Infect Immun 63:4535–4539

Stevenson B, Bono JL, Schwan TG, Rosa P (1998) *Borrelia burgdorferi* Erp proteins are immunogenic in mammals infected by tick bite, and their synthesis is inducible in cultured bacteria. Infect Immun 66:2648–2654

Stevenson B, von Lackum K, Wattier RL, McAlister JD, Miller JC, Babb K (2003) Quorum sensing by the Lyme disease spirochete. Microbes Infect 5:991–997

Stevenson B, von Lackum K, Riley SP, Cooley AE, Woodman ME, Bykowski T (2006) Evolving models of Lyme disease spirochete gene regulation. Wien Klin Wochenschr 118:643–652

Stewart PE, Wang X, Bueschel DM, Clifton DR, Grimm D, Tilly K, Carroll JA, Weis JJ, Rosa PA (2006) Delineating the requirement for the *Borrelia burgdorferi* virulence factor OspC in the mammalian host. Infect Immun 74:3547–3553

Surette MG, Miller MB, Bassler BL (1999) Quorum sensing in *Escherichia coli*, *Salmonella typhimurium*, and *Vibrio harveyi*: a new family of genes responsible for autoinducer production. Proc Natl Acad Sci USA 96:1639–1644

Sze CW, Li C (2011) Inactivation of bb0184, which encodes carbon storage regulator A, represses the infectivity of *Borrelia burgdorferi*. Infect Immun 79:1270–1279

Sze CW, Morado DR, Liu J, Charon NW, Xu H, Li C (2011) Carbon storage regulator A (CsrA (Bb)) is a repressor of *Borrelia burgdorferi* flagellin protein FlaB. Mol Microbiol 82:851–864

Taga ME, Miller ST, Bassler BL (2003) Lsr-mediated transport and processing of AI-2 in *Salmonella typhimurium*. Mol Microbiol 50:1411–1427

Tilly K, Fuhrman J, Campbell J, Samuels DS (1996) Isolation of *Borrelia burgdorferi* genes encoding homologues of DNA-binding protein HU and ribosomal protein S20. Microbiology 142:2471–2479

Tilly K, Grimm D, Bueschel DM, Krum JG, Rosa P (2004) Infectious cycle analysis of a *Borrelia burgdorferi* mutant defective in transport of chitobiose, a tick cuticle component. Vector-borne Zoonotic Dis 4:159–168

Tilly K, Krum JG, Bestor A, Jewett MW, Grimm D, Bueschel D, Byram R, Dorward D, Vanraden MJ, Stewart P, Rosa P (2006) *Borrelia burgdorferi* OspC protein is required exclusively in a crucial early stage of mammalian infection. Infect Immun 74:3554–3564

Tilly K, Bestor A, Jewett MW, Rosa P (2007) Rapid clearance of Lyme disease spirochetes lacking OspC from skin. Infect Immun 75:1517–1519

Tilly K, Bestor A, Dulebohn DP, Rosa PA (2009) OspC-independent infection and dissemination by host-adapted *Borrelia burgdorferi*. Infect Immun 77:2672–2682

Tilly K, Bestor A, Rosa PA (2013) Lipoprotein succession in *Borrelia burgdorferi*: similar but distinct roles for OspC and VlsE at different stages of mammalian infection. Mol Microbiol 89:216–227

Tokarz R, Anderton JM, Katona LI, Benach JL (2004) Combined effects of blood and temperature shift on *Borrelia burgdorferi* gene expression as determined by whole genome array. Infect Immun 72:5419–5432

Troxell B, Xu H, Yang XF (2012) *Borrelia burgdorferi*, a pathogen that lacks iron, encodes manganese-dependent superoxide dismutase essential for resistance to streptonigrin. J Biol Chem 287:19284–19293

Troxell B, Ye M, Yang Y, Carrasco SE, Lou Y, Yang XF (2013) Manganese and zinc regulate virulence determinants in *Borrelia burgdorferi*. Infect Immun 81:2743–2752

Troy EB, Lin T, Gao L, Lazinski DW, Lundt M, Camilli A, Norris SJ, Hu LT (2016) Global Tn-seq analysis of carbohydrate utilization and vertebrate infectivity of *Borrelia burgdorferi*. Mol Microbiol 101:1003–1023

Vakulskas CA, Potts AH, Babitzke P, Ahmer BM, Romeo T (2015) Regulation of bacterial virulence by Csr (Rsm) systems. Micribiol Mol Biol Rev 79:193–224

Van Laar TA, Lin YH, Miller CL, Karna SLR, Chambers JP, Seshu J (2012) Effects of levels of acetate on the mevalonate pathway of *Borrelia burgdorferi*. PLoS ONE 5:e38171

von Lackum K, Stevenson B (2005) Carbohydrate utilization by the Lyme borreliosis spirochete, *Borrelia burgdorferi*. FEMS Microbiol Lett 243:173–179

von Lackum K, Miller JC, Bykowski T, Riley SP, Woodman ME, Brade V, Kraiczy P, Stevenson B, Wallich R (2005) *Borrelia burgdorferi* regulates expression of complement regulator-acquiring surface protein 1 during the mammal-tick infection cycle. Infect Immun 73:7398–7405

von Lackum K, Babb K, Riley SP, Wattier RL, Bykowski T, Stevenson B (2006) Functionality of *Borrelia burgdorferi* LuxS: the Lyme disease spirochete produces and responds to the pheromone autoinducer-2, and lacks a complete activated-methyl cycle. Int J Med Microbiol 296(S1):92–102

Wang H, Wang F, Hua X, Ma T, Chen J, Xu X, Wang L, Tian B, Hua Y (2012a) Genetic and biochemical characteristics of the histone-like protein DR0199 in *Deinococcus radiodurans*. Microbiology 158:936–943

Wang P, Lutton A, Olesik J, Vali H, Li X (2012b) A novel iron- and copper-binding protein in the Lyme disease spirochaete. Mol Microbiol 86:1441–1451

Wang P, Dadhwal P, Cheng Z, Zianni MR, Rikihisa Y, Liang FT, Li X (2013) *Borrelia burgdorferi* oxidative stress regulator BosR directly represses lipoproteins primarily expressed in the tick during mammalian infection. Mol Microbiol 89:1140–1153

Wang P, Yu Z, Santangelo TJ, Olesik J, Wang Y, Heldwein E, Li X (2017) BosR is a novel Fur family member responsive to copper and regulating copper homeostasis in *Borrelia burgdorferi*. J Bacteriol

Xavier KB, Bassler BL (2003) LuxS quorum sensing: more than just a numbers game. Curr Opin Microbiol 6:191–197

Xavier KB, Miller ST, Lu W, Kim JH, Rabinowitz J, Pelczer I, Semmelhack MF, Bassler BL (2007) Phosphorylation and processing of the quorum-sensing molecule autoinducer-2 in enteric bacteria. ACS Chem Biol 2:128–136

Xu H, Caimano MJ, Lin T, He M, Radolf JD, Norris SJ, Gherardini F, Wolfe AJ, Yang XF (2010) Role of acetyl-phosphate in activation of the Rrp2-RpoN-RpoS pathway in *Borrelia burgdorferi*. PLoS Pathog 6:e1001104

Xu Q, Shi Y, Dadhwal P, Liang FT (2012) RpoS regulates essential virulence factors remaining to be identified in *Borrelia burgdorferi*. PLoS ONE 7:e53212

Yang X, Goldberg MS, Popova TG, Schoeler GB, Wikel SK, Hagman KE, Norgard MV (2000) Interdependence of environmental factors influencing reciprocal patterns of gene expression in virulent *Borrelia burgdorferi*. Mol Microbiol 37:1470–1479

Yang X, Popova TG, Goldberg MS, Norgard MV (2001) Influence of cultivation media on genetic regulatory patterns in *Borrelia burgdorferi*. Infect Immun 69:4159–4163

Yang XF, Alani SM, Norgard MV (2003a) The response regulator Rrp2 is essential for the expression of major membrane lipoproteins in *Borrelia burgdorferi*. Proc Natl Acad Sci USA 100:11001–11006

Yang XF, Hübner A, Popova TG, Hagman KE, Norgard MV (2003b) Regulation of expression of the paralogous Mlp family in *Borrelia burgdorferi*. Infect Immun 71:5012–5020

Ye M, Zhang JJ, Fang X, Lawlis GB, Troxell B, Zhou Y, Gomelsky M, Lou Y, Yang XF (2014) DhhP, a cyclic di-AMP phosphodiesterase of *Borrelia burgdorferi*, is essential for cell growth and virulence. Infect Immun 82:1840–1849

Yin Y, Yang Y, Xiang X, Wang Q, Yang ZN, Blevins J, Lou Y, Yang XF (2016) Insight into the dual functions of bacterial enhancer-binding protein Rrp2 of *Borrelia burgdorferi*. J Bacteriol 198:1543–1552

Zhang J-R, Norris SJ (1998a) Genetic variation of the *Borrelia burgdorferi* gene *vlsE* involves cassette-specific, segmental gene conversion. Infect Immun 66:3698–3704

Zhang J-R, Norris SJ (1998b) Kinetics and in vivo induction of genetic variation of *vlsE* in *Borrelia burgdorferi*. Infect Immun 66:3689–3697

Zhang J-R, Hardham JM, Barbour AG, Norris SJ (1997) Antigenic variation in Lyme disease borreliae by promiscuous recombination of VMP-like sequence cassettes. Cell 89:1–20

Zückert WR, Meyer J (1996) Circular and linear plasmids of Lyme disease spirochetes have extensive homology: characterization of a repeated DNA element. J Bacteriol 178:2287–2298

Genetic Manipulation of *Borrelia* Spp.

Dan Drecktrah and D. Scott Samuels

Abstract The spirochetes *Borrelia* (*Borreliella*) *burgdorferi* and *Borrelia hermsii*, the etiologic agents of Lyme disease and relapsing fever, respectively, cycle in nature between an arthropod vector and a vertebrate host. They have extraordinarily unusual genomes that are highly segmented and predominantly linear. The genetic analyses of Lyme disease spirochetes have become increasingly more sophisticated, while the age of genetic investigation in the relapsing fever spirochetes is just dawning. Molecular tools available for *B. burgdorferi* and related species range from simple selectable markers and gene reporters to state-of-the-art inducible gene expression systems that function in the animal model and high-throughput mutagenesis methodologies, despite nearly overwhelming experimental obstacles. This armamentarium has empowered borreliologists to build a formidable genetic understanding of the cellular physiology of the spirochete and the molecular pathogenesis of Lyme disease.

Contents

D. Drecktrah (✉) · D.S. Samuels (✉)
Division of Biological Sciences, University of Montana, Missoula, MT 59812, USA
e-mail: dan.drecktrah@mso.umt.edu

D.S. Samuels
e-mail: samuels@mso.umt.edu

Current Topics in Microbiology and Immunology (2018) 415:113–140
DOI 10.1007/82_2017_51

Published Online: 17 September 2017

1 Introduction

The maxim *spirochetes do it differently* has proved repeatedly to be true, from ultrastructure to motility to gene regulation (Charon and Goldstein 2002; Charon et al. 2012; Johnson 1977; Samuels and Radolf 2009; Samuels 2011), yet a few exceptions exist; one of these rule-proving exceptions regards genetically manipulating *Borrelia* spp. Except for the labyrinthine tactics required to grow these serpentine bacteria in culture and handle their Gordian genomes, which are exceedingly segmented and largely linear (Brisson et al. 2012; Fraser et al. 1997), the transformation and complementation of *Borrelia* (*Borreliella*) *burgdorferi* (and related species that cause Lyme disease) and *Borrelia hermsii* (and related species that cause relapsing fever) are not unlike that of model microorganisms (Battisti et al. 2008; Samuels 1995).

The awesome power of reverse genetics has been tapped to dissect the microbial physiology of *B. burgdorferi* and, to a lesser extent, *B. afzelii*, *B. garinii*, and *B. hermsii* as well as the molecular pathogenesis of Lyme disease and, more recently, relapsing fever (Brisson et al. 2012; Groshong and Blevins 2014; Raffel et al. 2014; Rosa et al. 2005). One good reason for the success of the genetic approach is the availability of robust animal models (Barthold et al. 2010); these experimental systems mimic the natural course of disease as well as the two-host enzootic cycles that include a vertebrate and an arthropod vector, usually a tick (Caimano et al. 2016; Piesman and Schwan 2010; Radolf et al. 2012). The genetic toolbox constructed since *Borrelia* was first transformed (Samuels et al. 1994a) includes genome sequences (Di et al. 2014; Fraser et al. 1997), selectable markers and gene reporters (see Sect. 3), shuttle vectors (see Sect. 6), transposon mutagenesis (see Sect. 7), inducible promoters (see Sect. 5), and a counterselectable marker (see Sect. 8). Borreliogists have mutated a multitude of genes and now molecularly probe the *Borrelia* lifestyle encompassing all things from metabolism to transcriptional regulation to host interactions.

2 Transformation

B. burgdorferi was first transformed in the early 1990s with an allele of the endogenous *gyrB* gene, encoding the B subunit of DNA gyrase, that confers resistance to coumarin antibiotics (Samuels et al. 1994a). The electrotransformation protocol is similar to that used for many bacteria (Nickoloff 1995), but low transformation efficiency and retention of the native suite of plasmids continues to challenge borreliologists. The basic protocol has remained relatively unchanged: competent cells are prepared by removing salts, purified DNA substrate is added, and a high-voltage electric current is applied to the cells. Electroporation transiently compromises the integrity of the outer membrane, visualized by electron microscopy to reveal dark-stained patches that likely represent breaches in the membrane (Fig. 1) (Samuels and Garon 1997). The membrane rapidly recovers (within 30 min) when the cells are placed in growth medium following electroporation. Substantial improvements in transformation efficiency have been wrought by increasing the amount of substrate DNA to tens of micrograms (Eggers et al. 2002; Elias et al. 2002; Hübner et al. 2001; Stewart et al. 2001; Tilly et al. 2000) and transforming with modified DNA (Chen et al. 2008; Jewett et al. 2007).

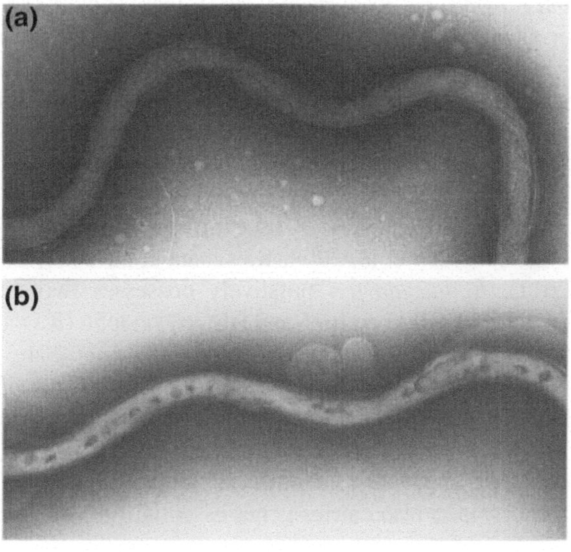

Fig. 1 *B. burgdorferi before* (**a**) and *after* (**b**) electroporation as visualized by negative staining and transmission electron microscopy. Note the darkly stained regions that are thought to be transient pores generated by the electric pulse that serve as a passageway for DNA to enter the spirochete. Reprinted with permission from Samuels and Garon (1997)

2.1 Competent Cells and Electroporation

Successful preparation of *B. burgdorferi* competent cells requires growth to mid- to late-log phase; cell density is probably the most vital parameter (Samuels et al. 1994a; Tilly et al. 2000). Certain low-passage and mutant strains are refractory to transformation due to inordinate clumping during competent cell preparation (Hübner et al. 2001; Samuels 1995; Tilly et al. 2000). Salts are removed from the growth medium by a series of cold, sterile washes and the *Borrelia* cells are resuspended in a glycerol-sucrose solution. Arcing will occur during electroporation if salts are not effectively removed, which will kill the vast majority of cells and, obviously, prevent transformation. Electrotransformation is most efficient with fresh competent cells, but cells can be stored at −80 °C and slow-thawed on ice, although there is about a 50% decrease in efficiency (Samuels 1995; Samuels and Garon 1997).

2.2 DNA Source, Form and Modification

Many different substrates have been transformed into *B. burgdorferi*, from short oligonucleotides (Samuels and Garon 1997) to an entire native 28-kb linear plasmid (Grimm et al. 2004a). The original allelic exchange transformation used PCR products to mutate the *gyrB* gene and confer antibiotic resistance (Samuels et al. 1994a). More recently, linear segments of cloned DNA containing a selectable marker are typically used for allelic exchange to generate mutants. In addition, circular suicide plasmids have been integrated into the genome by recombination (Stevenson et al. 1998). An important advance in the *Borrelia* genetic toolbox was the development of circular shuttle vectors that autonomously replicate in both *B. burgdorferi* and *Escherichia coli* (Eggers et al. 2002; Sartakova et al. 2000; Stewart et al. 2001). These plasmid vectors enhanced the number and sophistication of genetic experiments in *B. burgdorferi*, including *trans*-complementation, ectopic gene expression, and inducible gene expression. Linear shuttle vectors based on lp25, lp28-1, and lp17 have also been built using a similar strategy to the circular shuttle vector, although they have not been used frequently (Beaurepaire and Chaconas 2005; Stewart et al. 2003).

The low frequency of transformation is likely due to restriction-modification systems (Chen et al. 2008; Kawabata et al. 2004; Lawrenz et al. 2002; Rego et al. 2011). *B. burgdorferi* putatively possesses two type-IV restriction-modification enzymes: BBE02 on lp25 and BBQ67 on lp56 (Kawabata et al. 2004; Lawrenz et al. 2002). The gene products of *bbe02* and *bbq67* are predicted to methylate certain adenines and cleave foreign DNA lacking this modification. The presence of these two plasmids is correlated with low transformation efficiency of shuttle vectors isolated from *E. coli*. Transformants often lose lp25 and transformation efficiency increases about 40-fold in *bbe02* mutants (Kawabata et al. 2004). To complicate matters, lp25 also carries genes essential for vertebrate and tick infectivity (Grimm et al. 2004a; Labandeira-Rey and Skare 2001; Purser and Norris 2000; Purser et al. 2003; Revel et al. 2005; Strother and de Silva 2005). Thus, low-passage, infectious strains are more difficult to transform than high-passage, attenuated strains. BBE02 is not required for in vitro cultivation, so replacing *bbe02* with a selectable marker both increased transformation efficiency and provided a selection for lp25, allowing transformants to retain infectivity (Chan et al. 2015; Gilbert et al. 2007; Kasumba et al. 2015; Rego et al. 2011). Clever tactics to subvert the restriction-modification systems have been developed over the years. Transformation efficiency can be increased by first transforming the DNA into a strain carrying only one of the putative restriction-modification systems and using this partially methylated DNA for subsequent transformation (Jewett et al. 2007). Alternatively, in vitro CpG methylation of substrate DNA increases transformation efficiency (Chen et al. 2008). Both approaches have facilitated genetic manipulation of *B. burgdorferi*.

2.3 Relapsing Fever Borrelia

B. hermsii, one of the species of *Borrelia* that causes relapsing fever in humans (Barbour and Guo 2010), has also been successfully transformed (Battisti et al. 2008), but the history of genetic manipulation is considerably younger. The protocol for preparation of competent cells, electrotransformation and selection are basically the same as for *B. burgdorferi*. The notable changes include using large amounts of DNA for transformation (25 μg) and selecting for transformants in antibiotics for up to a week in batch before plating to isolate clones (Battisti et al. 2008; Fine et al. 2011; Raffel et al. 2014). The shuttle vector pBhSV2, based on the *B. burgdorferi* shuttle vector pBSV2, functions in *B. hermsii*, but only the antibiotics gentamicin and kanamycin have been effective for selection (Battisti et al. 2008). So far, three genes have been mutated: *vtp*, which encodes variable tick protein, an outer surface lipoprotein synthesized when the spirochete is resident in its tick vector (Battisti et al. 2008; Raffel et al. 2014), *vmp*, which is the expression locus for the genes encoding the variable major proteins that are synthesized in the mammalian host and allow the spirochete to evade the immune response (Raffel et al. 2014), and *dam*, which encodes DNA adenine methyltransferase (James et al. 2016). In addition, the green fluorescent protein (GFP) gene has been expressed in *B. hermsii* (Fine et al. 2011) and its cousin *Borrelia turicatae* (Krishnavajhala et al. 2017). With these advances in hand, we are beginning to see the dawn of the age of reverse genetics in relapsing fever spirochetes.

3 Selectable Markers and Assayable Reporters

Selectable markers allow for the preferential isolation of genetically altered microorganisms, including *B. burgdorferi*. Antibiotic resistance markers have proved to be remarkably utilitarian tools for negative selection, the inhibition of the unaltered parental cells. *B. burgdorferi* is susceptible to many antibiotics and the genes conferring resistance to antibiotics such as streptomycin, kanamycin, gentamicin, and erythromycin are paramount for genetic studies. The antibiotics ampicillin and tetracycline are used to treat patients with Lyme disease, so their use in genetic experiments is discouraged.

The challenges, both in cost and time, in culturing *B. burgdorferi* to study gene expression have made reporter genes also incredibly useful tools. Reporter gene products provide both a convenient readout to measure transcriptional regulation when fused to promoters and a signal for visualization of tagged proteins or entire cells in vitro and in vivo. Advances in imaging of live mice and ticks infected with *B. burgdorferi* have provided an unprecedented vista of the interaction of the spirochete with the host (Dunham-Ems et al. 2009; Moriarty et al. 2008).

3.1 Native Antibiotic Resistance

The first selectable marker used for genetic manipulation in *B. burgdorferi* was an allele of the *gyrB* gene conferring resistance to the coumarin antibiotics (Samuels et al. 1994a, b; Samuels 2006). A synthetic version of *gyrB* designed to minimize recombination into the native *gyrB* locus on the chromosome was used briefly as a selectable marker (Elias et al. 2003). Unfortunately, the *gyrB* mutations also affect DNA supercoiling in *B. burgdorferi* (Samuels and Garon 1993), causing pleiotropic effects such as increased expression of *groEL*, encoding a chaperonin (Alverson and Samuels 2002) and *ospC*, encoding an outer surface lipoprotein (Alverson et al. 2003; Drecktrah et al. 2013; Yang et al. 2005), which spurred the exploitation of antibiotic resistance markers from other microorganisms.

3.2 Foreign Antibiotic Resistance Genes

The efficiency of genetic manipulation of *B. burgdorferi* was enhanced when new selectable markers were developed by fusing endogenous promoters to nonnative antibiotic resistance genes (Bono et al. 2000). The strong, constitutive promoters of the flagellin and basal body rod genes *flaB* and *flgB*, respectively, were chosen to drive gene expression (Bono et al. 2000; Ge et al. 1997; Sohaskey et al. 1997). These promoters were first fused to the *aphI* gene from *Tn903*, resulting in *flaBp-aphI* and *flgBp-aphI*, which mediate resistance to kanamycin (Bono et al. 2000). Other markers were constructed soon after by the same strategy and conferred resistance to streptomycin using *aadA* from *Shigella flexneri* plasmid R100 (Frank et al. 2003) and resistance to gentamicin using *aacC1* from *Tn1696* (Elias et al. 2003). These antibiotic resistance markers function in both *E. coli* and *B. burgdorferi*, making them versatile and practical tools for genetic studies. Not all antibiotic resistance cassettes function in *B. burgdorferi*, including *cat* and *pac*, the genes encoding resistance to chloramphenicol and puromycin, respectively (Elias et al. 2003). The *cat* gene encodes a chloramphenicol acetyl transferase, but the acetylation of chloramphenicol is reversed by esterases present in the *B. burgdorferi* growth medium (Sohaskey and Barbour 1999), which might be overcome by modifying the medium. Some foreign antibiotic resistance genes driven by their own promoters function in *B. burgdorferi*, such as *erm* conferring erythromycin resistance (Sartakova et al. 2000). Others include *aph(3')-IIIa* conferring kanamycin resistance, *aad9* conferring spectinomycin resistance, *linA'* conferring lincomycin resistance, and *aac(3)*-VIa conferring sisomycin and gentamicin resistance, but they have not been widely used (Sartakova et al. 2003). Use of the *erm* gene has been restricted by some Institutional Biosafety Committees because erythromycin can be used to treat Lyme disease patients (Terekhova et al. 2002). Spectinomycin is not efficient at selecting for transformants containing the *aadA* (and probably the *aad9*) gene because *B. burgdorferi* readily becomes resistant to this antibiotic (Criswell et al. 2006). Lincomycin has not been utilized for selection in *B. burgdorferi*, but could be adapted to expand the toolbox for genetic manipulation. While the early years entailed plodding development of methodologies for genetic manipulation, borreliologists have refined their experimental approaches and now almost exclusively employ selectable markers conferring antibiotic resistance to kanamycin, streptomycin and gentamicin.

3.3 Mutant Isolation

There are two techniques for plating and selecting transformed *B. burgdorferi*. One is semi-solid agarose plating where transformed cells are allowed to recover overnight before plating in growth medium mixed with molten agarose (Kurtti et al. 1987; Rosa and Hogan 1992; Samuels et al. 1994a). *B. burgdorferi* is mixed with the molten agarose-growth medium containing the antibiotic so the colonies form within the mixture, not on the surface. After incubation for 5 days to 2 months in a CO_2 incubator, individual colonies appear as small white clouds in the plate and are then expanded in liquid culture to be screened for the desired genetic alteration. The second method is dilution plating in which the transformed cells are diluted in 100 ml growth medium and then distributed in four 96-well plates following overnight recovery (Yang et al. 2004). Plates are incubated in a CO_2 incubator until the medium in individual wells begins to turn from red to yellow, as metabolizing *B. burgdorferi* excretes lactic acid that is indicated by phenol red in the medium. Positive wells are expanded in liquid culture and screened.

3.4 Reporter Genes

The first reporter used in *B. burgdorferi* was the *cat* gene (Sohaskey et al. 1997). Although the gene product does not confer resistance to chloramphenicol in

B. burgdorferi, the reporter provided some of the first assays of promoter activity (Alverson et al. 2003; Sohaskey et al. 1997, 1999), just as the age of spirochete genomics was dawning (Fraser et al. 1997). More recently, the gene for β-galactosidase, *lacZ*, fused to the *ospC* promoter, was used to study the regulation of *ospC* expression in both liquid and semi-solid media (Hayes et al. 2010, 2014; Sarkar et al. 2011).

The luciferase (*luc*) gene from the firefly *Photinus pyralis* has been adapted as a transcriptional reporter to assay gene expression from endogenous promoters (Blevins et al. 2007; Ouyang et al. 2010). The *luc* reporter also provided the means to follow *B. burgdorferi* infection in vivo by bioluminescence imaging of infected mice (Chan et al. 2015; Hyde et al. 2011). *B. burgdorferi* infection and dissemination of mutant strains lacking the fibronectin-binding protein BBK32 or the decorin-binding proteins DbpA and DbpB (Hyde et al. 2011) were assayed as well as the pathogenokinetics of *ospC* expression (Skare et al. 2016). Recently, a dual *luc* reporter system has been successfully implemented to study gene expression in the tick vector. The second *luc* reporter, from *Renilla reniformis*, allows for normalization to *B. burgdorferi* number, while the *P. pyralis luc* gene product can then accurately measure transcriptional activity (Adams et al. 2017a). The *luc* reporter systems developed for in vivo work provide a more convenient and lower cost system to follow gene expression in the tick-mouse model.

The convenience and versatility of GFP has made it a popular reporter for studies of *B. burgdorferi* gene expression (Babb et al. 2004; Carroll et al. 2003; Eggers et al. 2002, 2004) and served to demonstrate the functionality of an inducible gene system (Whetstine et al. 2009). The GFP reporter has the advantage of allowing the analyses of single cells, which has highlighted the heterogeneity of *B. burgdorferi* populations (Srivastava and de Silva 2008). Additionally, the yellow fluorescent protein and cyan fluorescent protein have both been used as reporters (Eggers et al. 2002). The monomeric red fluorescent protein has proved useful for studying lipoprotein secretion to the cell surface (Schulze and Zückert 2006; Schulze et al. 2010) and was a crucial component of a FACS-based protein localization screen (Kumru et al. 2010). The GFP marker could not be directly visualized in these studies, as fluorescence was compromised in the periplasm (Schulze and Zückert 2006). Constitutively expressed *gfp* has allowed for visualization of *B. burgdorferi* in live-cell imaging studies. A strain of *B. burgdorferi* containing a plasmid expressing *gfp* from the *flaB* promoter allowed for exquisite examination of the spirochete's interaction with the mouse vasculature in vivo by live-cell intravital microscopy (Moriarty et al. 2008). Stable integration of a constitutively expressed *gfp* driven by the *flaB* promoter on cp26 has served as a fluorescent tag of *B. burgdorferi* to allow in vivo studies on morphology and population dynamics in ticks (Dunham-Ems et al. 2009).

Finally, GFP has been expressed in the relapsing fever agents *B. hermsii* (Fine et al. 2011) and *B. turicatae* (Krishnavajhala et al. 2017), and used to follow the persistence of the latter in the *Ornithodoros turicatae* tick midgut and salivary glands (Krishnavajhala et al. 2017). Visualization of GFP-relapsing fever *Borrelia* in vivo should aid in describing the spirochete lifestyle in both the tick vector and mammalian hosts of these understudied *Borrelia* species.

4 Mutagenesis

Using reverse genetics to reveal the function of a gene entails generating mutations in the gene, which is often suspected of producing a particular phenotype, and isolating the resulting mutants. Although transduction has been demonstrated in *Borrelia* (Eggers et al. 2001, 2016), mutagenesis currently relies exclusively on transformation (see Sect. 2) and selection with an antibiotic resistance marker (see Sect. 3). Besides pleiotropy and polar effects, a crucial caveat is off-target repercussions of genetic manipulation, which is particularly problematic in *B. burgdorferi* due to plasmid loss from the segmented genome (Kawabata et al. 2004; Labandeira-Rey and Skare 2001; Lawrenz et al. 2002; Purser and Norris 2000); therefore, mutant phenotypes should be confirmed by complementation or genetic reconstitution (see Sect. 6).

4.1 Gene Inactivation by Homologous Recombination

Almost every mutant in *Borrelia* has been generated by allelic exchange, requiring recombination of a synthetic DNA construct into the genome. The other mechanisms for mutagenesis, albeit not commonly utilized, are displacement of entire plasmids (Grimm et al. 2004a) and deletion of large segments of plasmids (Beaurepaire and Chaconas 2005; Bestor et al. 2010; Chaconas et al. 2001). Both *B. burgdorferi* (and other Lyme disease species) and *B. hermsii* (and other relapsing fever species) have antigenic variation systems that entail intra-genomic recombination (Nordstrand et al. 2000; Stoenner et al. 1982; Zhang et al. 1997) and a RecA homolog (Liveris et al. 2004; Putteet-Driver et al. 2004) that are likely hijacked to incorporate foreign DNA. VlsE antigenic variation in *B. burgdorferi* requires the branch migrase RuvAB (Dresser et al. 2009; Lin et al. 2009), which is also implicated in homologous recombination during genetic manipulation.

One curious phenomenon that was observed during allelic exchange (Brisson et al. 2012; Knight et al. 2000) and also documented during VlsE variation (Coutte et al. 2009) is the presence of mosaic recombination products, which implies that the crossover hops back and forth between the donor and recipient DNA strands, so the recombination machinery skips short stretches of sequence. This intermittent recombination generates further diversity during antigenic variation (Brisson et al. 2012; Coutte et al. 2009), but the caveat for genetic manipulation is that not all the expected changes may be present, so mutant and flanking sequences should be experimentally confirmed.

4.2 Site-Directed Mutagenesis

There have been only a handful of studies in which the sequence of a gene or regulatory element in its endogenous genomic location was specifically and subtly altered (as opposed to grossly deleting or inserting a gene) in *B. burgdorferi*. In the original report of transformation, a single codon of the *gyrB* gene was mutated to confer resistance to the antibiotic coumermycin A_1 (Samuels et al. 1994a; Samuels 2006). Several other genes have since been surgically manipulated in situ.

The C-terminal domain of the A subunit of DNA gyrase is uniquely synthesized as an independent DNA-binding protein termed Gac in *B. burgdorferi* (Knight and Samuels 1999). In order to disrupt *gac* expression without affecting the essential *gyrA* gene in which *gac* is embedded, the ribosome-binding site, the start codon and the second methionine codon of *gac* were mutated, which abrogated Gac production, although the mutant phenotype was disappointingly suppressed (Knight et al. 2000).

Rrp2 is an enhancer-binding protein and two-component system response regulator (Fraser et al. 1997; Samuels 2011). Exhaustive efforts failed to generate a null *rrp2* mutant, suggesting that Rrp2 had an essential function. Yang et al. (2003) changed a single residue in the putative activation domain, which eliminated expression of *rpoS* and the RpoS regulon, establishing the Rrp2-RpoN-RpoS pathway paradigm for gene regulation during the enzootic cycle (Burtnick et al. 2007; Caimano et al. 2007; Hübner et al. 2001; Samuels 2011), but retained the essential, and still unidentified, function of Rrp2.

OspC is a transmission factor required to establish *B. burgdorferi* infection in the vertebrate host (Dunham-Ems et al. 2012; Fingerle et al. 2007; Gilbert et al. 2007; Gilmore and Piesman 2000; Grimm et al. 2004b; Pal et al. 2004; Stewart et al. 2006; Tilly et al. 2006), but its molecular function has not been fully elucidated. Marconi and colleagues have mutated both a presumptive ligand-binding site (Earnhart et al. 2010) and a cysteine residue involved in quaternary structure (Earnhart et al. 2011): infectivity was attenuated when the binding site was mutated, although the specific ligand has not been defined, and oligomerization was prevented by eradicating the disulfide bond.

The regulation of *ospC* expression has also been probed by site-directed mutagenesis. OspC levels increase dramatically during transmission from the tick to the vertebrate (Fingerle et al. 2002; Gilmore and Piesman 2000; Leuba-Garcia et al. 1998; Montgomery et al. 1996; Ohnishi et al. 2001; Rathinavelu and de Silva 2001; Schwan et al. 1995; Schwan and Piesman 2000), but OspC synthesis is turned off after the vertebrate infection is established in order for the spirochete to evade adaptive immunity (Crother et al. 2004; Liang et al. 2002a, b; 2004; Tilly et al. 2006). We (Drecktrah et al. 2013) introduced site-directed mutations in situ into an inverted repeat that constitutes part of an *ospC* operator (Xu et al. 2007) and demonstrated that impairing the intra-complementarity of the inverted repeat reduced induction of *ospC* expression in response to environmental factors in vitro.

Site-directed mutations have also been incorporated in *trans* into the genome by transforming a mutated gene or regulatory sequence on a shuttle vector into a null background or fused to a reporter, respectively. Targets of these plasmid-borne dissections include the *ospC* operator/promoter (Eggers et al. 2004; Yang et al. 2005), the *ospE* and *ospF* promoters (Eggers et al. 2006), *ospA* (Schulze and Zückert 2006), *ospC* (Kumru et al. 2011a, b), *p13* (Kumru et al. 2011a), *bb0646* (Shaw et al. 2012), and *bbd18* (Hayes et al. 2014).

4.3 Promoter Fusion

Borrelia promoters have been fused to selectable markers (see Sect. 3.2), reporters (see Sect. 3.4), and transposons (see Sect. 7) as well as genes of interest to effect constitutive, high-level expression, often as an alternative approach when attempts to disrupt the gene fail. One of the first applications was the fusion of the *flaB* promoter to the *ospC* ORF, which corroborated that constitutive synthesis of OspC during vertebrate infection prevented *B. burgdorferi* from avoiding the immune response (Xu et al. 2006). In a more recent study, Tilly et al. (2013) swapped the promoters and ORFs for three outer membrane lipoprotein genes, including *ospC* and *vlsE*, revealing that OspC and VlsE have similar in vivo functions.

4.4 *Cre-*lox *and Telomere-Mediated Recombination*

Perhaps the most conspicuous genetic method missing from the *Borrelia* armamentarium is markerless mutagenesis, which is somewhat surprising because Bestor et al. (2010) have developed an elegant Cre-*lox* recombination system that functions in *B. burgdorferi*. Two *loxP* sites were inserted several kilobase pairs apart into one of the linear plasmids and the Cre recombinase was introduced on a shuttle vector, which resulted in the deletion of a substantial swath of genes (Bestor et al. 2010). Chaconas et al. (2001) fashioned a similar system utilizing a replicated telomere that is inserted into a linear plasmid and triggers a telomere resolvase-mediated deletion between the inserted telomere and the end of the linear plasmid opposite the origin of replication (Beaurepaire and Chaconas 2005). Genetic borreliology stoically awaits a tractable system to make markerless mutations.

5 Inducible Gene Expression

The ability to artificially regulate gene expression in *B. burgdorferi* has been immensely valuable for probing the function of essential genes and for assaying the kinetic requirements of gene expression in vivo (Samuels 2011). Failure to generate a null mutant after exhaustive attempts suggests that the targeted gene is essential

(Bandy et al. 2014; Chu et al. 2016; Dunn et al. 2015; Lenhart et al. 2012; Liang et al. 2010). In these instances, a conditional mutant, generated by replacing the endogenous promoter with an inducible promoter, allows for gradual reduction of gene expression, which is not possible with a null mutation. Essentiality is demonstrated if growth ceases in the absence of gene expression. In addition, the phenotype associated with decreased expression may provide clues to the function of the gene. Inducible systems can also be harnessed to experimentally overcome a growth phenotype even if the gene is not essential. For example, changes in morphology, metabolism or gene expression may be challenging or laborious to assess in a null mutant, but may be expedited using a conditional mutant where expression is induced for outgrowth and then stopped by removing the inducer.

5.1 lac *System*

A couple of strategies for inducible gene expression that have been workhorses in other bacteria have been adapted for *B. burgdorferi*. The first was based on the *lac* operon of *E. coli* where the repressor LacI binds to operators to repress gene expression (Jacob and Monod 1961). Repression is relieved by an inducer, such as isopropyl β-D-1-thiogalactopyranoside (IPTG), which binds LacI and causes an allosteric change that inhibits binding to the operators. In *B. burgdorferi*, the strong *flgB* promoter (Ge et al. 1997) fused to *lacI* was inserted in *bbe02* on lp25, resulting in high levels of LacI production (Gilbert et al. 2007). The inducible *flac* promoter was fashioned by integrating an optimized operator, *lacO*$_{\mathrm{id}}$ (Oehler et al. 1994), into the *flgB* promoter; the hybrid *flacp* is then fused to any gene (Gilbert et al. 2007). In vitro, IPTG induced OspC production in a *flacp-ospC* strain (Gilbert et al. 2007) and RpoS production in a *flacp-rpoS* strain (Caimano et al. 2007; Gilbert et al. 2007). Furthermore, using a mouse model in which IPTG was administered in the drinking water, OspC was required immediately upon injection of *B. burgdorferi* into the mouse as well as during the early stages of murine infection (Gilbert et al. 2007). *flacp* has also been fused to *bosR*, encoding a transcriptional regulator (Hyde et al. 2010), *bamA* and *tamB*, encoding the outer membrane protein transport machinery (Iqbal et al. 2016; Lenhart and Akins 2010), and *ftsH*, encoding a membrane protein protease (Chu et al. 2016).

Another inducible system based on the *lac* operon consists of the *flaB* promoter fused to *lacI* and the hybrid T5/*lac* promoter *pQE30* from plasmid QE30 (Blevins et al. 2007). IPTG induced expression of several genes fused to *pQE30*, including *bptA*, encoding an outer surface lipoprotein required for tick persistence (Blevins et al. 2007), *rrp2* (Groshong et al. 2012; Ouyang et al. 2014a), *csrA*, encoding an RNA-binding protein (Ouyang et al. 2014b), *dhhP*, encoding a cyclic-di-AMP phosphodiesterase (Ye et al. 2014), and *resT*, encoding the telomere resolvase (Bandy et al. 2014).

5.2 tet System

In addition to the *lac*-based gene regulation in *B. burgdorferi*, there have been two other inducible systems incorporating tetracycline resistance regulation from the *E. coli* transposon *Tn*10 (Cabello et al. 2006; Whetstine et al. 2009). The TetR repressor binds to the *tet* operator and the tetracycline analog anhydrotetracycline induces gene expression. In one version, the *flaB* promoter was fused to *tetR* and the *tet* operator was combined with the *B. burgdorferi ospA* promoter forming the hybrid P_{ost} (Whetstine et al. 2009), which was fused to *gfp* (Whetstine et al. 2009) as well as *bpaB* and *ebfC*, which encode small DNA-binding proteins (Jutras et al. 2012a, b). In the other *tet* system in *B. burgdorferi*, *tetR* was also driven by the *flaB* promoter and expression of *gfp* or *bmpA*, encoding an outer membrane protein, was controlled by fusing *tet* operators to the *bmpA* promoter (Cabello et al. 2006).

Thus, a few variations on inducible gene regulation have been developed for *B. burgdorferi*, although the *lacI* systems are more widely used and, notably, function in vivo in an animal model (Chu et al. 2016; Gilbert et al. 2007).

6 Complementation

The fulfillment of molecular Koch's postulates assigning a function to a gene product in a microbe requires complementation of an isogenic mutant (Falkow 1988). In bacteria, including *B. burgdorferi*, synthetically constructed mutants are typically complemented either in *cis* by genetically restoring the wild-type allele to the native locus or in *trans* by introducing the functional gene to a heterologous locus, often on a plasmid vector. Each approach has strengths and weaknesses.

6.1 Obstacles

The barriers to *B. burgdorferi* transformation, especially the restriction-modification systems (the *bbe02* and *bbq56* gene products) and loss of genomic elements, some of which are essential for host infection, are particularly pertinent to complementation. In addition to the aforementioned tactics of *bbe02* disruption and DNA methylation, various methods facilitate transformation of infectious mutant strains for complementation. A *bbe02* mutant has been deployed as the parental strain, in both wild-type and mutant *bbq56* backgrounds (Chan et al. 2015; Gilbert et al. 2007; Kasumba et al. 2015; Rego et al. 2011). This modification serves double duty by increasing transformation efficiency and furnishing a selection for lp25, which is readily lost during in vitro cultivation, but essential for host infection (Grimm et al. 2004a; Labandeira-Rey and Skare 2001; Purser and Norris 2000; Purser et al. 2003; Revel et al. 2005; Strother and de Silva 2005). Another method employed to ease complementation utilizes the *pncA* gene from lp25, which is necessary and sufficient to maintain mouse infectivity, carried on a shuttle vector, along with the complementing gene, and transformed into a strain that lacks lp25 (and lp56) (Seshu et al. 2006; Xu et al. 2005). This approach allows for vertebrate host studies, but not tick experiments as undefined genes from lp25, besides *pncA*, are necessary for persistence in the tick (Grimm et al. 2005; Revel et al. 2005; Strother and de Silva 2005). More recently, a suicide plasmid that recombines both the complementing gene and a selectable marker into the *bbe02* locus on lp25 has been established that is competent for in vivo experiments in the tick-mouse model (Kasumba et al. 2015). Thus, what was once an impediment to genetic manipulation and animal studies, a restriction-modification system on a plasmid unstable in vitro but required in vivo, has been exploited to study the role of gene products in *B. burgdorferi*.

6.2 Trans-*complementation and Shuttle Vectors*

B. burgdorferi sequences essential for plasmid partitioning and replication have been incorporated to construct several shuttle vectors that originally made complementation experiments possible. pBSV2 was the first, and likely remains the most popular, shuttle vector; it contains a 3.3-kb segment from the endogenous plasmid cp9 and is capable of autonomous replication in both *E. coli* and *B. burgdorferi* (Stewart et al. 2001). This region contains three open reading frames of a paralogous gene family (PGF) that is not homologous to any known replication or partitioning genes, but is found in many of the *B. burgdorferi* circular and linear plasmids (Casjens et al. 2000; Chaconas and Norris 2013; Zückert and Meyer 1996). In all, there are five different members in the PGF, with different combinations present on *B. burgdorferi* plasmids (Beaurepaire and Chaconas 2005; Chaconas and Norris 2013; Eggers et al. 2002; Zückert and Meyer 1996). pBSV2 (Stewart et al. 2001), and its derivatives containing alternative selectable markers (Elias et al. 2003; Frank et al. 2003), as well as the pCE320 series derived from cp32 (Eggers et al. 2002) are stably maintained both in vitro and in vivo. A number of shuttle vectors have been constructed using the PGFs from lp25, lp28-1, lp17, a cp32, and cp26 (Beaurepaire and Chaconas 2005; Byram et al. 2004; Eggers et al. 2002; Stewart et al. 2003; Tilly et al. 2012). Interestingly, all of these shuttle vectors are incompatible with their native plasmid from which the partitioning region has been cloned, as evidenced by transformation displacing the endogenous plasmid (Chaconas and Norris 2013), except cp26, which contains other essential genetic elements (Bandy et al. 2014; Byram et al. 2004).

Broad-host-range plasmids also have served as shuttle vectors to complement *B. burgdorferi* mutants, and to ectopically express genes. These plasmids can be replicated and maintained in a number of different bacterial hosts. The plasmid pJRS525 and related pGK12, both from *Lactococcus lactis*, have been used to complement *B. burgdorferi* mutants and express *gfp*, respectively (Hübner et al.

2001; Sartakova et al. 2000). pGK12 is less stable, without selection, and present at lower copy number than the more widely used pBSV2, potentially practical properties for plasmid curing and matching gene dosage during complementation (Sartakova et al. 2000).

6.3 Genetic Reconstitution

The other way to complement a mutant in *B. burgdorferi* is genetic reconstitution. This strategy involves allelic exchange at the original mutated locus to restore the wild-type gene (Hübner et al. 2001; Tilly et al. 2001). The restriction-modification systems do not pose a hurdle with this method as linear, non-autonomously replicating DNA substrates are used, but loss of native plasmids during this second transformation remains a concern. The advantage of genetic reconstitution is that returning the wild-type gene to its native locus restores the copy number, regulation, supercoiling state, and any small noncoding RNAs that may have been disrupted in the mutant. Disadvantages of genetic reconstitution include additional manipulation of the targeted loci, which may cause polar effects, thereby confounding comparison of the mutant and complemented strain phenotypes. Genetic reconstitution in a separate noncoding region of the genome may alleviate some of these concerns, but we now know that *B. burgdorferi* has an extensive, and incompletely characterized, sRNA transcriptome that must be considered when altering any region of the genome that is considered "non-coding" (Adams et al. 2017b; Arnold et al. 2016; Lybecker and Samuels 2017; Popitsch et al. 2017).

6.4 Caveats and Considerations

Occasionally, despite a plethora of painstaking attempts, complementing a *B. burgdorferi* mutant is simply not successful (Stewart et al. 2008). The reason is rarely clear, but some mutants have severe growth phenotypes, including retarded replication, elongated morphology and "clumping," which adversely affect the preparation of competent cells or efficient electrotransformation and recovery. In the case where a mutant is unable to be complemented, the last best course to assign a function to a mutated gene is to generate multiple independent clones from separate transformations. If all the clones have the same phenotype, then this is stronger evidence than relying on the phenotype of a single mutant. This line of reasoning is obviously not as robust as complementation of the mutant, but decreases the chance that an unrelated second-site mutation in the genome is responsible for the phenotype, as that change is unlikely to occur in two independent mutants, particularly if two different selectable markers are used. There are several analyses that should be undertaken if complementation does not restore the wild-type phenotype (Bugrysheva et al. 2005). Of course, the plasmid profile should be determined following all genetic manipulations, and transformants lacking an intact genome should not be further studied. Second, an alternative complementation construction (in *trans* or genetic reconstitution) should be attempted as both methods have their limitations. Third, the recombinant DNA in the mutant should be recovered from the complemented strain and sequenced to determine if any changes have occurred during genetic manipulation. Inadvertent mutations could be introduced during subcloning in *E. coli* or in *B. burgdorferi*. If the sequence is correct, then the level of gene expression should be compared between the complemented strain and the isogenic wild type by qRT-PCR or Northern blot analyses. Northern blot analysis of the mRNA has the advantage of detecting differential transcript processing between the two strains, which may affect stability or translation. A different complementation strategy would be prudent if the levels or kinetics of expression are significantly different between the complemented and wild-type strains. Finally, if all else fails, then polar effects from the mutation, second-site mutations and disruption of small noncoding RNAs should be considered. If the expression of adjacent genes has been altered in either the mutant or complemented strain, then another strategy such as site-directed mutagenesis may be warranted. Recently, *B. burgdorferi* has been found to have an extensive collection of noncoding sRNAs, nearly all of unknown function (Adams et al. 2017b; Arnold et al. 2016; Lybecker and Samuels 2017; Popitsch et al. 2017). These include antisense, intergenic, and intragenic (within a gene on the coding strand) RNAs. Mutant construction and complementation may alter these previously unappreciated sRNAs, resulting in failed complementation or confounding data interpretation. These genetic elements should be considered when designing both mutant and complementation strategies. Finally, the effect of extraneous second-site mutations on complementation may finally be possible to address as the cost of sequencing entire bacterial genomes continues to decrease. This may be especially applicable to the small genome ($\sim$1.5 Mb) of *B. burgdorferi* (Fraser et al. 1997). While there remain numerous challenges for genetically manipulating *B. burgdorferi*, particularly during complementation, there is now truly light at the end of the tunnel.

7 Transposon Mutagenesis and High-Throughput Genetics

Transposon mutagenesis is undoubtedly an exceedingly powerful genetic manipulation tool (Lin et al. 2014; Morozova et al. 2005; Stewart et al. 2004; Stewart and Rosa 2008). The most popular transposon system employed in *B. burgdorferi* entails a hyperactive allele of *Himar1*, a *mariner*-family transposable element from *Haematobia irritans*, fused to a strong, constitutive promoter and carried on a suicide vector; also included are the inverted terminal repeats that serve as binding sites for the transposon and flank a ColE1 origin as well as a selectable marker (Stewart et al. 2004; Stewart and Rosa 2008). The genome-wide mutagenesis technology has led to the generation of a signature-tagged mutant (STM) library in an infectious background (Botkin et al. 2006; Lin et al. 2012), the application of transposon sequencing (Tn-seq) (Troy et al. 2013, 2016), and the identification of scores of genes involved in diverse aspects of spirochete physiology as well as disease pathogenesis (Botkin et al. 2006; Khajanchi et al. 2016; Lin et al. 2009, 2012, 2015; Liu et al. 2009; Morozova et al. 2005; Ramsey et al. 2017; Stewart et al. 2004; Troy et al. 2013, 2016). A notable caveat is that STM and Tn-seq libraries may include mutants that either carry second-site mutations or are missing genomic elements; again, the segmented nature of the genome is a challenge for high-throughput methodologies in *Borrelia*. Prudence dictates that clean mutants be constructed before a function is ascribed to a gene. For example, the tour de force implementation of in vivo expression technology, another high-throughput genetic screen, inadvertently identified a red herring due to the loss of a plasmid (Ellis et al. 2014), although the system has proved insightful in defining an infectivity-associated sRNA transcriptome in the spirochete (Adams et al. 2017b).

8 Counterselection

Counterselectable markers (CSM) select for the loss of a DNA sequence (Reyrat et al. 1998) and can provide another avenue to generate markerless mutations. CSM generally confer susceptibility to a substance that is lethal to an organism. This scenario requires that the parental strain in the absence of the CSM is resistant to the substance used for negative selection, which limits employing counterselection.

Counterselection has been demonstrated using the *rpsL* gene, encoding the S12 ribosomal protein, in the spirochetes *Leptospira biflexa* (Picardeau et al. 2001) and *B. burgdorferi* (Drecktrah et al. 2010). A mutant *rpsL* gene located on the chromosome confers resistance to the antibiotic streptomycin (Criswell et al. 2006). The presence of a second, wild-type allele of *rpsL* in *trans* renders the merodiploid *B. burgdorferi* susceptible to streptomycin; although the mechanism has not been fully defined, susceptibility is dominant because translational errors by the sensitive ribosomes (Karimi and Ehrenberg 1994, 1996) will poison the cell. The addition of streptomycin was shown to select for the loss of the plasmid carrying the wild-type *rpsL* gene that confers susceptibility, thus providing the first functioning CSM in *Borrelia* (Drecktrah et al. 2010). Additionally, mutations to the *parC* gene, encoding a subunit of topoisomerase IV, confer resistance to fluoroquinolones in *B. burgdorferi* (Galbraith et al. 2005), providing another potential CSM.

9 Surrogate Genetics

Many barriers to genetic manipulation of *B. burgdorferi*, including low transformation efficiency, loss of endogenous plasmids during transformation, low GC content, slow growth, and expensive, complex media, have made surrogate genetics using *E. coli* an attractive and productive method to study spirochete gene function.

Numerous *B. burgdorferi* genes have been heterologously expressed in *E. coli* mutants to complement the phenotype in order to define the gene products: *grpE* and *dnaJ* (Tilly et al. 1993), *guaA* (Margolis et al. 1994), *lon* (Cloud et al. 1997; Coleman et al. 2009), *gac* (Knight and Samuels 1999), *oppA* (Lin et al. 2001), *pncA* (Purser et al. 2003), *recA* (Liveris et al. 2004; Putteet-Driver et al. 2004), *rpoS* (Eggers et al. 2004), rel_{Bbu} (Bugrysheva et al. 2005), and *hfq* (Lybecker et al. 2010). Conversely, the *E. coli hfq* gene has been heterologously expressed in a *B. burgdorferi hfq* mutant and partially complemented the phenotype, providing evidence that the spirochete has an Hfq-like RNA chaperone (Lybecker et al. 2010). Several endogenous *B. burgdorferi* promoters are capable of driving gene expression in *E. coli*. In fact, determining the consensus sequences for the sigma factors RpoS and RpoN was facilitated by studies in *E. coli* (Alverson et al. 2003; Burtnick et al. 2007; Caimano et al. 2004; Eggers et al. 2004, 2006). In addition, key experiments characterizing the promoter for the operon encoding the decorin-binding proteins (Ouyang et al. 2010) and describing the transcriptional regulator BosR (Boylan et al. 2003) were performed in *E. coli*.

Curiously, *B. burgdorferi* itself has served as a surrogate to heterologously express genes from less genetically tractable spirochetes, such as *Treponema pallidum*, the etiological agent of syphilis. *T. pallidum* lacks systems for both genetic manipulation and in vitro culture. The *T. pallidum* gene encoding the putative lipoprotein TP0435 was expressed in *B. burgdorferi* and shown to mediate adherence to host cells (Chan et al. 2016). *tp0435* expression was driven by its endogenous promoter, demonstrating that a *T. pallidum* promoter functions in *B. burgdorferi* (Chan et al. 2016). Using a similar system, the *T. pallidum* gene *tp0751*, encoding pallilysin, was also expressed in *B. burgdorferi* to complement an adhesin-attenuated mutant and restore interactions with the mouse vasculature in vivo (Kao et al. 2017). The promise of these surrogate systems to study surface-exposed adhesins provides a significant step toward the goal of developing *T. pallidum* vaccine candidates. *B. burgdorferi*, which has its own experimental hurdles, can, surprisingly, serve as a valuable tool to facilitate studies in other genetic-less members of the spirochete phylum.

10 Conclusions

The post-genomic era has ushered in powerful genetic approaches that, combined with robust animal models, have elevated spirochetology from its relegated plight as a microbiological curiosity into the limelight. The disciplines of *Borrelia* microbiology and Lyme disease pathogenesis have come to depend on genetic manipulation, and the study of relapsing fever spirochetes is not too far behind. Mutagenesis in these recalcitrant bacteria is still not for the faint-hearted, but most molecular laboratories are proficient at reverse genetics, although ignoring several considerations and caveats remains perilous. Certain genetic tricks, notably markerless mutations, are still absent from the borreliologist's quiver, but this just requires some attention and a bit of technical elbow grease. The state-of-the-art technologies for inducible gene expression in vivo and intravital microscopy are destined to reveal incredible insights into how these wily microbes interact with their hosts and are maintained in nature.

Acknowledgements We thank Ben Adler for thoughtful reading of the manuscript and Darrin Akins, Jim Battisti, Melissa Caimano, Sherwood Casjens, George Chaconas, Christian Eggers, Frank Gherardini, Mike Gilbert, Laura Hall, Chris Li, Meghan Lybecker, Rich Marconi, Motaleb, Steve Norris, Justin Radolf, Sandy Raffel, Patti Rosa, Tom Schwan, Jon Skare, Phil Stewart, Kit Tilly, and Frank Yang for useful discussions about the genetic manipulation of *Borrelia*. Our laboratory is supported by National Institutes of Health grant AI051486 (to D.S.S.).

References

Adams PP, Flores Avile C, Jewett MW (2017a) A dual luciferase reporter system for *B. burgdorferi* measures transcriptional activity during tick-pathogen interactions. Front Cell Infect Microbiol 7:225

Adams PP, Flores Avile C, Popitsch N, Bilusic I, Schroeder R, Lybecker M, Jewett MW (2017b) In vivo expression technology and 5′ end mapping of the *Borrelia burgdorferi* transcriptome identify novel RNAs expressed during mammalian infection. Nucleic Acids Res 45:775–792

Alverson J, Samuels DS (2002) *groEL* expression in *gyrB* mutants of *Borrelia burgdorferi*. J Bacteriol 184:6069–6072

Alverson J, Bundle SF, Sohaskey CD, Lybecker MC, Samuels DS (2003) Transcriptional regulation of the *ospAB* and *ospC* promoters from *Borrelia burgdorferi*. Mol Microbiol 48:1665–1677

Arnold WK, Savage CR, Brissette CA, Seshu J, Livny J, Stevenson B (2016) RNA-seq of *Borrelia burgdorferi* in multiple phases of growth reveals insights into the dynamics of gene expression, transcriptome architecture, and noncoding RNAs. PLoS ONE 11:e0164165

Babb K, McAlister JD, Miller JC, Stevenson B (2004) Molecular characterization of *Borrelia burgdorferi erp* promoter/operator elements. J Bacteriol 186:2745–2756

Bandy NJ, Salman-Dilgimen A, Chaconas G (2014) Construction and characterization of a *Borrelia burgdorferi* strain with conditional expression of the essential telomere resolvase, ResT. J Bacteriol 196:2396–2404

Barbour AG, Guo BP (2010) Pathogenesis of relapsing fever. In Samuels DS, Radolf JD (eds) *Borrelia*: Molecular biology, host interaction and pathogenesis. Norfolk, UK Caister Academic Press, pp 333–357

Barthold SW, Cadavid D, Philipp MT (2010) Animal models of borreliosis. In Samuels DS, Radolf JD (eds) *Borrelia*: Molecular biology, host interaction and pathogenesis. Norfolk, UK Caister Academic Press, pp 359–411

Battisti JM, Raffel SJ, Schwan TG (2008). A system for site-specific genetic manipulation of the relapsing fever spirochete *Borrelia hermsii*. In DeLeo FR, Otto M (eds) Bacterial pathogenesis: Methods and protocols. Totowa, New Jersey Humana Press, pp 69–84

Beaurepaire C, Chaconas G (2005) Mapping of essential replication functions of the linear plasmid lp17 of *B. burgdorferi* by targeted deletion walking. Mol Microbiol 57:132–142

Bestor A, Stewart PE, Jewett MW, Sarkar A, Tilly K, Rosa PA (2010) Use of the Cre-*lox* recombination system to investigate the lp54 gene requirement in the infectious cycle of *Borrelia burgdorferi*. Infect Immun 78:2397–2407

Blevins JS, Revel AT, Smith AH, Bachlani GN, Norgard MV (2007) Adaptation of a luciferase gene reporter and *lac* expression system to *Borrelia burgdorferi*. Appl Environ Microbiol 73:1501–1513

Bono JL, Elias AF, Kupko JJ III, Stevenson B, Tilly K, Rosa P (2000) Efficient targeted mutagenesis in *Borrelia burgdorferi*. J Bacteriol 182:2445–2452

Botkin DJ, Abbott AN, Stewart PE, Rosa PA, Kawabata H, Watanabe H, Norris SJ (2006) Identification of potential virulence determinants by *Himar1* transposition of infectious *Borrelia burgdorferi* B31. Infect Immun 74:6690–6699

Boylan JA, Posey JE, Gherardini FC (2003) *Borrelia* oxidative stress response regulator, BosR: a distinctive Zn-dependent transcriptional activator. Proc Natl Acad Sci USA 100:11684–11689

Brisson D, Drecktrah D, Eggers CH, Samuels DS (2012) Genetics of *Borrelia burgdorferi*. Annu Rev Genet 46:515–536

Bugrysheva JV, Bryksin AV, Godfrey HP, Cabello FC (2005) *Borrelia burgdorferi rel* is responsible for generation of guanosine-3′-diphosphate-5′-triphosphate and growth control. Infect Immun 73:4972–4981

Burtnick MN, Downey JS, Brett PJ, Boylan JA, Frye JG, Hoover TR, Gherardini FC (2007) Insights into the complex regulation of *rpoS* in *Borrelia burgdorferi*. Mol Microbiol 65:277–293

Byram R, Stewart PE, Rosa P (2004) The essential nature of the ubiquitous 26-kilobase circular replicon of *Borrelia burgdorferi*. J Bacteriol 186:3561–3569

Cabello FC, Dubytska L, Bryksin AV, Bugrysheva JV, Godfrey HP (2006). Genetic studies of the *Borrelia burgdorferi bmp* gene family. In Cabello FC, Hulinska D, Godfrey HP (eds) Molecular Biology of Spirochetes. Amsterdam, Netherlands IOS Press, pp 235–249

Caimano MJ, Eggers CH, Hazlett KRO, Radolf JD (2004) RpoS is not central to the general stress response in *Borrelia burgdorferi* but does control expression of one or more essential virulence determinants. Infect Immun 72:6433–6445

Caimano MJ, Iyer R, Eggers CH, Gonzalez C, Morton EA, Gilbert MA, Schwartz I, Radolf JD (2007) Analysis of the RpoS regulon in *Borrelia burgdorferi* in response to mammalian host signals provides insight into RpoS function during the enzootic cycle. Mol Microbiol 65: 1193–1217

Caimano MJ, Drecktrah D, Kung F, Samuels DS (2016) Interaction of the Lyme disease spirochete with its tick vector. Cell Microbiol 18:919–927

Carroll JA, Stewart PE, Rosa P, Elias AF, Garon CF (2003) An enhanced GFP reporter system to monitor gene expression in *Borrelia burgdorferi*. Microbiology 149:1819–1828

Casjens S, Palmer N, van Vugt R, Huang WM, Stevenson B, Rosa P, Lathigra R, Sutton G, Peterson J, Dodson RJ, Haft D, Hickey E, Gwinn M, White O, Fraser CM (2000) A bacterial genome in flux: the twelve linear and nine circular extrachromosomal DNAs in an infectious isolate of the Lyme disease spirochete *Borrelia burgdorferi*. Mol Microbiol 35:490–516

Chaconas G, Norris SJ (2013) Peaceful coexistence amongst *Borrelia* plasmids: getting by with a little help from their friends? Plasmid 70:161–167

Chaconas G, Stewart PE, Tilly K, Bono JL, Rosa P (2001) Telomere resolution in the Lyme disease spirochete. EMBO J 20:3229–3237

Chan K, Alter L, Barthold SW, Parveen N (2015) Disruption of *bbe02* by insertion of a luciferase gene increases transformation efficiency of *Borrelia burgdorferi* and allows live imaging in Lyme disease susceptible C3H mice. PLoS ONE 10:e0129532

Chan K, Nasereddin T, Alter L, Centurion-Lara A, Giacani L, Parveen N (2016) *Treponema pallidum* lipoprotein TP0435 expressed in *Borrelia burgdorferi* produces multiple surface/periplasmic isoforms and mediates adherence. Sci Rep 6:25593

Charon NW, Goldstein SF (2002) Genetics of motility and chemotaxis of a fascinating group of bacteria: the spirochetes. Annu Rev Genet 36:47–73

Charon NW, Cockburn A, Li C, Liu J, Miller KA, Miller MR, Motaleb MA, Wolgemuth CW (2012) The unique paradigm of spirochete motility and chemotaxis. Annu Rev Microbiol 66:349–370

Chen Q, Fischer JR, Benoit VM, Dufour NP, Youderian P, Leong JM (2008) In vitro CpG methylation increases the transformation efficiency of *Borrelia burgdorferi* strains harboring the endogenous linear plasmid lp56. J Bacteriol 190:7885–7891

Chu C-Y, Stewart PE, Bestor A, Hansen B, Lin T, Gao L, Norris SJ, Rosa PA (2016) Function of the *Borrelia burgdorferi* FtsH homolog is essential for viability both in vitro and in vivo and independent of HflK/C. MBio 7:e00404–16

Cloud JL, Marconi RT, Eggers CH, Garon CF, Tilly K, Samuels DS (1997) Cloning and expression of the *Borrelia burgdorferi lon* gene. Gene 194:137–141

Coleman JL, Katona LI, Kuhlow C, Toledo A, Okan NA, Tokarz R, Benach JL (2009) Evidence that two ATP-dependent (Lon) proteases in *Borrelia burgdorferi* serve different functions. PLoS Pathog 5:e1000676

Coutte L, Botkin DJ, Gao L, Norris SJ (2009) Detailed analysis of sequence changes occurring during *vlsE* antigenic variation in the mouse model of *Borrelia burgdorferi* infection. PLoS Pathog 5:e1000293

Criswell D, Tobiason VL, Lodmell JS, Samuels DS (2006) Mutations conferring aminoglycoside and spectinomycin resistance in *Borrelia burgdorferi*. Antimicrob Agents Chemother 50: 445–452

Crother TR, Champion CI, Whitelegge JP, Aguilera R, Wu X-Y, Blanco DR, Miller JN, Lovett MA (2004) Temporal analysis of the antigenic composition of *Borrelia burgdorferi* during infection in rabbit skin. Infect Immun 72:5063–5072

Di L, Pagan PE, Packer D, Martin CL, Akther S, Ramrattan G, Mongodin EF, Fraser CM, Schutzer SE, Luft BJ, Casjens SR, Qiu W-G (2014) *BorreliaBase*: a phylogeny-centered browser of *Borrelia* genomes. BMC Bioinf 15:233

Drecktrah D, Douglas JM, Samuels DS (2010) Use of *rpsL* as a counterselectable marker in *Borrelia burgdorferi*. Appl Environ Microbiol 76:985–987

Drecktrah D, Hall LS, Hoon-Hanks LL, Samuels DS (2013) An inverted repeat in the *ospC* operator is required for induction in *Borrelia burgdorferi*. PLoS ONE 8:e68799

Dresser AR, Hardy PO, Chaconas G (2009) Investigation of the genes involved in antigenic switching at the *vlsE* locus in *Borrelia burgdorferi*: an essential role for the RuvAB branch migrase. PLoS Pathog 5:e1000680

Dunham-Ems SM, Caimano MJ, Pal U, Wolgemuth CW, Eggers CH, Balic A, Radolf JD (2009) Live imaging reveals a biphasic mode of dissemination of *Borrelia burgdorferi* within ticks. J Clin Invest 119:3652–3665

Dunham-Ems SM, Caimano MJ, Eggers CH, Radolf JD (2012) *Borrelia burgdorferi* requires the alternative sigma factor RpoS for dissemination within the vector during tick-to-mammal transmission. PLoS Pathog 8:e1002532

Dunn JP, Kenedy MR, Iqbal H, Akins DR (2015) Characterization of the β-barrel assembly machine accessory lipoproteins from *Borrelia burgdorferi*. BMC Microbiol 15:70

Earnhart CG, LeBlanc DV, Alix KE, Desrosiers DC, Radolf JD, Marconi RT (2010) Identification of residues within ligand-binding domain 1 (LBD1) of the *Borrelia burgdorferi* OspC protein required for function in the mammalian environment. Mol Microbiol 76:393–408

Earnhart CG, Rhodes DV, Marconi RT (2011) Disulfide-mediated oligomer formation in *Borrelia burgdorferi* outer surface protein C, a critical virulence factor and potential Lyme disease vaccine candidate. Clin Vaccine Immunol 18:901–906

Eggers CH, Kimmel BJ, Bono JL, Elias A, Rosa P, Samuels DS (2001) Transduction by ϕBB-1, a bacteriophage of *Borrelia burgdorferi*. J Bacteriol 183:4771–4778

Eggers CH, Caimano MJ, Clawson ML, Miller WG, Samuels DS, Radolf JD (2002) Identification of loci critical for replication and compatibility of a *Borrelia burgdorferi* cp32 plasmid and use of a cp32-based shuttle vector for expression of fluorescent reporters in the Lyme disease spirochaete. Mol Microbiol 43:281–296

Eggers CH, Caimano MJ, Radolf JD (2004) Analysis of promoter elements involved in the transcriptional initiation of RpoS-dependent *Borrelia burgdorferi* genes. J Bacteriol 186: 7390–7402

Eggers CH, Caimano MJ, Radolf JD (2006) Sigma factor selectivity in *Borrelia burgdorferi*: RpoS recognition of the *ospE*/*ospF*/*elp* promoters is dependent on the sequence of the −10 region. Mol Microbiol 59:1859–1875

Eggers CH, Gray CM, Preisig AM, Glenn DM, Pereira J, Ayers RW, Alshahrani M, Acabbo C, Becker MR, Bruenn KN, Cheung T, Jendras TM, Shepley AB, Moeller JT (2016) Phage-mediated horizontal gene transfer of both prophage and heterologous DNA by ϕBB-1, a bacteriophage of *Borrelia burgdorferi*. Pathog Dis 74:ftw107

Elias AF, Stewart PE, Grimm D, Caimano MJ, Eggers CH, Tilly K, Bono JL, Akins DR, Radolf JD, Schwan TG, Rosa P (2002) Clonal polymorphism of *Borrelia burgdorferi* strain B31 MI: implications for mutagenesis in an infectious strain background. Infect Immun 70:2139–2150

Elias AF, Bono JL, Kupko JJ 3rd, Stewart PE, Krum JG, Rosa PA (2003) New antibiotic resistance cassettes suitable for genetic studies in *Borrelia burgdorferi*. J Mol Microbiol Biotechnol 6:29–40

Ellis TC, Jain S, Linowski AK, Rike K, Bestor A, Rosa PA, Halpern M, Kurhanewicz S, Jewett MW (2014) Correction: In vivo expression technology identifies a novel virulence factor critical for *Borrelia burgdorferi* persistence in mice. PLoS Pathog 10:e1004260

Falkow S (1988) Molecular Koch's postulates applied to microbial pathogenicity. Rev Infect Dis 10(Suppl 2):S274–276

Fine LM, Earnhart CG, Marconi RT (2011) Genetic transformation of the relapsing fever spirochete *Borrelia hermsii*: stable integration and expression of green fluorescent protein from linear plasmid 200. J Bacteriol 193:3241–3245

Fingerle V, Rauser S, Hammer B, Kahl O, Heimerl C, Schulte-Spechtel U, Gern L, Wilske B (2002) Dynamics of dissemination and outer surface protein expression of different European *Borrelia burgdorferi* sensu lato strains in artificially infected *Ixodes ricinus* nymphs. J Clin Microbiol 40:1456–1463

Fingerle V, Goettner G, Gern L, Wilske B, Schulte-Spechtel U (2007) Complementation of a *Borrelia afzelii* OspC mutant highlights the crucial role of OspC for dissemination of *Borrelia afzelii* in *Ixodes ricinus*. Int J Med Microbiol 297:97–107

Frank KL, Bundle SF, Kresge ME, Eggers CH, Samuels DS (2003) *aadA* confers streptomycin-resistance in *Borrelia burgdorferi*. J Bacteriol 185:6723–6727

Fraser CM, Casjens S, Huang WM, Sutton GG, Clayton R, Lathigra R, White O, Ketchum KA, Dodson R, Hickey EK, Gwinn M, Dougherty B, Tomb J-F, Fleischmann RD, Richardson D, Peterson J, Kerlavage AR, Quakenbush J, Salzberg S, Hanson M, van Vugt R, Palmer N, Adams MK, Gocayne J, Weidman J, Utterback T, Watthey L, McDonald L, Artiach P, Bowman C, Garland S, Fujii C, Cotton MD, Horst K, Roberts K, Hatch B, Smith HO, Venter JC (1997) Genomic sequence of a Lyme disease spirochete, *Borrelia burgdorferi*. Nature 390:580–586

Galbraith KM, Ng AC, Eggers BJ, Kuchel CR, Eggers CH, Samuels DS (2005) *parC* mutations in fluoroquinolone-resistant *Borrelia burgdorferi*. Antimicrob Agents Chemother 49:4354–4357

Ge Y, Old IG, Saint Girons I, Charon NW (1997) Molecular characterization of a large *Borrelia burgdorferi* motility operon which is initiated by a consensus σ^{70} promoter. J Bacteriol 179:2289–2299

Gilbert MA, Morton EA, Bundle SF, Samuels DS (2007) Artificial regulation of *ospC* expression in *Borrelia burgdorferi*. Mol Microbiol 63:1259–1273

Gilmore RD Jr, Piesman J (2000) Inhibition of *Borrelia burgdorferi* migration from the midgut to the salivary glands following feeding by ticks on OspC-immunized mice. Infect Immun 68:411–414

Grimm D, Eggers CH, Caimano MJ, Tilly K, Stewart PE, Elias AF, Radolf JD, Rosa PA (2004a) Experimental assessment of the roles of linear plasmids lp25 and lp28-1 of *Borrelia burgdorferi* throughout the infectious cycle. Infect Immun 72:5938–5946

Grimm D, Tilly K, Byram R, Stewart PE, Krum JG, Bueschel DM, Schwan TG, Policastro PF, Elias AF, Rosa PA (2004b) Outer-surface protein C of the Lyme disease spirochete: a protein induced in ticks for infection of mammals. Proc Natl Acad Sci USA 101:3142–3147

Grimm D, Tilly K, Bueschel DM, Fisher MA, Policastro PF, Gherardini FC, Schwan TG, Rosa PA (2005) Defining plasmids required by *Borrelia burgdorferi* for colonization of tick vector *Ixodes scapularis* (Acari: Ixodidae). J Med Entomol 42:676–684

Groshong AM, Blevins JS (2014) Insights into the biology of *Borrelia burgdorferi* gained through the application of molecular genetics. Adv Appl Microbiol 86:41–143

Groshong AM, Gibbons NE, Yang XF, Blevins JS (2012) Rrp2, a prokaryotic enhancer-like binding protein, is essential for viability of *Borrelia burgdorferi*. J Bacteriol 194:3336–3342

Hayes BM, Jewett MW, Rosa PA (2010) *lacZ* reporter system for use in *Borrelia burgdorferi*. Appl Environ Microbiol 76:7407–7412

Hayes BM, Dulebohn DP, Sarkar A, Tilly K, Bestor A, Ambroggio X, Rosa PA (2014) Regulatory protein BBD18 of the Lyme disease spirochete: essential role during tick acquisition? MBio 5: e01017–14

Hübner A, Yang X, Nolen DM, Popova TG, Cabello FC, Norgard MV (2001) Expression of *Borrelia burgdorferi* OspC and DbpA is controlled by a RpoN-RpoS regulatory pathway. Proc Natl Acad Sci USA 98:12724–12729

Hyde JA, Shaw DK, Smith R III, Trzeciakowski JP, Skare JT (2010) Characterization of a conditional *bosR* mutant in *Borrelia burgdorferi*. Infect Immun 78:265–274

Hyde JA, Weening EH, Chang M, Trzeciakowski JP, Höök M, Cirillo JD, Skare JT (2011) Bioluminescent imaging of *Borrelia burgdorferi* in vivo demonstrates that the fibronectin-binding protein BBK32 is required for optimal infectivity. Mol Microbiol 82:99–113

Iqbal H, Kenedy MR, Lybecker M, Akins DR (2016) The TamB ortholog of *Borrelia burgdorferi* interacts with the β-barrel assembly machine (BAM) complex protein BamA. Mol Microbiol 102:757–774

Jacob F, Monod J (1961) Genetic regulatory mechanisms in the synthesis of proteins. J Mol Biol 3:318–356

James AE, Rogovskyy AS, Crowley MA, Bankhead T (2016) Characterization of a DNA adenine methyltransferase gene of *Borrelia hermsii* and its dispensability for murine infection and persistence. PLoS ONE 11:e0155798

Jewett MW, Lawrence K, Bestor AC, Tilly K, Grimm D, Shaw P, VanRaden M, Gherardini F, Rosa PA (2007) The critical role of the linear plasmid lp36 in the infectious cycle of *Borrelia burgdorferi*. Mol Microbiol 64:1358–1374

Johnson RC (1977) The spirochetes. Annu Rev Microbiol 31:89–106

Jutras BL, Bowman A, Brissette CA, Adams CA, Verma A, Chenail AM, Stevenson B (2012a) EbfC (YbaB) is a new type of bacterial nucleoid-associated protein and a global regulator of gene expression in the Lyme disease spirochete. J Bacteriol 194:3395–3406

Jutras BL, Verma A, Adams CA, Brissette CA, Burns LH, Whetstine CR, Bowman A, Chenail AM, Zückert WR, Stevenson B (2012b) BpaB and EbfC DNA-binding proteins regulate production of the Lyme disease spirochete's infection-associated Erp surface proteins. J Bacteriol 194:778–786

Kao W-CA, Pětrošová H, Ebady R, Lithgow KV, Rojas P, Zhang Y, Kim Y-E, Kim Y-R, Odisho T, Gupta N, Moter A, Cameron CE, Moriarty TJ (2017) Identification of Tp0751 (pallilysin) as a *Treponema pallidum* vascular adhesin by heterologous expression in the Lyme disease spirochete. Sci Rep 7:1538

Karimi R, Ehrenberg M (1994) Dissociation rate of cognate peptidyl-tRNA from the A-site of hyper-accurate and error-prone ribosomes. Eur J Biochem 226:355–360

Karimi R, Ehrenberg M (1996) Dissociation rates of peptidyl-tRNA from the P-site of *E. coli* ribosomes. EMBO J 15:1149–1154

Kasumba IN, Bestor A, Tilly K, Rosa PA (2015) Use of an endogenous plasmid locus for stable in *trans* complementation in *Borrelia burgdorferi*. Appl Environ Microbiol 81:1038–1046

Kawabata H, Norris SJ, Watanabe H (2004) BBE02 disruption mutants of *Borrelia burgdorferi* B31 have a highly transformable, infectious phenotype. Infect Immun 72:7147–7754

Khajanchi BK, Odeh E, Gao L, Jacobs MB, Philipp MT, Lin T, Norris SJ (2016) Phosphoenolpyruvate phosphotransferase system components modulate gene transcription and virulence of *Borrelia burgdorferi*. Infect Immun 84:754–764

Knight SW, Samuels DS (1999) Natural synthesis of a DNA-binding protein from the C-terminal domain of DNA gyrase A in *Borrelia burgdorferi*. EMBO J 18:4875–4881

Knight SW, Kimmel BJ, Eggers CH, Samuels DS (2000) Disruption of the *Borrelia burgdorferi gac* gene, encoding the naturally synthesized GyrA C-terminal domain. J Bacteriol 182: 2048–2051

Krishnavajhala A, Wilder HK, Boyle WK, Damania A, Thornton JA, Pérez de León AA, Teel PD, Lopez JE (2017) Imaging of *Borrelia turicatae* producing the green fluorescent protein reveals persistent colonization of the *Ornithodoros turicata* midgut and salivary glands from nymphal acquisition through transmission. Appl Environ Microbiol 83:e02503–16

Kumru OS, Schulze RJ, Slusser JG, Zückert WR (2010) Development and validation of a FACS-based lipoprotein localization screen in the Lyme disease spirochete *Borrelia burgdorferi*. BMC Microbiol 10:277

Kumru OS, Bunikis I, Sorokina I, Bergström S, Zückert WR (2011a) Specificity and role of the *Borrelia burgdorferi* CtpA protease in outer membrane protein processing. J Bacteriol 193:5759–5765

Kumru OS, Schulze RJ, Rodnin MV, Ladokhin AS, Zückert WR (2011b) Surface localization determinants of *Borrelia* OspC/Vsp family lipoproteins. J Bacteriol 193:2814–2825

Kurtti TJ, Munderloh UG, Johnson RC, Ahlstrand GG (1987) Colony formation and morphology in *Borrelia burgdorferi*. J Clin Microbiol 25:2054–2058

Labandeira-Rey M, Skare JT (2001) Decreased infectivity in *Borrelia burgdorferi* strain B31 is associated with loss of linear plasmid 25 or 28-1. Infect Immun 69:446–455

Lawrenz MB, Kawabata H, Purser JE, Norris SJ (2002) Decreased electroporation efficiency in *Borrelia burgdorferi* containing linear plasmids lp25 and lp56: impact on transformation of infectious *B. burgdorferi*. Infect Immun 70:4798–4804

Lenhart TR, Akins DR (2010) *Borrelia burgdorferi* locus BB0795 encodes a BamA orthologue required for growth and efficient localization of outer membrane proteins. Mol Microbiol 75:692–709

Lenhart TR, Kenedy MR, Yang X, Pal U, Akins DR (2012) BB0324 and BB0028 are constituents of the *Borrelia burgdorferi* β-barrel assembly machine (BAM) complex. BMC Microbiol 12:60

Leuba-Garcia S, Martinez R, Gern L (1998) Expression of outer surface proteins A and C of *Borrelia afzelii* in *Ixodes ricinus* ticks and in the skin of mice. Zentralbl Bakteriol 287:475–484

Liang FT, Jacobs MB, Bowers LC, Philipp MT (2002a) An immune evasion mechanism for spirochetal persistence in Lyme borreliosis. J Exp Med 195:415–422

Liang FT, Nelson FK, Fikrig E (2002b) Molecular adaptation of *Borrelia burgdorferi* in the murine host. J Exp Med 196:275–280

Liang FT, Yan J, Mbow ML, Sviat SL, Gilmore RD, Mamula M, Fikrig E (2004) *Borrelia burgdorferi* changes its surface antigenic expression in response to host immune responses. Infect Immun 72:5759–5767

Liang FT, Xu Q, Sikdar R, Xiao Y, Cox JS, Doerrler WT (2010) BB0250 of *Borrelia burgdorferi* is a conserved and essential inner membrane protein required for cell division. J Bacteriol 192:6105–6115

Lin B, Short SA, Eskildsen M, Klempner MS, Hu LT (2001) Functional testing of putative oligopeptide permease (Opp) proteins of *Borrelia burgdorferi*: a complementation model in *opp*$^-$ *Escherichia coli*. Biochim Biophys Acta 1499:222–231

Lin T, Gao L, Edmondson DG, Jacobs MB, Philipp MT, Norris SJ (2009) Central role of the Holliday junction helicase RuvAB in *vlsE* recombination and infectivity of *Borrelia burgdorferi*. PLoS Pathog 5:e1000679

Lin T, Gao L, Zhang C, Odeh E, Jacobs MB, Coutte L, Chaconas G, Philipp MT, Norris SJ (2012) Analysis of an ordered, comprehensive STM mutant library in infectious *Borrelia burgdorferi*: insights into the genes required for mouse infectivity. PLoS ONE 7:e47532

Lin T, Troy EB, Hu LT, Gao L, Norris SJ (2014) Transposon mutagenesis as an approach to improved understanding of *Borrelia* pathogenesis and biology. Front Cell Infect Microbiol 4:63

Lin T, Gao L, Zhao X, Liu J, Norris SJ (2015) Mutations in the *Borrelia burgdorferi* flagellar type III secretion system genes *fliH* and *fliI* profoundly affect spirochete flagellar assembly, morphology, motility, structure, and cell division. MBio 6:e00579–15

Liu J, Lin T, Botkin DJ, McCrum E, Winkler H, Norris SJ (2009) Intact flagellar motor of *Borrelia burgdorferi* revealed by cryo-electron tomography: evidence for stator ring curvature and rotor/C-ring assembly flexion. J Bacteriol 191:5026–5036

Liveris D, Mulay V, Schwartz I (2004) Functional properties of *Borrelia burgdorferi recA*. J Bacteriol 186:2275–2280

Lybecker MC, Abel CA, Feig AL, Samuels DS (2010) Identification and function of the RNA chaperone Hfq in the Lyme disease spirochete *Borrelia burgdorferi*. Mol Microbiol 78: 622–635

Lybecker MC, Samuels DS (2017) Small RNAs of *Borrelia burgdorferi*: characterizing functional regulators in a sea of sRNAs. Yale J Biol Med 90:317–323

Margolis N, Hogan D, Tilly K, Rosa PA (1994) Plasmid location of *Borrelia* purine biosynthesis gene homologs. J Bacteriol 176:6427–6432

Montgomery RR, Malawista SE, Feen KJ, Bockenstedt LK (1996) Direct demonstration of antigenic substitution of *Borrelia burgdorferi* ex vivo: exploration of the paradox of the early immune response to outer surface proteins A and C in Lyme disease. J Exp Med 183:261–269

Moriarty TJ, Norman MU, Colarusso P, Bankhead T, Kubes P, Chaconas G (2008) Real-time high resolution 3D imaging of the Lyme disease spirochete adhering to and escaping from the vasculature of a living host. PLoS Pathog 4:e1000090

Morozova OV, Dubytska LP, Ivanova LB, Moreno CX, Bryksin AV, Sartakova ML, Dobrikova EY, Godfrey HP, Cabello FC (2005) Genetic and physiological characterization of 23S rRNA and *ftsJ* mutants of *Borrelia burgdorferi* isolated by mariner transposition. Gene 357:63–72

Nickoloff JA (1995) Electroporation protocols for microorganisms. In Walker JM (ed) *Methods in Molecular Biology*. Totowa, New Jersey: Humana Press, p 372

Nordstrand A, Barbour AG, Bergström S (2000) *Borrelia* pathogenesis research in the post-genomic and post-vaccine era. Curr Opin Microbiol 3:86–92

Oehler S, Amouyal M, Kolkhof P, von Wilcken-Bergmann B, Müller-Hill B (1994) Quality and position of the three *lac* operators of *E. coli* define efficiency of repression. EMBO J 13: 3348–3355

Ohnishi J, Piesman J, de Silva AM (2001) Antigenic and genetic heterogeneity of *Borrelia burgdorferi* populations transmitted by ticks. Proc Natl Acad Sci USA 98:670–675

Ouyang Z, Haq S, Norgard MV (2010) Analysis of the *dbpBA* upstream regulatory region controlled by RpoS in *Borrelia burgdorferi*. J Bacteriol 192:1965–1974

Ouyang Z, Zhou J, Norgard MV (2014a) Synthesis of RpoS is dependent on a putative enhancer binding protein Rrp2 in *Borrelia burgdorferi*. PLoS ONE 9:e96917

Ouyang Z, Zhou J, Norgard MV (2014b) CsrA (BB0184) is not involved in activation of the RpoN-RpoS regulatory pathway in *Borrelia burgdorferi*. Infect Immun 82:1511–1522

Pal U, Yang X, Chen M, Bockenstedt LK, Anderson JF, Flavell RA, Norgard MV, Fikrig E (2004) OspC facilitates *Borrelia burgdorferi* invasion of *Ixodes scapularis* salivary glands. J Clin Invest 113:220–230

Picardeau M, Brenot A, Saint Girons I (2001) First evidence for gene replacement in *Leptospira* spp. Inactivation of *L. biflexa flaB* results in non-motile mutants deficient in endoflagella. Mol Microbiol 40:189–199

Piesman J, Schwan TG (2010) Ecology of borreliae and their arthropod vectors. In Samuels DS, Radolf JD (eds) *Borrelia*: Molecular biology, host interaction and pathogenesis. Norfolk, UK Caister Academic Press, pp 251–278

Popitsch N, Bilusic I, Rescheneder P, Schroeder R, Lybecker M (2017) Temperature-dependent sRNA transcriptome of the Lyme disease spirochete. BMC Genom 18:28

Purser JE, Norris SJ (2000) Correlation between plasmid content and infectivity in *Borrelia burgdorferi*. Proc Natl Acad Sci USA 97:13865–13870

Purser JE, Lawrenz MB, Caimano MJ, Howell JK, Radolf JD, Norris SJ (2003) A plasmid-encoded nicotinamidase (PncA) is essential for infectivity of *Borrelia burgdorferi* in a mammalian host. Mol Microbiol 48:753–764

Putteet-Driver AD, Zhong J, Barbour AG (2004) Transgenic expression of RecA of the spirochetes *Borrelia burgdorferi* and *Borrelia hermsii* in *Escherichia coli* revealed differences in DNA repair and recombination phenotypes. J Bacteriol 186:2266–2274

Radolf JD, Caimano MJ, Stevenson B, Hu LT (2012) Of ticks, mice and men: understanding the dual-host lifestyle of Lyme disease spirochaetes. Nat Rev Microbiol 10:87–99

Raffel SJ, Battisti JM, Fischer RJ, Schwan TG (2014) Inactivation of genes for antigenic variation in the relapsing fever spirochete *Borrelia hermsii* reduces infectivity in mice and transmission by ticks. PLoS Pathog 10:e1004056

Ramsey ME, Hyde JA, Medina-Perez DN, Lin T, Gao L, Lundt ME, Li X, Norris SJ, Skare JT, Hu LT (2017) A high-throughput genetic screen identifies previously uncharacterized *Borrelia burgdorferi* genes important for resistance against reactive oxygen and nitrogen species. PLoS Pathog 13:e1006225

Rathinavelu S, de Silva AM (2001) Purification and characterization of *Borrelia burgdorferi* from feeding nymphal ticks (*Ixodes scapularis*). Infect Immun 69:3536–3541

Rego ROM, Bestor A, Rosa PA (2011) Defining the plasmid-borne restriction-modification systems of the Lyme disease spirochete *Borrelia burgdorferi*. J Bacteriol 193:1161–1171

Revel AT, Blevins JS, Almazán C, Neil L, Kocan KM, de la Fuente J, Hagman KE, Norgard MV (2005) *bptA* (*bbe16*) is essential for the persistence of the Lyme disease spirochete, *Borrelia burgdorferi*, in its natural tick vector. Proc Natl Acad Sci USA 102:6972–6977

Reyrat J-M, Pelicic V, Gicquel B, Rappuoli R (1998) Counterselectable markers: untapped tools for bacterial genetics and pathogenesis. Infect Immun 66:4011–4017

Rosa PA, Hogan DM (1992) Colony formation by *Borrelia burgdorferi* in solid medium: clonal analysis of *osp* locus variants. In Munderloh UG, Kurtti TJ (eds) First international conference on tick-borne pathogens at the host-vector interface: an agenda for research. St. Paul University of Minnesota, pp 95–103

Rosa PA, Tilly K, Stewart PE (2005) The burgeoning molecular genetics of the Lyme disease spirochaete. Nat Rev Microbiol 3:129–143

Samuels DS (1995) Electrotransformation of the spirochete *Borrelia burgdorferi*. In Nickoloff JA (ed) Electroporation protocols for microorganisms. Totowa, New Jersey Humana Press, pp 253–259

Samuels DS (2006). Antibiotic resistance in *Borrelia burgdorferi*: applications for genetic manipulation and implications for evolution. In Cabello FC, Hulinska D, Godfrey HP (eds) Molecular biology of spirochetes. Amsterdam, Netherlands IOS Press, pp 56–70

Samuels DS (2011) Gene regulation in *Borrelia burgdorferi*. Annu Rev Microbiol 65:479–499

Samuels DS, Garon CF (1993) Coumermycin A_1 inhibits growth and induces relaxation of supercoiled plasmids in *Borrelia burgdorferi*, the Lyme disease agent. Antimicrob Agents Chemother 37:46–50

Samuels DS, Garon CF (1997) Oligonucleotide-mediated genetic transformation of *Borrelia burgdorferi*. Microbiology 143:519–522

Samuels DS, Mach KE, Garon CF (1994a) Genetic transformation of the Lyme disease agent *Borrelia burgdorferi* with coumarin-resistant *gyrB*. J Bacteriol 176:6045–6049

Samuels DS, Radolf JD (2009) Who is the BosR around here anyway? Mol Microbiol 74:1295–1299

Samuels DS, Marconi RT, Huang WM, Garon CF (1994b) *gyrB* mutations in coumermycin A_1-resistant *Borrelia burgdorferi*. J Bacteriol 176:3072–3075

Sarkar A, Hayes BM, Dulebohn DP, Rosa PA (2011) Regulation of the virulence determinant OspC by *bbd18* on linear plasmid lp17 of *Borrelia burgdorferi*. J Bacteriol 193:5365–5373

Sartakova M, Dobrikova E, Cabello FC (2000) Development of an extrachromosomal cloning vector system for use in *Borrelia burgdorferi*. Proc Natl Acad Sci USA 97:4850–4855

Sartakova ML, Dobrikova EY, Terekhova DA, Devis R, Bugrysheva JV, Morozova OV, Godfrey HP, Cabello FC (2003) Novel antibiotic-resistance markers in pGK12-derived vectors for *Borrelia burgdorferi*. Gene 303:131–137

Schulze RJ, Zückert WR (2006) *Borrelia burgdorferi* lipoproteins are secreted to the outer surface by default. Mol Microbiol 59:1473–1484

Schulze RJ, Chen S, Kumru OS, Zückert WR (2010) Translocation of *Borrelia burgdorferi* surface lipoprotein OspA through the outer membrane requires an unfolded conformation and can initiate at the C-terminus. Mol Microbiol 76:1266–1278

Schwan TG, Piesman J (2000) Temporal changes in outer surface proteins A and C of the Lyme disease-associated spirochete, *Borrelia burgdorferi*, during the chain of infection in ticks and mice. J Clin Microbiol 38:382–388

Schwan TG, Piesman J, Golde WT, Dolan MC, Rosa PA (1995) Induction of an outer surface protein on *Borrelia burgdorferi* during tick feeding. Proc Natl Acad Sci USA 92:2909–2913

Seshu J, Esteve-Gassent MD, Labandeira-Rey M, Kim JH, Trzeciakowski JP, Höök M, Skare JT (2006) Inactivation of the fibronectin-binding adhesin gene *bbk32* significantly attenuates the infectivity potential of *Borrelia burgdorferi*. Mol Microbiol 59:1591–1601

Shaw DK, Hyde JA, Skare JT (2012) The BB0646 protein demonstrates lipase and haemolytic activity associated with *Borrelia burgdorferi*, the aetiological agent of Lyme disease. Mol Microbiol 83:319–334

Skare JT, Shaw DK, Trzeciakowski JP, Hyde JA (2016) In vivo imaging demonstrates that *Borrelia burgdorferi ospC* is uniquely expressed temporally and spatially throughout experimental infection. PLoS ONE 11:e0162501

Sohaskey CD, Barbour AG (1999) Esterases in serum-containing growth media counteract chloramphenicol acetyltransferase activity in vitro. Antimicrob Agents Chemother 43:655–660

Sohaskey CD, Arnold C, Barbour AG (1997) Analysis of promoters in *Borrelia burgdorferi* by use of a transiently expressed reporter gene. J Bacteriol 179:6837–6842

Sohaskey CD, Zückert WR, Barbour AG (1999) The extended promoters for two outer membrane lipoprotein genes of *Borrelia* spp. uniquely include a T-rich region. Mol Microbiol 33:41–51

Srivastava SY, de Silva AM (2008) Reciprocal expression of *ospA* and *ospC* in single cells of *Borrelia burgdorferi*. J Bacteriol 190:3429–3433

Stevenson B, Bono JL, Elias A, Tilly K, Rosa P (1998) Transformation of the Lyme disease spirochete *Borrelia burgdorferi* with heterologous DNA. J Bacteriol 180:4850–4855

Stewart PE, Rosa PA (2008). Transposon mutagenesis of the Lyme disease agent *Borrelia burgdorferi*. In DeLeo FR, Otto M (eds) Bacterial pathogenesis: methods and protocols. Totowa, New Jersey Humana Press, pp 85–95

Stewart PE, Chaconas G, Rosa P (2003) Conservation of plasmid maintenance functions between linear and circular plasmids in *Borrelia burgdorferi*. J Bacteriol 185:3202–3209

Stewart P, Thalken R, Bono J, Rosa P (2001) Isolation of a circular plasmid region sufficient for autonomous replication and transformation of infectious *Borrelia burgdorferi*. Mol Microbiol 39:714–721

Stewart PE, Wang X, Bueschel DM, Clifton DR, Grimm D, Tilly K, Carroll JA, Weis JJ, Rosa PA (2006) Delineating the requirement for the *Borrelia burgdorferi* virulence factor OspC in the mammalian host. Infect Immun 74:3547–3553

Stewart PE, Bestor A, Cullen JN, Rosa PA (2008) A tightly regulated surface protein of *Borrelia burgdorferi* is not essential to the mouse-tick infectious cycle. Infect Immun 76:1970–1978

Stewart PE, Hoff J, Fischer E, Krum JG, Rosa PA (2004) Genome-wide transposon mutagenesis of *Borrelia burgdorferi* for identification of phenotypic mutants. Appl Environ Microbiol 70:5973–5979

Stoenner HG, Dodd T, Larsen C (1982) Antigenic variation of *Borrelia hermsii*. J Exp Med 156:1297–1311

Strother KO, de Silva A (2005) Role of *Borrelia burgdorferi* linear plasmid 25 in infection of *Ixodes scapularis* ticks. J Bacteriol 187:5776–5781

Terekhova D, Sartakova ML, Wormser GP, Schwartz I, Cabello FC (2002) Erythromycin resistance in *Borrelia burgdorferi*. Antimicrob Agents Chemother 46:3637–3640

Tilly K, Hauser R, Campbell J, Ostheimer GJ (1993) Isolation of *dnaJ*, *dnaK*, and *grpE* homologues from *Borrelia burgdorferi* and complementation of *Escherichia coli* mutants. Mol Microbiol 7:359–369

Tilly K, Elias AF, Bono JL, Stewart P, Rosa P (2000) DNA exchange and insertional inactivation in spirochetes. J Mol Microbiol Biotechnol 2:433–442

Tilly K, Elias AF, Errett J, Fischer E, Iyer R, Schwartz I, Bono JL, Rosa P (2001) Genetics and regulation of chitobiose utilization in *Borrelia burgdorferi*. J Bacteriol 183:5544–5553

Tilly K, Krum JG, Bestor A, Jewett MW, Grimm D, Bueschel D, Byram R, Dorward D, Vanraden MJ, Stewart P, Rosa P (2006) *Borrelia burgdorferi* OspC protein required exclusively in a crucial early stage of mammalian infection. Infect Immun 74:3554–3564

Tilly K, Checroun C, Rosa PA (2012) Requirements for *Borrelia burgdorferi* plasmid maintenance. Plasmid 68:1–12

Tilly K, Bestor A, Rosa PA (2013) Lipoprotein succession in *Borrelia burgdorferi*: similar but distinct roles for OspC and VlsE at different stages of mammalian infection. Mol Microbiol 89:216–227

Troy EB, Lin T, Gao L, Lazinski DW, Camilli A, Norris SJ, Hu LT (2013) Understanding barriers to *Borrelia burgdorferi* dissemination during infection using massively parallel sequencing. Infect Immun 81:2347–2457

Troy EB, Lin T, Gao L, Lazinski DW, Lundt M, Camilli A, Norris SJ, Hu LT (2016) Global Tn-seq analysis of carbohydrate utilization and vertebrate infectivity of *Borrelia burgdorferi*. Mol Microbiol 101:1003–1023

Whetstine CR, Slusser JG, Zückert WR (2009) Development of a single-plasmid-based regulatable gene expression system for *Borrelia burgdorferi*. Appl Environ Microbiol 75:6553–6558

Xu Q, Seemanapalli SV, Lomax L, McShan K, Li X, Fikrig E, Liang FT (2005) Association of linear plasmid 28-1 with an arthritic phenotype of *Borrelia burgdorferi*. Infect Immun 73:7208–7215

Xu Q, Seemanapalli SV, McShan K, Liang FT (2006) Constitutive expression of outer surface protein C diminishes the ability of *Borrelia burgdorferi* to evade specific humoral immunity. Infect Immun 74:5177–5184

Xu Q, McShan K, Liang FT (2007) Identification of an *ospC* operator critical for immune evasion of *Borrelia burgdorferi*. Mol Microbiol 64:220–231

Yang XF, Alani SM, Norgard MV (2003) The response regulator Rrp2 is essential for the expression of major membrane lipoproteins in *Borrelia burgdorferi*. Proc Natl Acad Sci USA 100:11001–11006

Yang XF, Pal U, Alani SM, Fikrig E, Norgard MV (2004) Essential role for OspA/B in the life cycle of the Lyme disease spirochete. J Exp Med 199:641–648

Yang XF, Lybecker MC, Pal U, Alani SM, Blevins J, Revel AT, Samuels DS, Norgard MV (2005) Analysis of the *ospC* regulatory element controlled by the RpoN-RpoS regulatory pathway in *Borrelia burgdorferi*. J Bacteriol 187:4822–4829

Ye M, Zhang J-J, Fang X, Lawlis GB, Troxell B, Zhou Y, Gomelsky M, Lou Y, Yang XF (2014) DhhP, a cyclic di-AMP phosphodiesterase of *Borrelia burgdorferi*, is essential for cell growth and virulence. Infect Immun 82:1840–1849

Zhang J-R, Hardham JM, Barbour AG, Norris SJ (1997) Antigenic variation in Lyme disease borreliae by promiscuous recombination of VMP-like sequence cassettes. Cell 89:275–285

Zückert WR, Meyer J (1996) Circular and linear plasmids of Lyme disease spirochetes have extensive homology: characterization of a repeated DNA element. J Bacteriol 178:2287–2298

Toolbox of Molecular Techniques for Studying *Leptospira* Spp.

Mathieu Picardeau

Abstract This chapter covers the progress made in the *Leptospira* field since the application of mutagenesis techniques and how they have allowed the study of virulence factors and, more generally, the biology of *Leptospira*. The last decade has seen advances in our ability to perform molecular genetic analysis of *Leptospira*. Major achievements include the generation of large collections of mutant strains and the construction of replicative plasmids, enabling complementation of mutations. However, there are still no practical tools for routine genetic manipulation of pathogenic *Leptospira* strains, slowing down advances in pathogenesis research. This review summarizes the status of the molecular genetic toolbox for *Leptospira* species and highlights new challenges in the nascent field of *Leptospira* genetics.

Contents

M. Picardeau (✉)
Biology of Spirochetes Unit, Institut Pasteur, 28 Rue Du Docteur Roux, 75724 Paris Cedex 15, France
e-mail: mathieu.picardeau@pasteur.fr

Current Topics in Microbiology and Immunology (2018) 415:141–162
DOI 10.1007/82_2017_45

Published Online: 29 August 2017

1 Introduction

The genetic tools available to study species of the genus *Leptospira*, which comprises both saprophytic and pathogenic members, are limited relative to those for other bacterial species. Prior to 2000, genetic analysis of *Leptospira* was impeded by the lack of methods for the introduction of DNA into leptospiral cells. In the first genetic studies carried out in the 1990s, several *Leptospira* genes were isolated and characterized by the functional complementation of *E. coli* mutants. This method led to the identification of several amino acid biosynthesis genes, including *asd*, *aroD*, *dapD*, *metX*, *metY*, *trpE*, *proA*, and *leuB* (Baril et al. 1992; Belfaiza et al. 1998; Richaud et al. 1990; Yelton and Cohen 1986). Similarly, the *L. biflexa recA* gene was found to confer functional complementation of *recA* mutations in *E. coli* (Stamm et al. 1991). Transformation of *Leptospira* spp. was first achieved using a plasmid shuttle vector which was constructed using the replication origin of a lysogenic leptophage (Saint Girons et al. 1990, 2000). Today, genetic manipulation of these spirochetes, in particular pathogenic strains, is still cumbersome and not yet routine due to the paucity of genetic tools. The genomic sequences of hundreds of *Leptospira* strains have been determined, including representatives of each of the *Leptospira* species (Fouts et al. 2016), and efforts are underway to take advantage of this information to learn more about the biology and pathogenesis of these bacteria. However, new genetic tools such as tools for gene knockout in pathogens need to be developed to take full advantage of the wealth of information contained in these genomes.

1.1 Traditional Genetic Strategies

The genus *Leptospira* is composed of 22 species that are divided into more than 300 serovars. *Leptospira* species are phylogenetically clustered into three distinct clades: 10 pathogenic species (*L. interrogans, L. kirschneri, L. borgpetersenii, L. santarosai, L. noguchii, L. weilii, L. alexanderi, L. kmetyi, L. alstonii,* and *L. mayottensis*), which are able to infect and cause disease in humans and animals, five intermediates (*L. inadai, L. broomii, L. fainei, L. wolffii, and L. licerasiae*), which have been isolated from humans and animals and may cause a variety of mild clinical manifestations, and seven saprophytes (*L. biflexa, L. wolbachii, L. meyeri, L. vanthielii, L. terpstrae, L. yanagawae,* and *L. idonii*), which do not cause disease

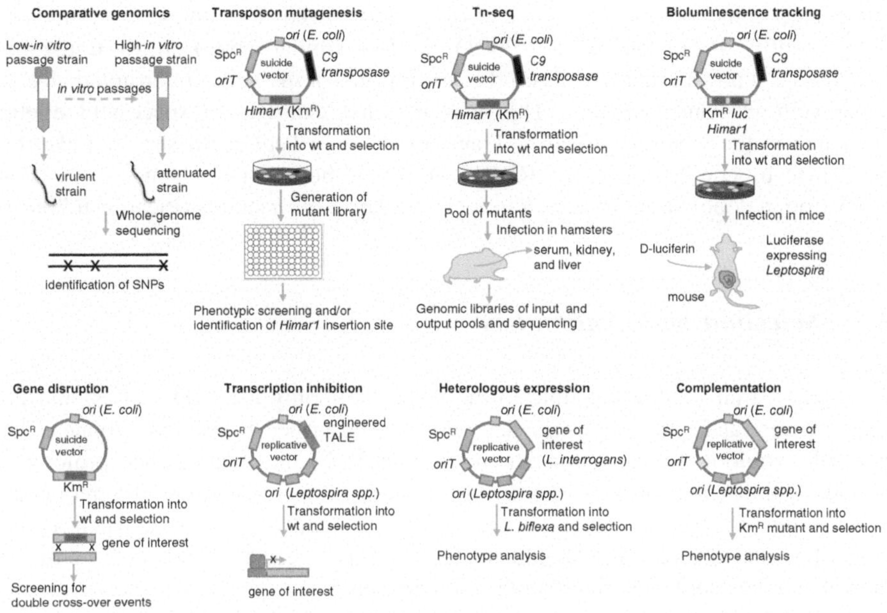

Fig. 1 Genetic toolbox for *Leptospira* spp.

(Picardeau 2014). *Leptospira* cells in vitro grow in a complex and nutrient-rich medium at an optimal temperature of 30 °C; minimal media for the growth of *Leptospira* are not available. Efforts to develop a system for genetic analysis in *Leptospira* spp. have focused primarily on the saprophytic strain *L. biflexa*, which is a fast-growing *Leptospira* with a generation time of five h, compared to up to 20 h for pathogenic strains (or 1 week vs. 1 month to obtain colonies on solid medium). The current molecular genetic toolbox includes methods that enable random transposon mutagenesis followed by screening for mutants that display a specific phenotype, approaches for targeted mutagenesis, and shuttle vectors that enable complementation of mutations and heterologous expression (Fig. 1).

1.2 Transformation Methods

Leptospira strains are not naturally competent bacteria; their genomes do not contain the DNA-uptake machinery of naturally competent bacteria. In 2000, Saint Girons et al. (2000) reported the first successful transformation of *Leptospira* spp. In this study, they introduced a replicative plasmid DNA into the saprophyte *L. biflexa* by electroporation. For electroporation, cells are grown to mid-log phase and washed extensively with water to eliminate all salts. To electroporate DNA into cells, washed *L. biflexa* cells are mixed with the DNA to be transformed. The competent cells cannot be stored frozen and saved for future experiments. Subsequently, Picardeau et al. (2001) used a similar approach to transform *L. biflexa* with a suicide plasmid carrying the *flaB* gene interrupted by a kanamycin-resistance cassette. This was the first example of successful allelic exchange in the genus *Leptospira*. Transformation in *Leptospira* spp. can also be performed by conjugation using RP4-based, broad-host-range plasmids to transfer DNA from a donor strain of *E. coli* to recipient *Leptospira* strains (Picardeau 2008).

1.3 Selection with Antibiotics

Leptospira strains are susceptible to the most commonly used antibiotics, and no acquired resistance to antibiotics has yet been described in clinical isolates. An optimal antibiotic for selection should be stable in the medium for prolonged periods of time (plate incubation may last more than one month), and the frequency of spontaneous resistant mutants must be low. In addition, antibiotics such as the penicillins and doxycycline should not be used as selectable markers because of their therapeutic use in patients with leptospirosis.

Antibiotic resistance markers from other bacteria, including kanamycin (*Enterococcus faecalis* cassette), spectinomycin (*Staphylococcus aureus* cassette), and gentamicin (*Enterobacter cloacae* cassette), fused to strong spirochetal promoters, have been used as selectable markers in *Leptospira* spp. (Bauby et al. 2003; Bourhy et al. 2005a; Poggi et al. 2010; Saint Girons et al. 2000). However, it is possible to select rare spontaneous spectinomycin-resistant strains resulting from mutations in the 16S rRNA gene (Poggi et al. 2010). The commonly used counterselectable marker *sacB* was found to be unsuitable in *L. biflexa*, but the *rpsL* gene, which encodes the S12 ribosomal protein, has been used as an efficient counterselectable marker in a *L. biflexa* streptomycin-resistant strain (Picardeau et al. 2001).

1.4 Promoters for Heterologous Expression and Inducible Expression Systems

Early studies demonstrating expression of foreign resistance cassettes in *Leptospira* made use of the *flgB* promoter of *Borrelia burgdorferi* (Bono et al. 2000; Saint Girons et al. 2000). This promoter was also used to drive strong and constitutive expression of LigA and LigB proteins in *L. biflexa* serovar Patoc (Figueira et al. 2011).

Recent determination of primary transcription start sites (pTSS) for *L. interrogans* genes has enabled accurate identification and delineation of most of the promoter regions (Zhukova et al. 2017). An unusually high number of pTSSs were located between 0 and 10 nucleotides from the translational start site, suggesting that leaderless transcripts are a common feature of the leptospiral translational landscape (Zhukova et al. 2017). In the absence of clear consensus sequences, 200–300 bp of DNA sequence immediately upstream from the translation initiation codon are generally cloned.

Several strong leptospiral promoters have been used for heterologous gene expression in *Leptospira* spp., such as the promoter of *lipL32* (Toma et al. 2014). The most abundant proteins in *L. interrogans* include lipoproteins LipL32 (38,000 proteins per cell), LipL41 (30,000), LipL36 (14,000), and Loa22 (10,500), the flagellar protein FlaB1 (12,000), the chaperone GroEL (17,500), the elongation factor Tu Ef-Tu (14,000), and a transcriptional regulator of the ArsR family (12,000) (Malmström et al. 2009). The promoters of these genes are thus good candidates to drive high gene expression levels in transformed cells.

Codon usage in *E. coli* is different from that of *Leptospira* spp., and several rare codons in *Leptospira* spp. are more frequently used in *E. coli*. Significantly different codon usage of the target protein from that of the expression host may lead to low expression levels or toxicity. This is particularly true when the gene is overexpressed. Optimizing codon usage for the expression host is therefore required for optimal protein expression.

Inducible promoters are useful for the maintenance of toxic genes or timely expression of essential genes. The *E. coli lacI* and *B. burgdorferi*-codon optimized *lacI* gene, encoding the LacI repressor protein, can be used in *L. biflexa* to control the expression of genes of interest after modification of their promoters by insertion of the *lac* operator (*lacO*) (Aviat et al. 2010; Slamti et al. 2011). The LacI repressor-based system can be simply regulated using the lactose analog, IPTG. In the absence of IPTG, the LacI regulator binds to *lacO* and represses transcription of the gene. When introduced to the cells, IPTG causes a conformational change in the repressor, which cannot then bind the operator, allowing RNA polymerase to gain access to the promoter and initiate transcription. This strategy was used to generate an *L. biflexa* strain that conditionally expresses the endogenous *mreB*, an essential gene, demonstrating its role in cell morphogenesis (Slamti et al. 2011).

1.5 E. coli-Leptospira *Shuttle Plasmids*

The genome of *Leptospira* spp. is composed of two chromosomes and until recently no plasmids had been identified (Picardeau et al. 2008; Wang et al. 2015; Zhu et al. 2014). Phages or chromosomal prophage regions of *Leptospira* spp. have been used for the generation of replicative plasmids in *Leptospira* spp. The putative replication origins can be determined by identifying the partitioning locus, consisting of the *parA* and *parB* genes, which is usually close to the replication origin. Saint Girons et al. (1990) isolated and characterized three bacteriophages from sewage water, of which replication was limited to the saprophyte *L. biflexa*. One of these phages, LE1, was shown to replicate as a circular plasmid in *L. biflexa* and cloning of the predicted *parA-parB-rep* locus and selection markers conferring antibiotic resistance in an *E. coli* plasmid enabled the generation of the first *E. coli-Leptospira* plasmid shuttle vector (Saint Girons et al. 2000). The partition locus confers stability to plasmids which can persist in bacteria for several generations, without the presence of a selective antibiotic (Saint Girons et al. 2000). Cloned versions of the putative replication origins, which consist of the putative partition locus downstream of a *rep*-like gene encoding a DNA-binding protein, of the *L. biflexa* replicon p74 (Picardeau et al. 2008) and a phage-related genomic island from *L. interrogans* (Bourhy et al. 2007) have also been used to achieve autonomous replication in *L. biflexa*, but until recently, there has been no replicative plasmid vector available for pathogenic *Leptospira* strains. Two recent studies reported plasmid replication in pathogenic *Leptospira* strains by cloning the replication origins of extrachromosomal replicons, including an inducible prophage, into *L. interrogans* serovar Linhai (Zhu et al. 2015) and a prophage-like element into *L. mayottensis* (Pappas et al. 2015). *E. coli-Leptospira* shuttle vectors described to date are low-copy number plasmids. It is generally accepted that plasmids containing the same origin of replication are incompatible, whereas plasmids containing different origins of replication are generally compatible and can coexist together in a cell. The use of compatible leptospiral vectors with distinct antibiotic resistance markers should enable, for example, the expression of multiprotein complexes.

Replicative vectors enable complementation of mutants (Pappas et al. 2015) and the heterologous expression of pathogen-specific genes in the saprophyte *L. biflexa* (Figueira et al. 2011; Matsunaga and Coutinho 2012; Toma et al. 2014; Zhang et al. 2012). The pathogen-specific LigB protein belongs to the superfamily of bacterial immunoglobulin (Ig)-like (Big) repeat domain proteins, including virulence determinants that mediate pathogen–host cell interactions (Matsunaga et al. 2003). The expression of LigB on the surface of *L. biflexa* increased adhesion to fibronectin, fibrinogen, and cultured cells (Choy et al. 2011; Figueira et al. 2011), survival in human serum (Castiblanco-Valencia et al. 2016), and phagocytic uptake (Toma et al. 2014), suggesting that LigB is a multifunctional protein. The faster growing saprophyte *L. biflexa* may therefore represent a good surrogate host for the characterization of the role of key virulence factors that contribute to leptospirosis.

1.6 Targeted Mutagenesis

Targeted mutagenesis was first achieved in *L. biflexa* using a suicide plasmid delivering an inactivated *flaB* allele to the targeted chromosomal gene. However, treatment of the DNA used for transformation by UV irradiation or alkaline denaturation before electroporation is essential for the generation of large numbers of recombinants, including mutants displaying allelic exchange (Picardeau et al. 2001). The proportion of allelic exchange mutants is low, and single crossover recombination (plasmid integration) is the general rule, but the use of counterselectable *rpsL* in a *L. biflexa* streptomycin-resistant strain can permit the positive selection of double crossover recombinants (Picardeau et al. 2001). Several chromosomal genes have been disrupted by allelic exchange in saprophytic strains, including *flaB, trpE, metY, metX, metW, hemH*, and *recA* (Louvel and Picardeau 2007). In contrast, transformation efficiencies achieved in pathogenic *Leptospira* strains are low. The first demonstration of site-directed homologous recombination in pathogenic *Leptospira* was reported by Croda et al. (2008), with the disruption of the *ligB* gene in *L. interrogans*. Although challenging, other examples of successful allelic exchanges have been reported, including the *L. interrogans* genes *fcpA* (Wunder et al. 2016), *mce* (Zhang et al. 2012), *colA* (Kassegne et al. 2014), *chpIK* or *mazEF* (Komi et al. 2015), a phospholipase C (Zhao et al. 2013), and *fliY* (Liao et al. 2009).

1.7 Random Mutagenesis

Random genetic alterations can be introduced into genomes by various methods, such as insertional transposition, UV irradiation, and exposure to DNA-damaging chemicals.

Mutagenesis was first performed by exposing *Leptospira* cells to nitrosoguanidine, followed by isolation of approximately 500 colonies and screening for mutants with altered colony morphologies, indicative of motility deficiencies (Bromley and Charon 1979). At that time, whole-genome sequencing was not available and the molecular factors responsible for the phenotype could not been identified.

A system for random mutagenesis using the *Himar1 mariner* transposon has been developed in both saprophytic and pathogenic *Leptospira* strains (Bourhy et al. 2005b; Louvel et al. 2005; Murray et al. 2009a). Thousands of random mutants can be easily obtained in *L. biflexa*, thereby generating extensive libraries of mutants that could be screened for phenotypes affecting diverse aspects of metabolism and physiology. For example, six heme-requiring mutants, with insertions in genes that were predicted to encode iron transporters and the regulator of a two-component system, were identified by screening a library of $\approx$2000 random transposon mutants in *L. biflexa* on media with or without hemin supplementation (Louvel et al. 2006, 2008). Similarly, 29 mutants were selected for further analysis by screening a library of $\approx$5000 random mutants in *L. biflexa* for increased susceptibility to the DNA intercalating agent, ethidium bromide (EtBr). Genes involved in transport (including efflux pumps) and regulation (two-component systems, anti-sigma factor antagonists, etc.) were overrepresented, demonstrating that these genes are major contributors to EtBr tolerance (Pětrošová and Picardeau 2014). The *Himar1 mariner* transposon system is also applicable to pathogenic *Leptospira* spp., but the transformation efficiencies are two to three orders of magnitude lower than those for *L. biflexa* (Bourhy et al. 2005b). An avirulent mutant of the *L. interrogans* serovar Lai was obtained via disruption of *loa22*, a gene that encodes an outer membrane protein containing an OmpA domain (Ristow et al. 2007). The resulting mutant did not express Loa22 and showed attenuated virulence in hamster models of leptospirosis, whereas genetic complementation restored Loa22 expression and virulence. Indeed, Loa22 is the first genetically defined virulence factor in *Leptospira* species (Ristow et al. 2007). Three years of transformation experiments in different *Leptospira* pathogenic strains, performed simultaneously in two different laboratories, have generated a library of 1000 distinct *L interrogans* mutants (Murray et al. 2009a). Consistent with *mariner*-based mutagenesis systems tested in other bacterial species, analysis of transposon integration sites showed that all *Himar1* insertions in *Leptospira* spp. occurred at a TA dinucleotide with no other target site preference (Murray et al. 2009a). Mapping of the insertions on the *L. interrogans* genome showed that they were scattered throughout the genome, and no hot spots could be detected, apart from genes encoding the 16S and 23S rRNAs. A bias toward low insertion-site density in the genomic region that contains essential genes, such as the 100 kb-lipopolysaccharide (LPS) and heme biosynthesis loci, was also identified (Murray et al. 2009a). Of the 1000 insertion sites, 721 of the mutations identified affected the protein-coding regions of 551 different genes. Today, a library of $\approx$2000 random mutants of *L. interrogans* serovar Manilae strain L495 is now available (our unpublished data). This strain was chosen because of its high virulence and relatively high transformation efficiency relative to other pathogenic strains. A set of mutants with transposon insertions in genes encoding putative virulence factors has been selected for further in vitro and in vivo studies.

Transposon mutants of genes encoding the bacterial homolog of the molecular chaperones Hsp90, and ClpB have shown attenuated virulence. The *clpB*, but not *hsp90* mutant, exhibited in vitro growth defects and increased susceptibility to oxidative stress (King et al. 2014; Lourdault et al. 2011). A *katE* mutant was attenuated in its virulence and showed reduced survival under oxidative stress, further demonstrating the requirement for a functional ROS (reactive oxygen species) resistance mechanism for full leptospiral pathogenesis (Eshghi et al. 2012a). A transposon mutant in a gene encoding a putative HtpX-like protein, related to the M48 family of metalloproteases, produced an iron precipitate and formed outer membrane vesicles under high iron conditions (Henry et al. 2013). Mutation of the *lipL32* and *lipL41* genes, which encode the first and third most abundant OM lipoproteins, respectively, did not result in virulence attenuation in animal models (King et al. 2013; Murray et al. 2009b). In contrast, a transposon mutant in another lipoprotein-encoded gene, *lruA*, showed partial attenuation of virulence (Zhang et al. 2013). Transposon mutants in genes encoding the lipid A and O-antigen of the LPS were avirulent in animal infection models (Eshghi et al. 2015a; Matsunaga and Coutinho 2012; Srikram et al. 2011). Other attenuated mutants include a mutant with a transposon insertion in a putative regulatory locus, containing genes encoding a sensor protein with a phosphatase domain and an anti-sigma factor antagonist, eliciting pleiotropic deregulation of genes involved in motility/chemotaxis and signal transduction, as well as genes encoding predicted extracellular proteins (Eshghi et al. 2014). Transposon insertion in *sph2* resulted in reduced hemolytic and sphingomyelinase activities (Narayanavari et al. 2015). Transposon mutants of *flaA1* and *flaA2* assemble periplasmic flagella, but their motility, virulence, and flagellar and cellular morphologies were affected, depending on whether FlaA1 or both FlaA proteins were expressed (Lambert et al. 2012). A more detailed analysis of mutations affecting virulence is given in the chapter by Bulach and Adler in this volume.

Hamsters can be infected with pools of up to 10 *Himar1* mutants to reduce the number of animals used in virulence studies. The composition of the surviving members of mutant pools was assessed in the kidneys and blood after in vivo growth, five days after challenge (Marcsisin et al. 2013). The authors showed that the phenotype of some mutants can be different in acute (hamsters) and chronic (mice) models of infection, suggesting that different factors are required in acute and chronic leptospiral infection (Marcsisin et al. 2013). More recently, transposon sequencing (Tn-Seq) was used to analyze a pool of 42 *L. interrogans* transposon mutants in hamsters (Lourdault et al. 2016). Deep sequencing of the DNA regions flanking the transposons make it possible to determine the frequencies of mutants recovered from tissues and relate them to their frequencies in the input pool. With this technique, gene fitness profiles were generated by simultaneously assessing the effect of the absence of each gene product on the ability of the bacteria to colonize specific organs and their growth kinetics (Lourdault et al. 2016).

1.8 Studying Protein Localization

Proteins exposed on the surface of *Leptospira* constitute an important set of virulence factors that may promote interaction with host cell receptors or other macromolecules (Haake and Matsunaga 2010). The appropriate surface localization of leptospiral proteins, mostly lipoproteins, is a prerequisite for the characterization of their adhesion activity. This has proven to be challenging: One of the most highly studied lipoproteins, LipL32, was believed to be a surface lipoprotein, but the recent use of several complementary methods now indicates that LipL32 is a subsurface lipoprotein, located in the periplasmic leaflet of the outer membrane (Pinne and Haake 2013).

Protein localization in *Leptospira* spp. has been predicted in silico and analyzed by subcellular localization by immunostaining and/or cell fractionation. Genome analysis of *L. interrogans* revealed approximately 145 putative lipoproteins (Setubal et al. 2006), of which only a small fraction is predicted to be surface-exposed (Gamberini et al. 2005; Viratyosin et al. 2008). Yang et al. used a filtering approach, combining in silico analysis, comparative genome hybridization, and microarray methods to identify 226 leptospiral putative surface proteins (Yang et al. 2006). Two-dimensional gel electrophoresis (2-DE) of outer membrane proteins (Cullen et al. 2002; Nally et al. 2005b, 2007), surfaceome studies (Cullen et al. 2005), and localization of surface-exposed proteins by immunofluorescence and/or Triton X-114 solubilization and phase partitioning studies (Cullen et al. 2004) have led to the identification of several outer membrane proteins, including OmpL1, Qlp42, LipL32, LipL41, LipL36, and Loa22. The separation of culture supernatants from whole bacteria has enabled the identification of 325 exoproteins in *L. interrogans* by label-free quantitative mass spectrometry, from those that are 1.5-fold more abundant in supernatants than in whole cells to those detected only in supernatants (Eshghi et al. 2015b). Of the 325 identified exoproteins, 89 ($\sim$27%) contained an N-terminal signal sequence and 23 ($\sim$7%) were predicted to be exported through a nonclassical protein export system, suggesting that *Leptospira* predominantly exports proteins via a mechanism yet to be characterized (Eshghi et al. 2015b). Expression of tagged proteins, using a FLAG or polyhistidine tag either at the C- or N-terminus, can also facilitate the determination of protein localization (our unpublished data).

1.9 Reporter Genes

A β-galactosidase reporter system has been developed to measure the promoter activity of leptospiral genes in *L. biflexa* (Matsunaga and Coutinho 2012). The fluorescent alleles *gfp* and mRFP1 were transferred into both saprophytic and pathogenic *Leptospira* strains. Generation of transcriptional fusions placing GFP expression under the control of a bacterial promoter is a valuable tool to examine gene expression patterns during *Leptospira* infection or under in vitro conditions (Aviat et al. 2010; Cerqueira et al. 2011). However, fluorescent leptospires were not sufficiently bright to be clearly visible in cell cultures (Aviat et al. 2010). The firefly luciferase-derived *luc* gene, driven by the *L. interrogans flgB* promoter, was successfully used to produce luminescent leptospires in live mice in the presence of luciferin. These bioluminescent leptospires are novel tools that will be useful to track the dissemination of mutants in laboratory animals and to test the efficacy of antibiotic treatments or vaccines against leptospirosis (Ratet et al. 2014). More recently, the *luxCDABE* cassette from *Photorhabdus luminescens* has been used to generate luciferase-expressing strains of *Leptospira spp.* This cassette does not require addition of an exogenous substrate, in contrast to the *luc* system, and allowed tracking of bacterial dissemination and important colonization sites in hamsters (Ozuru et al. 2017). This technology was also proposed for the enumeration of viable cells (Murray et al. 2010).

2 Alternative Strategies

2.1 Targeted Mutagenesis Using TALE

Allelic exchange in pathogenic leptospires remains challenging due to poor transformation efficiency and/or inefficient homologous recombination machinery. Thus, a new strategy for creating targeted gene knockdowns in *Leptospira* spp. has been developed. Transcription activator-like effectors (TALEs) are a group of repressors which modify transcriptional activity in prokaryotic and eukaryotic cells by binding directly to a targeted sequence within the host genome. A recent study showed that TALE is effective at reducing the expression of targeted genes in saprophytic and pathogenic strains of *Leptospira* spp. (Pappas and Picardeau 2015). The recent mapping of TSS in the *L. interrogans* genome will be helpful for designing TALEs for targeted genes in the future (Zhukova et al. 2017) as the role of TALE proteins is to repress transcription by binding directly to DNA within the promoter region of a gene (which in turn inhibits promoter recognition by RNA polymerase or by abrogating transcription initiation).

2.2 Comparative Genomics and Pathogenomics

Genome sequences aid in the global understanding of the biology of living organisms by revealing complete gene repertoires. However, the functional analysis of these proteins in *Leptospira* spp. has been limited by the lack of tools for molecular genetic manipulation (Cerqueira et al. 2011).

All members of the *Leptospira* genus that have been analyzed carry at least two circular replicons (Zuerner 1991), yielding a total genome size of more than 3.9 Mb. In 2003, *L. interrogans* serovar Lai was the first *Leptospira* genome to be sequenced, revealing 4727 coding sequences (Ren et al. 2003). However, after re-sequencing, reannotation, and proteomic analysis, the total number of coding sequences was reduced to 3718 (Zhong et al. 2011). Today, the genome sequences of hundreds of *Leptospira* strains have been determined, including representative of each of the 20 *Leptospira* species (Fouts et al. 2016). Examination of the 1100 genes of the core genome of *Leptospira* spp. reveals that it has a robust metabolism with complete sets of enzymes for glycolysis and the TCA cycle and may be able synthesize all 20 amino acids de novo. Other housekeeping genes include ATP-binding cassette (ABC) transporters ($\approx$80 genes), lipoproteins ($\approx$150 genes), and flagellar and chemotaxis genes ($\approx$80 genes). The pan-genome of *Leptospira* spp. is open, suggesting that the entire gene repertoire has not yet been elucidated (Xu et al. 2016). Inspection of the noncore genome has revealed important differences between species and strains that may modulate their virulence such as (i) the expansion of pathogen-specific gene families, such as virulence modifying (VM) (PF07598 paralogs) proteins, (ii) pathogen-specific adhesins, and (iii) proteins involved in resistance to oxidative stress (Fouts et al. 2016; Xu et al. 2016).

A large number of *Leptospira* pathogen-specific genes encode proteins of completely unknown function, suggesting the presence of pathogenic mechanisms unique to *Leptospira* (Adler et al. 2011; Picardeau et al. 2008). Genomic analysis has revealed a remarkable reduction of metabolic capacities in the strictly parasitic pathogen *L. borgpetersenii* and the presence of numerous pseudogenes and mobile elements (Bulach et al. 2006). The intermediate *L. licerasiae* shares more genes with pathogens than saprophytes, suggesting that intermediate species are more closely related to pathogenic than saprophytic *Leptospira* species (Ricaldi et al. 2012a).

The saprophyte *L. biflexa* has many more genes encoding environmental sensing and metabolic proteins than pathogenic species, reflecting its capacity to survive in diverse environments (Picardeau et al. 2008).

Leptospira strains can accumulate mutations during repeated in vitro passage. Identification of point mutations responsible for specific phenotypic changes can be identified by whole-genome sequencing of the parental and mutant strains. Subtle genomic variations include the substitution of one nucleotide with another, referred to as a single-nucleotide polymorphism (SNP) or single-nucleotide variant (SNV), or the insertion or deletion of a single nucleotide. Single-nucleotide substitution in a DNA sequence that encodes a protein may produce either a synonymous (silent) codon mutation, and thus no change in the encoded amino acid, or a nonsynonymous mutation, resulting in an amino acid change that may have an impact on the function of the encoded protein. Single-nucleotide insertions or deletions in protein-coding sequences result in a frameshift leading to significant alteration of the encoded protein, including protein truncation due to creation of an early stop codon.

The full-genome sequences of a low- and high-passage strain obtained after 67 in vitro passages of *L. interrogans* serovar Manilae resulted in a 51-nucleotide difference in the size of the genome (Satou et al. 2015). A high-passage, virulence-attenuated isolate of *L. interrogans* serovar Lai was sequenced and the sequence compared with that of a nonisogenic virulent isolate, revealing mutations (insertions, deletions, and single-nucleotide variations) in 101 genes, representing 2.7% of total gene content (Zhong et al. 2011). In another study, a virulent isolate of *L. interrogans* serovar Lai was serially passaged in vitro 18 times to identify mutations associated with the disappearance of virulence. Through a process of aging for eight months, corresponding to approximately 800 generations, 41 nonsynonymous SNVs, resulting in dominant alleles and corresponding to a total of 35 protein-coding genes, were identified and shown to be associated with attenuation (Lehmann et al. 2013). The process of accumulation of mutations resulting in attenuation was further determined by assessing the SNV allele frequency between the polyclonal populations of low- and high-*L. interrogans* serovar Lai strains (Lehmann et al. 2016). Approximately, 100 SNVs were identified as having undergone a significant change in allele frequency after passaging the parental strain for eight subcultures (Lehmann et al. 2016). It has to be noted that few mutations were shared among the three culture-attenuated strains of *L. interrogans* serovar Lai (Lehmann et al. 2013, 2016; Zhong et al. 2011), suggesting that the in vitro accumulation of SNV is rather stochastic. Culture attenuation experiments would be therefore more informative if whole-genome sequencing data from several independent attenuated lineages are compared.

Spontaneous nonmotile strains displaying altered cell morphology and attenuated virulence have been recently isolated after plating fresh isolates of *L. interrogans* onto solid medium. Whole-genome sequencing of two spontaneous mutants allowed the identification of an insertion and a deletion of one nucleotide in *fcpA* (Wunder et al. 2016) and *fliM* (Fontana et al. 2016), respectively, which introduced a frameshift and resulted in a premature stop codon. Complementation of the mutations with wild-type genes restored the wild-type morphology, motility, and virulence phenotypes (Fontana et al. 2016; Wunder et al. 2016). Spontaneous mutations in *fliM* (Fontana et al. 2016) and *fcpA* (Wunder et al. 2016) may have arisen during in vitro culture isolation.

2.3 Transcriptomics

Different aspects of pathogenesis can be studied at the transcription level of leptospires. Transcriptome of pathogenic *Leptospira* spp. can be obtained from simple in vitro cultures and ex vivo cell cultures to in vivo animal models. Each model has specific advantages and disadvantages. For example, in vitro or ex vivo models can mimic distinct stages of infection, but lack the complexity of a living host. *Leptospira* is usually grown in nutrient-rich media of low osmolarity (67 mOsm, in comparison with the physiological level of 300 mOsm), and the optimal cultivation temperature is 30 °C. The in vitro culture, conditions are therefore very different from those found inside the host. Whole-genome microarrays have been used to determine global changes in transcript levels of *L. interrogans* in response to changes in temperature (Lo et al. 2006; Qin et al. 2006) and osmolarity (Matsunaga et al. 2007), iron depletion (Lo et al. 2010), and serum exposure (Patarakul et al. 2010), which are relevant to changes that occur during acute and chronic disease. RNA-sequencing (RNA-Seq) of the wild-type, LB139 (Eshghi et al. 2014), and *perR* (Lo et al. 2010) mutants has enabled investigation of the regulon of these genes in *L. interrogans*. Transcriptional profiling of *L. interrogans* during its interaction with phagocytic cells was also studied (Xue et al. 2010). More recently, the transcriptome of *L. interrogans* cultivated inside dialysis membrane chambers (DMCs) implanted intraperitoneally into rats has provided an invaluable first glimpse into the conditions inside the reservoir host during adaptation, as rats are not susceptible to leptospirosis infection (Caimano et al. 2014). One major advantage of growing the bacteria inside DMCs is to separate host and pathogen RNA as the bacteria are often in low number within infected tissues.

2.4 Proteomics and Protein Microarrays

While it is possible to gain much information from the transcriptome, the proteome is the final product of gene expression, and therefore, changes in the proteome are amplified relative to those in the transcriptome. Thus, only 25% of proteins that differed in level in *L. interrogans* also differed at the transcriptional level in response to changes in temperature (Lo et al. 2009). This may be the result of RNA and protein turnover and post-transcriptional regulation by noncoding small RNA (sRNA).

The *L. interrogans* genome encodes approximately 3700 proteins, including many of unknown function. In standard leptospiral culture medium (67 mOsm, 30 °C, exponential phase culture), some proteins, including LigA/LigB (Matsunaga et al. 2003), LipL36 (Haake et al. 1998), and heat shock proteins (Nally et al. 2001), are not detectable or only at low levels. Protein mass spectrometry of *L. interrogans* grown in culture in regular medium, or in media depleted of iron or supplemented with serum, showed that *L. interrogans* exposed to conditions that mimic those of an in vivo infection showed altered protein levels (Cao et al. 2010; Eshghi et al. 2009). Direct secretome analysis by mass spectrometry of *L. interrogans* grown in vitro at various temperatures or osmolarity showed that most exoproteins are conserved in the non-pathogenic species *L. biflexa* and annotated as being involved in metabolic processes, some of which may be moonlighting proteins (Eshghi et al. 2015b).

An approach combining cryo-electron tomography and quantitative mass spectrometry took advantage of the long, thin structure of leptospires to determine protein abundance per *Leptospira* cell and identified protein complexes localized to specific cellular compartments (Beck et al. 2009; Malmström et al. 2009). This approach provided concentration estimates for $\sim$1800 proteins of *L. interrogans,* representing 51% of the coding sequences of the genome, and showed, for example, that LipL32 is the most abundant protein, with 38,000 copies per cell (Malmström et al. 2009). Visual proteomics has also been used to produce 3D snapshots of the cellular distribution of large protein complexes in responses to heat shock, antibiotic treatment, and starvation conditions (Beck et al. 2009).

Analyzing the proteome of the pathogen in the host can yield indispensable insight into the disease process. The proteome of *L. interrogans* isolated from the tissues and/or urine of acutely and chronically infected animal models has been determined using standard proteomic methodologies, including two-dimensional gel electrophoresis, immunoblotting, and mass spectrometry (Monahan et al. 2008; Nally et al. 2005a, b, 2007, 2011), showing, for example, that Loa22 levels were higher under conditions of acute disease than in in vitro cultivated leptospires, whereas the lipopolysaccharide content was down-regulated during acute infection (Nally et al. 2005a, 2007). Proteomics can also identify which host proteins in lung tissues (Schuller et al. 2015) and urine (Nally et al. 2015) are targeted by leptospires during infection, providing crucial information about the mechanisms of host–microbe interactions.

Methylation, acetylation, and phosphorylation are common post-translational modifications (PTM) of proteins in both saprophytic and pathogenic *Leptospira* spp. (Cao et al. 2010; Eshghi et al. 2012b; Stewart et al. 2016; Witchell et al. 2014). Methylated and phosphorylated proteins were predominant in the membrane-associated fractions of in vitro cultures (Cao et al. 2010; Eshghi et al. 2009; Stewart et al. 2016), but also in in vivo-isolated *L. interrogans* (Witchell et al. 2014). Proteomic analysis also suggests that nonulosonic acids, which include sialic acids, are synthesized by *L. interrogans* and used for modification of surface-exposed lipoproteins (Ricaldi et al. 2012b). These PTMs may have an important role in protein function and/or host cell interactions, although their functional significance remains to be determined.

Protein microarrays have been widely used to evaluate the humoral immune response to many infectious agents (Liang and Felgner 2015). An *L. interrogans* microarray containing the protein products of 91% (3359) of all leptospiral predicted genes was thus used to identify *L. interrogans* protein-reactive antibodies in sera from naturally infected humans (Lessa-Aquino et al. 2015). Protein microarrays can also be used for high-throughput identification of host ligand-binding proteins, making it possible, for example, to screen for interactions between leptospiral OMPs and components of the extracellular matrix (Pinne et al. 2012).

3 Future Prospects

Our understanding of the genetics and molecular basis of the pathogenesis of leptospires is in its infancy relative to other bacterial species. Further tools and approaches will need to be developed to improve the characterization of individual leptospiral genes, including the ability to knockout genes in pathogenic *Leptospira* spp. Genetic manipulation of the spirochete *B. burgdorferi* was initially hindered by endogenous restriction-modification systems (Kawabata et al. 2004). Indeed, low transformation efficiency of pathogenic *Leptospira* spp. may be attributed to the presence of restriction-modification systems, and the use of strains lacking the restriction-modification systems by gene disruption may increase transformation efficiency. The increasing ease of genetic manipulation for fastidious organisms, including CRISPR/Cas9 technology, may also soon facilitate the identification of factors important for infection.

Although the number of essential genes in bacteria varies (Zhang and Lin 2009), the estimation that approximately 350 genes are essential for the growth of *L. interrogans* suggests that there are transposon insertions in 35% of the ~3300 genes that may be amenable to mutagenesis in our library in *L. interrogans* (our unpublished data). Completion of the mutant collection by systematic random transposon mutagenesis is necessary, even though assembling a much larger library of mutants will also result in a substantial increase in mutant redundancy and intergenic insertions. This is particularly important, since we now have a powerful tool, Tn-Seq, for screening large pools of mutants in animal models and identifying genes that are essential for virulence and the colonization of specific organs (Lourdault et al. 2016).

The refinement of genome annotation is required for accurate studies by the *Leptospira* scientific community. This will provide the basis for understanding key cellular processes, such as metabolism and transcriptional regulation at the genome scale. Further improvement of the accuracy of genome annotation can be achieved by extensive human curation, taking into account transcriptome and proteome datasets which can allow the identification of novel genes, noncoding RNAs, and genes encoding small proteins (<100 amino acids), as well as the correction of misannotated start codons.

Although technically challenging, future studies should investigate the genomic evolution of the pathogen during the course of infection by whole-genome sequencing directly from the biological samples as well as analyze the host and pathogen responses at the transcriptional and translational levels, as both must dynamically respond to changing conditions during the course of an infection.

4 Conclusions

Slow growth, complex nutritional requirements, poor transformation efficiency, and genetic redundancy are inherent to *Leptospira* spp. Further complexity is added by the large diversity of strains, serovars, and species in the genus *Leptospira*. However, a better understanding of the pathogenesis of *Leptospira* is now possible due to recent significant advancements in the field, allowing, for example, the fulfillment of the molecular version of Koch's postulates by demonstrating a causal relationship between a bacterial gene, its product, and a defined phenotype (Falkow 1988). Further progress in our understanding of the virulence of *Leptospira* will contribute to the development of novel strategies for the diagnosis and prevention of this neglected emerging disease.

References

Adler B, Lo M, Seemann T, Murray GL (2011) Pathogenesis of leptospirosis: the influence of genomics. Vet Microbiol 153:73–81

Aviat F, Slamti L, Cerqueira GM, Lourdault K, Picardeau M (2010) Expanding the genetic toolbox for *Leptospira* species by generation of fluorescent bacteria. Appl Environ Microbiol 76:8135–8142

Baril C, Richaud C, Fournie E, Baranton G, Saint Girons I (1992) Cloning of *dapD*, *aroD* and *asd* of *Leptospira interrogans* serovar icterohaemorrhagiae, and nucleotide sequence of the *asd* gene. J Gen Microbiol 138:47–53

Bauby H, Saint Girons I, Picardeau M (2003) Construction and complementation of the first auxotrophic mutant in the spirochaete *Leptospira meyeri*. Microbiology 149:689–693

Beck M, Malmström JA, Lange V, Schmidt A, Deutsch EW, Aebersold R (2009) Visual proteomics of the human pathogen *Leptospira interrogans*. Nat Methods 6:817–823

Belfaiza J, Martel A, Margarita D, Saint Girons I (1998) Direct sulfhydrylation for methionine biosynthesis in *Leptospira meyeri*. J Bacteriol 180:250–255

Bono JL, Elias AF, Kupko JJ, Stevenson B, Tilly K, Rosa P (2000) Efficient targeted mutagenesis in *Borrelia burgdorferi*. J Bacteriol 182:2445–2452

Bourhy P, Frangeul L, Couve E, Glaser P, Saint Girons I, Picardeau M (2005a) Complete nucleotide sequence of the LE1 prophage from the spirochete *Leptospira biflexa* and characterization of its replication and partition functions. J Bacteriol 187:3931–3940

Bourhy P, Louvel H, Saint Girons I, Picardeau M (2005b) Random insertional mutagenesis of *Leptospira interrogans*, the agent of leptospirosis, using a mariner transposon. J Bacteriol 187:3255–3258

Bourhy P, Salaün L, Lajus A, Médigue C, Boursaux-Eude C, Picardeau M (2007) A genomic island of the pathogen *Leptospira interrogans* serovar Lai can excise from its chromosome. Infect Immun 75:677–683

Bromley DB, Charon NW (1979) Axial filament involvement in the motility of *Leptospira interrogans*. J Bacteriol 137:1406–1412

Bulach DM, Zuerner RL, Wilson P, Seemann T, McGrath A, Cullen PA, Davis J, Johnson M, Kuczek E, Alt DP, Peterson-Burch B, Coppel RL, Rood JI, Davies JK, Adler B (2006) Genome reduction in *Leptospira borgpetersenii* reflects limited transmission potential. Proc Natl Acad Sci U S A 103:14560–14565

Caimano MJ, Sivasankaran SK, Allard A, Hurley D, Hokamp K, Grassmann AA, Hinton JC, Nally JE (2014) A model system for studying the transcriptomic and physiological changes associated with mammalian host-adaptation by *Leptospira interrogans* serovar Copenhageni. PLoS Pathog 10:e1004004

Cao XJ, Dai J, Xu H, Nie S, Chang X, Hu BY, Sheng QH, Wang LS, Ning ZB, Li YX, Guo XK, Zhao GP, Zeng R (2010) High-coverage proteome analysis reveals the first insight of protein modification systems in the pathogenic spirochete *Leptospira interrogans*. Cell Res 20:197–210

Castiblanco-Valencia MM, Fraga TR, Breda LC, Vasconcellos SA, Figueira CP, Picardeau M, Wunder E, Ko AI, Barbosa AS, Isaac L (2016) Acquisition of negative complement regulators by the saprophyte *Leptospira biflexa* expressing LigA or LigB confers enhanced survival in human serum. Immunol Lett 173:61–68

Cerqueira GM, Souza NM, Araújo ER, Barros AT, Morais ZM, Vasconcellos SA, Nascimento AL (2011) Development of transcriptional fusions to assess *Leptospira interrogans* promoter activity. PLoS ONE 6:e17409

Choy HA, Kelley MM, Croda J, Matsunaga J, Babbitt JT, Ko AI, Picardeau M, Haake DA (2011) The multifunctional LigB adhesin binds homeostatic proteins with potential roles in cutaneous infection by pathogenic *Leptospira interrogans*. PLoS One 6(2):e16879

Croda J, Figueira CP, Wunder EAJ, Santos CS, Reis MG, Ko AI, Picardeau M (2008) Targeted mutagenesis in pathogenic *Leptospira*: disruption of the ligB gene does not affect virulence in animal models of leptospirosis. Infect Immun 76:5826–5833

Cullen PA, Cordwell SJ, Bulach DM, Haake DA, Adler B (2002) Global analysis of outer membrane proteins from *Leptospira interrogans* serovar Lai. Infect Immun 70:2311–2318

Cullen PA, Haake DA, Adler B (2004) Outer membrane proteins of pathogenic spirochetes. FEMS Microbiol Rev 28:291–318

Cullen PA, Xu X, Matsunaga J, Sanchez Y, Ko AI, Haake DA, Adler B (2005) Surfaceome of *Leptospira* spp. Infect Immun 73:4853–4863

Eshghi A, Cullen PA, Cowen L, Zuerner RL, Cameron CE (2009) Global proteome analysis of *Leptospira interrogans*. J Proteome Res 8:4564–4578

Eshghi A, Lourdault K, Murray GL, Bartpho T, Sermswan RW, Picardeau M, Adler B, Snarr B, Zuerner RL, Cameron CE (2012a) *Leptospira interrogans* catalase is required for resistance to H_2O_2 and for virulence. Infect Imm 80:3892–3899

Eshghi A, Pinne M, Haake DA, Zuerner RL, Frank A, Cameron CE (2012b) Methylation and in vivo expression of the surface-exposed *Leptospira interrogans* outer-membrane protein OmpL32. Microbiology 158:622–635

Eshghi A, Becam J, Lambert A, Sismeiro O, Dillies MA, Jagla B, Wunder EA Jr, Ko AI, Coppee JY, Goarant C, Picardeau M (2014) A putative regulatory genetic locus modulates virulence in the pathogen *Leptospira interrogans*. Infect Immun 82:2542–2552

Eshghi A, Henderson J, Trent MS, Picardeau M (2015a) *Leptospira interrogans lpxD* homologue is required for thermal acclimatization and virulence. Infect Immun 83:4314–4321

Eshghi A, Pappalardo E, Hester S, Thomas B, Pretre G, Picardeau M (2015b) Pathogenic *Leptospira interrogans* exoproteins are primarily involved in heterotrophic processes. Infect Imm 83:3061–3073

Falkow S (1988) Molecular Koch's postulates applied to microbial pathogenicity. Rev Infect Dis 10:274–276

Figueira CP, Croda J, Choy HA, Haake DA, Reis MG, Ko AI, Picardeau M (2011) Heterologous expression of pathogen-specific genes ligA and ligB in the saprophyte *Leptospira biflexa* confers enhanced adhesion to cultured cells and fibronectin. BMC Microbiol 11:129

Fontana C, Lambert A, Benaroudj N, Gasparini D, Gorgette O, Cachet N, Bomchil N, Picardeau M (2016) Analysis of a spontaneous non-motile and avirulent mutant shows that FliM is Required for full endoflagella assembly in *Leptospira interrogans*. PLoS ONE 11:e0152916

Fouts DE, Matthias MA, Adhikarla H, Adler B, Berg DE, Bulach D, Buschiazzo A, Chang Y-F, Galloway RL, Haake DA, Haft DH, Hartskeerl R, Ko A, Levett P, Matsunaga J, Mechaly AE, Monk J, Nascimento A, Nelson KE, Palsson A, Peacock SJ, Picardeau M, Ricaldi JN, Thaipandungpanit J, Wunder E, Yang F, Zhang JJ, Vinetz JM (2016) What makes a bacterial species pathogenic?: Comparative genomic analysis of the genus *Leptospira*. PLoS Negl Trop Dis 10:e0004403

Gamberini M, Gómez RM, Atzingen MV, Martins EA, Vasconcellos SA, Romero EC, Leite LC, Ho PL, Nascimento AL (2005) Whole-genome analysis of *Leptospira* interrogans to identify potential vaccine candidates against leptospirosis. FEMS Microbiol Lett 244(2):305–313

Haake DA, Martinich C, Summers TA, Shang ES, Pruetz JD, McCoy AM, Mazel MK, Bolin CA (1998) Characterization of leptospiral outer membrane lipoprotein LipL36: downregulation associated with late-log-phase growth and mammalian infection. Infect Immun 66:1579–1587

Haake DA, Matsunaga J (2010) *Leptospira*: a spirochaete with a hybrid outer membrane. Mol Microbiol 77:805–814

Henry R, Lo M, Khoo C, Zhang H, Boysen RI, Picardeau M, Murray GL, Bulach DM, Adler B, print] MEao (2013). Iron precipitation on the surface of *Leptospira interrogans* is associated with mutation of the stress-response metalloprotease HtpX. Appl Environ Microbiol 79:4653–4660

Kassegne K, Hu W, Ojcius DM, Sun D, Ge Y, Zhao J, Yang XF, Li L, Yan J (2014) Identification of collagenase as a critical virulence factor for invasiveness and transmission of pathogenic leptospira species. J Infect Dis 209:1105–1115

Kawabata H, Norris SJ, Watanabe H (2004) BBE02 disruption mutants of *Borrelia burgdorferi* B31 have a highly transformable, infectious phenotype. Infect Imm 72:7147–7154

King AM, Bartpho T, Sermswan RW, Bulach DM, Eshghi A, Picardeau M, Adler B, Murray GL (2013) Leptospiral outer membrane protein LipL41 is not essential for acute leptospirosis, but requires a small chaperone, Lep, for stable expression. Infect Immun 81:2768–2776

King AM, Pretre G, Bartpho T, Sermswan RW, Toma C, Suzuki T, Eshghi A, Picardeau M, Adler B, Murray GL (2014) High-temperature protein G is an essential virulence factor of *Leptospira interrogans*. Infect Immun 82:1123–1131

Komi KK, Ge YM, Xin XY, Ojcius DM, Sun D, Hu WL, Zhao X, Lin X, Yan J (2015) ChpK and MazF of the toxin-antitoxin modules are involved in the virulence of *Leptospira interrogans* during infection. Microbes Infect 17:34–37

Lambert A, Picardeau M, Haake DA, Sermswan RW, Srikram A, Adler B, Murray GA (2012) FlaA proteins in *Leptospira interrogans* are essential for motility and virulence but are not required for formation of the flagellum sheath. Infect Immun 80:2019–2025

Lehmann JS, Fouts DE, Haft DH, Cannella AP, Ricaldi JN, Brinkac L, Harkins D, Durkin S, Sanka R, Sutton G, Moreno A, Vinetz JM, Matthias MA (2013) Pathogenomic inference of virulence-associated genes in *Leptospira interrogans*. PLoS Negl Trop Dis 7:e2468

Lehmann JS, Corey VC, Ricaldi JN, Vinetz JM, Winzeler EA, Matthias MA (2016) Whole genome shotgun sequencing shows selection on *Leptospira* regulatory proteins during in vitro culture attenuation. Am J Trop Med Hyg 94:302–313

Lessa-Aquino C, Wunder EA Jr, Lindow JC, Rodrigues CB, Pablo J, Nakajima R, Jasinskas A, Liang L, Reis MG, Ko AI, Medeiros MA, Felgner PL (2015) Proteomic features predict seroreactivity against leptospiral antigens in leptospirosis patients. J Proteome Res 14:549–556

Liang L, Felgner PL (2015) A systems biology approach for diagnostic and vaccine antigen discovery in tropical infectious diseases. Curr Opin Infect Dis 28:438–445

Liao S, Sun A, Ojcius DM, Wu S, Zhao J, Yan J (2009) Inactivation of the fliY gene encoding a flagellar motor switch protein attenuates mobility and virulence of *Leptospira interrogans* strain Lai. BMC Microbiol 9:253

Lo M, Bulach DM, Powell DR, Haake DA, Matsunaga J, Paustian ML, Zuerner RL, Adler B (2006) Effects of temperature on gene expression patterns in *Leptospira interrogans* serovar Lai as assessed by whole-genome microarrays. Infect Immun 74:848–859

Lo M, Cordwell SJ, Bulach DM, Adler B (2009) Comparative transcriptional and translational analysis of leptospiral outer membrane protein expression in response to temperature. PLoS Negl Trop Dis 3:e560

Lo M, Murray GL, Khoo CA, Haake DA, Zuerner RL, Adler B (2010) Transcriptional response of *Leptospira interrogans* to iron limitation and characterization of a PerR homolog. Infect Immun 78:4850–4859

Lourdault K, Cerqueira GM, Wunder EA, Picardeau M (2011). Inactivation of clpB in the pathogen *Leptospira interrogans* reduces virulence and resistance to stress conditions. Infect Immun 79:3711–3717

Lourdault K, Matsunaga J, Haake DA (2016) High-Throughput parallel sequencing to measure fitness of *Leptospira interrogans* transposon insertion mutants during acute infection. PLoS Negl Trop Dis 10:e0005117

Louvel H, Saint Girons I, Picardeau M (2005) Isolation and characterization of FecA- and FeoB-mediated iron acquisition systems of the spirochete *Leptospira biflexa* by random insertional mutagenesis. J Bacteriol 187:3249–3254

Louvel H, Bommezzadri S, Zidane N, Boursaux-Eude C, Creno S, Magnier A, Rouy Z, Medigue C, Girons IS, Bouchier C, Picardeau M (2006) Comparative and functional genomic analyses of iron transport and regulation in *Leptospira* spp. J Bacteriol 188:7893–7904

Louvel H, Picardeau M (2007) Genetic manipulation of *Leptospira biflexa*. Wiley, Hoboken

Louvel H, Betton JM, Picardeau M (2008) Heme rescues a two-component system *Leptospira biflexa* mutant. BMC Microbiol 8:25

Malmström J, Beck M, Schmidt A, Lange V, Deutsch EW, Aebersold R (2009) Proteome-wide cellular protein concentrations of the human pathogen *Leptospira interrogans*. Nature 460:762–765

Marcsisin RA, Bartpho T, Bulach DM, Srikram A, Sermswan RW, Adler B, Murray GL (2013) Use of a high-throughput screen to identify *Leptospira* mutants unable to colonize the carrier host or cause disease in the acute model of infection. J Med Microbiol 62:1601–1608

Matsunaga J, Coutinho ML (2012) Positive regulation of *Leptospira interrogans* kdp expression by KdpE as demonstrated with a novel β-galactosidase reporter in *Leptospira biflexa*. Appl Environ Microbiol 78:5699–5707

Matsunaga J, Barocchi MA, Croda J, Young TA, Sanchez Y, Siqueira I, Bolin CA, Reis MG, Riley LW, Haake DA, Ko AI (2003) Pathogenic *Leptospira* species express surface-exposed proteins belonging to the bacterial immunoglobulin superfamily. Mol Microbiol 49:929–945

Matsunaga J, Lo M, Bulach DM, Zuerner RL, Adler B, Haake DA (2007) Response of *Leptospira interrogans* to physiologic osmolarity: relevance in signaling the environment-to-host transition. Infect Immun 75:2864–2874

Monahan AM, Callanan JJ, Nally JE (2008) Proteomic analysis of *Leptospira interrogans* shed in urine of chronically infected hosts. Infect Immun 76:4952–4958

Murray GL, Morel V, Cerqueira GM, Croda J, Srikram A, Henry R, Ko AI, Dellagostin OA, Bulach DM, Sermswan R, Adler B, Picardeau M (2009a) Genome-wide transposon mutagenesis in pathogenic *Leptospira* spp. Infect Immun 77:810–816

Murray GL, Srikram A, Hoke DE, Wunder EAJ, Henry R, Lo M, Zhang K, Sermswan RW, Ko AI, Adler B (2009b) The major surface protein LipL32 is not required for either acute or chronic infection with *Leptospira interrogans*. Infect Immun 77:952–958

Murray GL, King AM, Srikram A, Sermswan RW, Adler B (2010) Use of luminescent *Leptospira interrogans* for enumeration in biological assays. J Clin Microbiol 48:2037–2042

Nally JE, Timoney JF, Stevenson B (2001) Temperature-regulated protein synthesis by *Leptospira interrogans*. Infect Immun 69:400–404

Nally JE, Chow E, Fishbein MC, Blanco DR, Lovett MA (2005a) Changes in lipopolysaccharide O antigen distinguish acute versus chronic *Leptospira interrogans* infections. Infect Immun 73:3251–3260

Nally JE, Whitelegge JP, Aguilera R, Pereira MM, Blanco DR, Lovett MA (2005b) Purification and proteomic analysis of outer membrane vesicles from a clinical isolate of *Leptospira interrogans* serovar Copenhageni. Proteomics 5:144–152

Nally JE, Whitelegge JP, Bassilian S, Blanco DR, Lovett MA (2007) Characterization of the outer membrane proteome of *Leptospira interrogans* expressed during acute lethal infection. Infect Immun 75:766–773

Nally JE, Monahan AM, Miller IS, Bonilla-Santiago R, Souda P, Whitelegge JP (2011) Comparative proteomic analysis of differentially expressed proteins in the urine of reservoir hosts of leptospirosis. PLoS ONE 6:e26046

Nally JE, Mullen W, Callanan JJ, Mischak H, Albalat A (2015) Detection of urinary biomarkers in reservoir hosts of leptospirosis by capillary electrophoresis-mass spectrometry. Proteomics Clin Appl 9:543–551

Narayanavari SA, Lourdault K, Sritharan M, Haake DA, Matsunaga J (2015) Role of sph2 gene regulation in hemolytic and sphingomyelinase activities produced by *Leptospira interrogans*. PLoS Negl Trop Dis 9:e0003952

Ozuru R, Saito M, Kanemaru T, Miyahara S, Villanueva SY, Murray GL, Adler B, Fujii J, Yoshida SI (2017) Adipose tissue is the first colonization site of *Leptospira interrogans* in subcutaneously infected hamsters. PLoS ONE 12:e0172973

Pappas CJ, Picardeau M (2015) Control of gene expression in *Leptospira* spp. by transcription activator-like effectors (TALEs) demonstrates a potential role for LigA and LigB in virulence in *L. interrogans*. Appl Environ Microbiol 81:7888–7892

Pappas CJ, Benaroudj N, Picardeau M (2015) A replicative plasmid vector allows efficient complementation of pathogenic *Leptospira* strains. Appl Environ Microbiol 81:3176–3781

Patarakul K, Lo M, Adler B (2010) Global transcriptomic response of *Leptospira interrogans* serovar Copenhageni upon exposure to serum. BMC Microbiol 10:31

Pětrošová H, Picardeau M (2014) Screening of a *Leptospira biflexa* mutant library to identify genes involved in ethidium bromide tolerance. Appl Environ Microbiol 80:6091–6103

Picardeau M (2008) Conjugative transfer between *Escherichia coli* and *Leptospira* spp. as a new genetic tool. Appl Environ Microbiol 74:319–322

Picardeau M (2014) Family Leptospiraceae. In: Rosenberg E (ed) The prokaryotes. Springer, Berlin, pp 711–729

Picardeau M, Brenot A, Saint Girons I (2001) First evidence for gene replacement in *Leptospira* spp. Inactivation of *L. biflexa flaB* results in non-motile mutants deficient in endoflagella. Mol Microbiol 40:189–199

Picardeau M, Bulach DM, Bouchier C, Zuerner RL, Zidane N, Wilson PJ, Creno S, Kuczek ES, Bommezzadri S, Davis JC, McGrath A, Johnson MJ, Boursaux-Eude C, Seemann T, Rouy Z, Coppel RL, Rood JI, Lajus A, Davies JK, Médigue C, Adler B (2008) Genome sequence of the saprophyte *Leptospira biflexa* provides insights into the evolution of *Leptospira* and the pathogenesis of leptospirosis. PLoS ONE 3:e1607

Pinne M, Matsunaga J, Haake DA (2012) Leptospiral outer membrane protein microarray, a novel approach to identification of host ligand-binding proteins. J Bacteriol 194:6074–6087

Pinne M, Haake DA (2013) LipL32 Is a Subsurface Lipoprotein of *Leptospira* interrogans: presentation of new data and reevaluation of previous studies. PLoS ONE 8:e51025

Poggi D, Oliveira de Giuseppe P, Picardeau M (2010) Antibiotic resistance markers for genetic manipulations of *Leptospira* spp. Appl Environ Microbiol 76:4882–4885

Qin JH, Sheng YY, Zhang ZM, Shi YZ, He P, Hu BY, Yang Y, Liu SG, Zhao GP, Guo XK (2006) Genome-wide transcriptional analysis of temperature shift in *Leptospira. interrogans* serovar lai strain 5660. BMC Microbiol 6:51

Ratet G, Veyrier FJ, Fanton d'Andon M, Kammerscheit X, Nicola MA, Picardeau M, Boneca IG, Werts C (2014) Live imaging of bioluminescent *Leptospira interrogans* in mice reveals renal colonization as a stealth escape from the blood defenses and antibiotics. PLoS Negl Trop Dis 8: e3359

Ren S, Fu G, Jiang X, Zeng R, Xiong H, Lu G, Jiang HQ, Miao Y, Xu H, Zhang Y, Guo X, Shen Y, Qiang BQ, Xia Q, Danchin A, SaintGirons I, Somerville RL, Weng YM, Shi M, Chen Z, Xu JG, Zhao GP (2003) Unique and physiological and pathogenic features of *Leptospira interrogans* revealed by whole genome sequencing. Nature 422:888–893

Ricaldi JN, Fouts DE, Selengut JD, Harkins DM, Moreno A, Lehmann JS, Purushe J, Sanka R, Torres M, Webster NJ, Vinetz JM, Matthias MA (2012a) Whole genome analysis of *Leptospira licerasiae* provides insight into Leptospiral evolution and pathogenicity. PLoS Negl Trop Dis 6:e1853

Ricaldi JN, Matthias MA, Vinetz JM, Lewis AL (2012b) Expression of sialic acids and other nonulosonic acids in *Leptospira*. BMC Microbiol 12:161

Richaud C, Margarita D, Baranton G, Saint Girons I (1990) Cloning of genes required for amino acid biosynthesis from *Leptospira interrogans* serovar icterohaemorrhagiae. J Gen Microbiol 136:651–656

Ristow P, Bourhy P, da Cruz McBride FW, Figueira CP, Huerre M, Ave P, Girons IS, Ko AI, Picardeau M (2007) The OmpA-like protein Loa22 is essential for leptospiral virulence. PLoS Pathog 3:e97

Saint Girons I, Margarita D, Amouriaux P, Baranton G (1990) First isolation of bacteriophages for a spirochaete: potential genetic tools for *Leptospira*. Res Microbiol 141:1131–1138

Saint Girons I, Bourhy P, Ottone C, Picardeau M, Yelton D, Hendrix RW, Glaser P, Charon N (2000) The LE1 bacteriophage replicates as a plasmid within *Leptospira biflexa*: construction of an *L. biflexa–Escherichia coli* shuttle vector. J Bacteriol 182:5700–5705

Satou K, Shimoji M, Tamotsu H, Juan A, Ashimine N, Shinzato M, Toma C, Nohara T, Shiroma A, Nakano K, Teruya K, Terabayashi Y, Ohki S, Koizumi N, Okano S, Suzuki T, Hirano T (2015) Complete genome sequences of low-passage virulent and high-passage avirulent variants of pathogenic *Leptospira interrogans* serovar manilae strain UP-MMC-NIID, originally isolated from a patient with severe leptospirosis, determined using PacBio single-molecule real-time technology. Genome Announc 3:e00815–e00882

Setubal JC, Reis MG, Matsunaga J, Haake DA (2006) Lipoprotein computational prediction in spirochaetal genomes. Microbiology 152:113–121

Schuller S, Sergeant K, Renaut J, Callanan JJ, Scaife C, Nally JE (2015) Comparative proteomic analysis of lung tissue from guinea pigs with leptospiral pulmonary haemorrhage syndrome (LPHS) reveals a decrease in abundance of host proteins involved in cytoskeletal and cellular organization. J Proteomics 122:55–72

Slamti L, de Pedro MA, Guichet E, Picardeau M (2011) Deciphering morphological determinants of the helix-shaped *Leptospira*. J Bacteriol 193:6266–6275

Srikram A, Zhang K, Bartpho T, Lo M, Hoke DE, Sermswan RW, Adler B, Murray GL (2011) Cross-protective immunity against leptospirosis elicited by a live, attenuated lipopolysaccharide mutant. J Infect Dis 203:870–879

Stamm LV, Parrish EA, Gherardini FC (1991) Cloning of the *recA* gene from a free-living leptospire and distribution of RecA-like protein among spirochetes. Appl Environ Microbiol 57:183–189

Stewart PE, Carroll JA, Olano LR, Sturdevant DE, Rosa PA (2016) Multiple posttranslational modifications of *Leptospira biflexa* proteins as revealed by proteomic analysis. Appl Environ Microbiol 82:1183–1195

Toma C, Murray GL, Nohara T, Mizuyama M, Koizumi N, Adler B, Suzuki T (2014) Leptospiral outer membrane protein LMB216 is involved in enhancement of phagocytic uptake by macrophages. Cell Microbiol

Viratyosin W, Ingsriswang S, Pacharawongsakda E, Palittapongarnpim P (2008) Genome-wide subcellular localization of putative outer membrane and extracellular proteins in Leptospira interrogans serovar Lai genome using bioinformatics approaches. BMC Genom 9:181

Wang Y, Zhuang X, Zhong Y, Zhang C, Zhang Y, Zeng L, Zhu Y, He P, Dong K, Pal U, Guo X, Qin J (2015) Distribution of plasmids in distinct *Leptospira* pathogenic species. PLoS Negl Trop Dis 9:e0004220

Witchell TD, Eshghi A, Nally JE, Hof R, Boulanger MJ, Wunder EAJ, Ko AI, Haake DA, Cameron CE (2014) Post-translational modification of LipL32 during *Leptospira interrogans* infection. PLoS Negl Trop Dis 8:e3280

Wunder EA Jr, Figueira CP, Benaroudj N, Hu BY, Tong BA, Trajtenberg F, Liu J, Reis MG, Charon NW, Buschiazzo A, Picardeau M, Ko AI (2016) A novel flagellar sheath protein,

FcpA, determines filament coiling, translational motility and virulence for the *Leptospira* spirochete. Mol Microbiol 101:457–470

Xu Y, Zhu Y, Wang Y, Chang YF, Zhang Y, Jiang X, Zhuang X, Zhu Y, Zhang J, Zeng L, Yang M, Li S, Wang S, Ye Q, Xin X, Zhao G, Zheng H, Guo X, Wang J (2016) Whole genome sequencing revealed host adaptation-focused genomic plasticity of pathogenic *Leptospira*. Sci Rep 6:20020

Xue F, Dong H, Wu J, Wu Z, Hu W, Sun A, Troxell B, Yang XF, Yan J (2010) Transcriptional responses of *Leptospira interrogans* to host innate immunity: significant changes in metabolism, oxygen tolerance, and outer membrane. PLoS Negl Trop Dis 7:e2477

Yang HL, Zhu YZ, Qin JH, He P, Jiang XC, Zhao GP, Guo XK (2006) In silico and microarray-based genomic approaches to identifying potential vaccine candidates against *Leptospira interrogans*. BMC Genom 7:293

Yelton DB, Cohen RA (1986) Analysis of cloned DNA from *Leptospira biflexa* serovar patoc which complements a deletion of the *Escherichia coli trpE* gene. J Bacteriol 165:41–46

Zhang R, Lin Y (2009) DEG 5.0, a database of essential genes in both prokaryotes and eukaryotes. Nucleic Acids Res 37:D455–D458

Zhang L, Zhang C, Ojcius DM, Sun D, Zhao J, Lin X, Li L, Li L, Yan J (2012) The mammalian cell entry (Mce) protein of pathogenic *Leptospira* species is responsible for RGD motif-dependent infection of cells and animals. Mol Microbiol 83:1006–1023

Zhang K, Murray GL, Seemann T, Srikram A, Bartpho T, Sermswan RW, Adler B, Hoke DE (2013) Leptospiral LruA is required for virulence and modulates an interaction with mammalian Apolipoprotein A-I. Infect Imm 81:3872–3879

Zhao JF, Chen HH, Ojcius DM, Zhao X, Sun D, Ge YM, Zheng LL, Lin X, Li LJ, Yan J (2013) Identification of *Leptospira interrogans* phospholipase C as a novel virulence factor responsible for intracellular free calcium ion elevation during macrophage death. PLoS ONE 8:e75652

Zhong Y, Chang X, Cao XJ, Zhang Y, Zheng H, Zhu YZ, Cai C, Cui Z, Zhang Y, Li YY, Jiang XG, Zhao GP, Wang S, Li Y, Zeng R, Li X, Guo XK (2011) Comparative proteogenomic analysis of the *Leptospira interrogans* virulence-attenuated strain IPAV against the pathogenic strain 56601. Cell Res 21:1210–1229

Zhu WN, Huang LL, Zeng LB, Zhuang XR, Chen CY, Wang YZ, Qin JH, Zhu YZ, Guo XK (2014) Isolation and characterization of two novel plasmids from pathogenic *Leptospira interrogans* serogroup canicola serovar canicola strain gui44. PLoS Negl Trop Dis 8:e3103

Zhu W, Wang J, Zhu Y, Tang B, Zhang Y, He P, Zhang Y, Liu B, Guo X, Zhao G, Qin J (2015) Identification of three extra-chromosomal replicons in *Leptospira* pathogenic strain and development of new shuttle vectors. BMC Genom 16:90

Zhukova A, Fernandes LV, Hugon P, Pappas C, Sismeiro O, Coppée JY, Becavin C, Malabat C, Eshghi A, Zhang JJ, Yang FX, Picardeau M (2017) Genome-wide transcriptional start site mapping and sRNA identification in the pathogen *Leptospira interrogans*. Front Cel Infect Microbiol 7:10

Zuerner RL (1991) Physical mapping of chromosomal and plasmid DNA comprising the genome of *Leptospira interrogans*. Nucleic Acids Res 19:4857–4860

Interaction of *Leptospira* with the Innate Immune System

Catherine Werts

Abstract Innate immunity encompasses immediate host responses that detect and respond to microbes. Besides recognition by the complement system (see the chapter by A. Barbosa, this volume), innate immunity concerns cellular responses. These are triggered through recognition of conserved microbial components (called MAMPs) by pattern recognition receptors (PRRs), leading, through secretion of cytokines, antimicrobial peptides, and immune mediators, to cellular recruitment and phagocytosis. *Leptospira spp.* are successful zoonotic pathogenic bacteria that obviously overcome the immune system of their hosts. The first part of this chapter summarizes what is known about leptospires recognition and interaction with phagocytes and other innate immune cells, and the second part describes specific interactions of leptospiral MAMPs with PRRs from the TLR and NLR families. On the one hand, pathogenic leptospires appear to escape macrophage and neutrophil phagocytosis. On the other hand, studies about PRR sensing of leptospires remain very limited, but suggest that pathogenic leptospires escape some of the PRRs in a host-specific manner, due to peculiar cell wall specificities or post-translational modifications that may impair their recognition. Further studies are necessary to clarify the mechanisms and consequences of leptospiral escape on phagocytic functions and hopefully give clues to potential therapeutic strategies aimed at restoring the defective activation of PRRs by pathogenic *Leptospira spp*.

Abbreviations

BMM	Bone marrow-derived macrophages
GLP	Glycolipoprotein
Ig	Immunoglobulins
LPS	Lipopolysaccharide
LRR	Leucine reach repeat
MAMP	Microbe-associated molecular patterns
Na/K-ATPase	Potassium pump

C. Werts (✉)
Unité Biologie et Génétique de La Paroi Bactérienne, Institut Pasteur, Paris, France
e-mail: cwerts@pasteur.fr

Current Topics in Microbiology and Immunology (2018) 415:163–188
DOI 10.1007/82_2017_46

Published Online: 17 October 2017

NLR	NOD-like receptor
PBMC	Peripheral blood mononuclear cell
PRR	Pattern recognition receptor
TLR	Toll-like receptor

Contents

1 Introduction

Microbe-associated molecular patterns (MAMPs) are conserved components of bacteria, essential for their survival, but not present in the host. They are recognized by the host pattern recognition receptors (PRRs), which trigger the innate response. The recognition of MAMPs at the cell surface or in endosomes by the Toll-like receptors (TLRs), or in the cytosol by the NOD-like receptors (NLRs), initiates a signaling cascade leading to the activation of MAP kinases and transcription factors (NF-κB, IRF3, or IRF7) that together activate the expression of effectors of innate immunity, such as pro-inflammatory cytokines and chemokines that attract leukocytes to the infection site, antimicrobial peptides, and co-stimulation molecules required for antigen presentation (reviewed in Werts et al. 2006). The inflammatory response is thus a requisite to attract the phagocytes, but it can also be excessive and harmful.

Leptospirosis provokes inflammation, with increased levels of pro-inflammatory cytokines in blood and organs of humans and animals (Chassin et al. 2009; Matsui et al. 2011; Papa and Kotrotsiou 2015). It is associated with fever, pulmonary hemorrhages (Truong and Coburn 2011), and renal lesions (Yang 2007).

Leptospira infection has recently been visualized in mice. Intraperitoneal infection with a pathogenic bioluminescent strain of *L. interrogans serovar* Manilae provokes a biphasic disease. In the first week, leptospires disseminate rapidly and replicate in blood, then disappear from the circulation, and progressively reappear in the kidney, where they establish a lifelong colonization (Ratet et al. 2014). Although some mechanisms of escape of *Leptospira* from complement are known (see chapter XXX), how leptospires escape the phagocytic innate response and colonize the kidney remains obscure.

Myeloid cells, such as monocytes, macrophages, and dendritic cells recognize microbes through numerous PRRs. Although not phagocytic, epithelial cells play an important role in innate immunity, since they also express PRRs and can produce cytokines. As an example, uropathogenic *Escherichia coli* is recognized in mouse kidneys by multiple PRRs, allowing for neutrophil recruitment and bacterial phagocytosis (Bens et al. 2014; Tourneur et al. 2013). Kidney epithelial cell lines also recognize and respond to leptospires, through MAP kinases and NF-κB activation, resulting in chemokine and cytokine production (Hung et al. 2006a). Hence, the kidney participates actively in the innate defense against bacteria and raises the question as to how pathogenic leptospires circumvent the renal innate defense to colonize and settle in the proximal region of the renal tubule.

The aim of this review is to summarize what is known about interaction of *Leptospira spp.* with phagocytes and which PRRs are involved in recognition of leptospires.

2 Innate Recognition of *Leptospira*

2.1 Macrophages

Macrophages are mononuclear leukocytes that engulf and destroy microbes and infected or dying cells by phagocytosis. These innate sensors recognize bacteria through different PRRs of the TLR, NLR, and C-lectin (CLR) families and secrete pro-(TNF, IL-6, RANTES, IL-1ß …) and anti-(IL-10, TGFß) inflammatory cytokines, important for the recruitment of other leukocytes to sites of infection and for healing of tissues. Uptake of microbes occurs after their binding to plasma membrane receptors, which can be opsonic (such as complement receptors, or receptors of conserved domains of IgG antibodies), or non-opsonic (scavenger receptors or lectin-like receptors). This binding triggers internalization of the membrane engulfing the microbe into a phagosome, through actin cytoskeleton rearrangements. During phagocytosis, macrophages produce oxygen- and nitrogen-reactive species such as ROS and NO, and antimicrobial peptides which are bactericidal. The phagosome matures into an acidic oxidative compartment, enriched in proteases that hydrolyze the microbe. Peptides in association with MHC class II are presented to naïve T cells, thereby initiating the adaptive immune response. Macrophages present in different tissues have various names such as Kupffer cells in the liver or microglial cells in the brain and may also differentiate from blood monocytes (Gordon 2016). Compared to other spirochetes like *Borrelia* species (see chapter XXX), relatively little is known about phagocytosis of leptospires.

Immortalized human (THP-1) and mouse (J774 and RAW264.7) macrophage cell lines have been used extensively as surrogates to study macrophage functions of phagocytosis and cytokine secretion, as well as defining the PRRs involved in sensing of leptospires (see Recognition of Leptospiral MAMPs by TLRs and NLRs). A murine microglial cell line is also available as a model for recognition of leptospires by brain macrophages (Blasi et al. 2007).

Ex vivo stimulation of primary cells constitutes a complementary approach supposed to mimic more closely than cell lines the host macrophage's response. Hence, stimulation with *L. interrogans* of mouse peritoneal macrophages (Nahori et al. 2005; Xue et al. 2013), bone marrow-derived macrophages (BMM) (Lacroix-Lamande et al. 2012), and human monocytes from peripheral blood (PBMC) (Goris et al. 2011; Nahori et al. 2005; Xue et al. 2013) has clearly established the reactivity of macrophages to leptospires or their components, such as LPS, resulting in secretion of murine TNF, IL-6, RANTES, IL-1ß cytokines and also TNF and IL-8 in human cells, among other cytokines. Live and heat-killed, virulent, and culture-adapted strains all stimulate cytokine production. In a comparative array study, it was observed that virulent *L. interrogans* serovar Lai elicited more IL-10, RANTES, and IL-12-p40 cytokines in murine peritoneal macrophages and more IL-6, IL-10, TNF, and RANTES in human PBMC than the saprophytic *L. biflexa* Patoc strain (Xue et al. 2013). Heat-killed bacteria generally induced less cytokines in macrophages than live bacteria (Goris et al. 2011; Lacroix-Lamande et al. 2012; Werts et al. 2001). Interestingly, virulent live *L. interrogans* serovar Icterohaemorrhagiae strain Verdun and serovar Bataviae (Goris et al. 2011; Werts et al. 2001) have been shown to be less effective at activating THP-1 production of TNF than their highly passaged culture counterparts, which could be interpreted as an adaptation to immune evasion. However, both host- and culture-adapted *L. interrogans* serovar Lai strains induced the same TNF responses and also displayed almost equivalent response whether heat killed or alive (Goris et al. 2011). Thus, it is difficult to draw general conclusions that could be relevant for all the serovars. Moreover, differences of reactivity toward *Leptospira* observed between THP-1 and PBMC suggest that studies in cell lines should be confirmed by ex vivo studies with host macrophages (Goris et al. 2011).

Early studies in the 1960 by Faine and co-workers determined that peritoneal macrophages from guinea pigs, rabbits, and mice all rapidly and efficiently internalized virulent or avirulent *Leptospira icterohaemorrhagiae* strain AB (Faine et al. 1964). This internalization was faster when leptospires were incubated before infection with specific antiserum. Silver staining revealed intracellular granular spherical structures resembling *Leptospira*, killed by heating at 50 °C for 5 min, and therefore most probably corresponding to dead bacteria that had lost their shape (Faine et al. 1964). A study in the murine J774.4 cell line showed that the highly virulent *L. interrogans* serovar Lai also invaded the nucleus of cells, as opposed to the less pathogenic *L. interrogans* serovar Luo that was only found in the cytosol within phagosomes (Li et al. 2007).

Although internalization of *Leptospira* in macrophages was confirmed unambiguously in several studies in guinea pigs, mouse and human primary and cell line macrophages using FITC or CFSE-labeled *Leptospira* and immunofluorescence, flow cytometry, and confocal microscopy (Cinco et al. 1981; Li et al. 2007; Liu et al. 2014; Toma et al. 2011), the question of intracellular survival or killing of *Leptospira* remains obscure and may depend on the leptospiral serovar, and/or on the host and type of macrophage. Suspensions of elicited guinea pig peritoneal macrophages did not kill *Leptospira* serovar Copenhageni PB3 within a 2-h infection period, as measured by serial dilutions and colony plating of whole suspensions or washed macrophages treated with saponin to release internalized bacteria (Banfi et al. 1982). Human monocytes and derived macrophages were not able to internalize or kill *Leptospira* serogroup Icterohaemorrhagiae, even in the presence of 10% serum as a complement source, unless the bacteria were incubated with specific anti-*Leptospira* antibodies (Wang et al. 1984b). This suggests that pathogenic leptospires are resistant to the bactericidal effect of non-immune serum (see chapter by Barbosa). Interestingly, Yan's group showed that despite the fact that the virulent *L. interrogans* serovar Lai entered both murine and human cell lines and primary macrophages and equally induced ROS production (Hu et al. 2013), the fate of intracellular *Leptospira* is different in mouse versus human macrophages (Li et al. 2010). In mouse macrophages, leptospires were observed in late phagosomes and did not survive, whereas they were free in the cytosol, survived, and replicated in human macrophages (Li et al. 2010). However, another study in murine bone marrow-derived macrophages showed that a small proportion of internalized *L. interrogans* serovar Manilae survived and replicated within membrane bound compartments 24 h post-infection, although *L. biflexa* was killed rapidly (Toma et al. 2011).

How leptospires enter macrophages is not clear. In contrast to other spirochetes, such as *Treponema* and *Borrelia* species that enter macrophages mostly through coiling phagocytosis, which involves long pseudopodia forming spirals around the microbe, it was shown that different serovars of pathogenic *Leptospira* entered PBMC mostly through conventional phagocytosis (Rittig et al. 1998). Cytochalasin, a potent inhibitor of actin polymerization, impaired the internalization of opsonized *L. interrogans* serovar Lai in murine RAW264.7 macrophages, suggesting a classical actin-dependent phagocytosis (Liu et al. 2014). However, in murine J774.7 cells, cytochalasin did not modify the uptake of the virulent *L. interrogans* serovar icterohaemorraghiae strain Verdun, although nocodazole, known to depolymerize microtubules, and monodansylcadaverine, which inhibits clathrin-coated vesicles, decreased the number of internalized bacteria (Merien et al. 1997), suggesting that *Leptospira* could enter macrophages through clathrin-dependent endocytosis as shown for other invasive bacteria in epithelial cells (Veiga and Cossart 2006).

Pathogenic *Leptospira* spp. possess numerous adhesins (Adler 2014), playing a role in pathogenesis, but only a few leptospiral proteins are involved in binding to, and internalization by, macrophages. Compared to a highly virulent *L. interrogans* serovar Manilae strain, the uptake of a high-passage, culture-attenuated derivative was decreased in BMM (Toma et al. 2014), and a microarray transcriptional screen identified two surface proteins, the adhesin LigB and the predicted lipoprotein LMB216, both down-regulated in the culture-attenuated strain. Both proteins were shown to bind fibronectin. The LMB216 mutant showed decreased uptake in macrophages, and heterologous expression in *L. biflexa* of LigB or LMB216 conferred enhanced uptake and phagocytosis (Toma et al. 2014). The mutant was not tested for ability to cause acute disease in hamsters, but retained the ability to colonize mouse kidneys (Marcsisin et al. 2013). In addition, homologous to the mammalian cell entry protein of mycobacteria, mce, an outer membrane protein with an RGD adhesion motif binding to integrins α5ß1 and αvß3 was found crucial for adhesion and internalization of *L. interrogans* serovar Lai in murine J774A.1 macrophages. Indeed, the *mce Leptospira* mutant, notably obtained by homologous recombination, lost the ability to invade macrophages and was attenuated in hamsters, since the number of *mce* mutant bacteria in urine was lower than WT *Leptospira* or the complemented mutant 15 days post-intraperitoneal infection (Zhang et al. 2012), suggesting that the ability of leptospires to traffic through macrophages could be important for virulence. This has recently been confirmed using zebra fish embryos that are transparent and allowed imaging of infection with fluorescent *L. interrogans* serovar Copenhageni strain Fiocruz L1-130 (Davis et al. 2009). It was observed that intravenously injected leptospires were rapidly ingested in phagocytes with macrophagic appearance, but not killed, and survived within phagocytes for 48 h. Interestingly, the infected phagocytes migrated on the dorso-lateral part of dorsal aorta of embryos, suggesting that infected macrophages could participate in leptospiral dissemination in organs (Davis et al. 2009).

Leptospires injected in the forebrain attracted phagocytes and induced morphological changes with blebs and vesicles resembling apoptotic bodies found in microglia (Davis et al. 2009). Of note, several studies reported that infection with pathogenic *Leptospira* serovars Icterohaemorraghiae and Lai, but not with the saprophytic *L. biflexa* Patoc, induced apoptosis in human and mouse cell lines, primary peritoneal mouse macrophages, and PBMCs (Jin et al. 2009; Li et al. 2010; Merien et al. 1997). *L. interrogans* serovar Lai has been further shown to induce cytoplasm vacuolization and chromatin condensation, hallmarks of apoptosis in macrophages. This *Leptospira*-induced apoptosis occurred through activation of caspases 8, 3, and 6 and also through other pathways independent of caspase 9 (Hu et al. 2013; Jin et al. 2009). However, no cell death (measured by lactate dehydrogenase release) occurred after infection of mouse bone marrow macrophages with *L. interrogans* serovar Manilae (Toma et al. 2011).

The lack of murine models deficient for macrophages hindered studies that could have clarified the contribution of macrophages as efficient phagocytes or as cargo ships for *Leptospira*. Only one study addressed the in vivo role of macrophages during leptospirosis in mice by depletion of macrophages, with intravenous injection of crystalline silica (Isogai et al. 1986). Compared to non-treated mice, silica-treated mice presented higher leptospiral loads in blood on the first day post-intravenous infection with *L. interrogans* serovar Copenhageni, suggesting that macrophages could play a role in leptospiral clearance at the acute phase, in naïve mice (Isogai et al. 1986). However, this depletion technique is not specific to macrophages, since all phagocytes including neutrophils may have ingested the very toxic silica dust particles that altered their functions and killed them.

2.2 Neutrophils

Polymorphonuclear neutrophils (PMN) are the most abundant subset of leukocytes in blood, characterized by their granulocytic nucleus. They migrate toward the site of infection by chemotaxis, attracted by inflammatory components such as IL-8 or IFN-γ. These highly motile cells extravasate from blood vessel endothelia and reach inflamed tissues to very efficiently destroy pathogens, mostly by phagocytosis and generation of reactive oxygen species (ROS). When in contact with pathogens, they also degranulate their vesicles and release antimicrobial components. In addition, neutrophils were recently shown to generate neutrophil extracellular traps (NETs), a mixture of chromatin- and protease-forming nets that capture and kill pathogens. Neutrophils are short-lived cells, presumably to avoid damaging host cells (Segal 2005).

Human leptospirosis is known to induce neutrophilia, especially in severe cases (De Silva et al. 2014; Lindow et al. 2016; Raffray et al. 2016). In addition, a recent Australian retrospective study on 210 patients with leptospirosis showed that their increased counts of PMN at admission differed according to the infecting serovar, with the higher counts of neutrophils found in patients infected with *Leptospira interrogans* serovars considered more pathogenic than *L. borgpeterseni* and *L.weili* (Craig et al. 2013). However, the significance of this finding is not clear.

Morphological observations by light microscopy in the 1960s suggested that mice, guinea pigs, and rabbit neutrophils could phagocytose leptospires. Engulfment of leptospires was more efficient upon opsonization with a specific antiserum (Faine et al. 1964). It was shown that human and guinea pig PMN did not ingest nor kill pathogenic *L. interrogans*, even in presence of 10% serum, unless leptospires are opsonized by specific antibodies (McGrath et al. 1984; Wang et al. 1984a). Isogai's group isolated peritoneally elicited PMNs from rats and showed that live, but not killed, leptospires induced PMN chemoluminescence, which is the hallmark of oxidative bursts occurring upon phagocytosis (Isogai et al. 1989). The killing of bacteria was boosted in the presence of complement and specific antiserum. Interestingly, *L. interrogans* elicited less chemoluminescence compared to non-pathogenic strains (Isogai et al. 1989), confirming studies which showed that the saprophytic *L. biflexa* Patoc strains were easily phagocytized even in the absence of specific antiserum (Banfi et al. 1982). Interestingly, Gomez's group showed recently that infection of mice with *L. interrogans* triggered the early formation of NETs (Scharrig et al. 2015), and that ex vivo infection of human PMNs with *L. interrogans* triggered NETs that killed the bacteria (Scharrig et al. 2015). However, the PMN defense through NETosis does not seem very efficient in mice since depletion of PMN with specific antibodies only slightly decreased the leptospiral loads in blood at the acute phase, but did not modify the overall course of the disease (Scharrig et al. 2015; Werts 2010). Bovine PMNs infected ex vivo with different serovars of *Leptospira* produced cytokines and ROS and also slightly induced NET formation. However, this bovine PMN response did not affect the leptospiral viability (Wilson-Welder et al. 2016). Of note, Raffray and co-workers recently showed that despite the increased levels of PMNs in patients with leptospirosis, expression markers of granulocyte activation as well as IL-8 secretion were strikingly not triggered by leptospirosis (Raffray et al. 2016). Therefore, all these studies in different animals suggest that neutrophils, like macrophages, would not be very proficient to combat leptospirosis in non-immune patients, despite their recruitment in humans.

2.3 Dendritic Cells/DC-SIGN

Dendritic cells (DCs) are professional antigen-presenting cells, linking the innate recognition of pathogens to the triggering of adaptive immunity. Immature DCs of the skin or mucosa may be the first line to recognize, capture, and process leptospires in antigenic peptides, then migrate to lymph nodes to present the processed antigens to naïve T cells. Only one study addressed the recognition of *Leptospira* by DCs (Gaudart et al. 2008). Different live virulent or attenuated strains of *L. interrogans* serovars Pyrogenes and Autumnalis induced in vitro maturation of monocyte-derived DCs (Mo-DCs), obtained from healthy blood donors. Of note, the different serovars induced different levels of IL-12p70 and TNF cytokines (Gaudart et al. 2008), two cytokines crucial for further differentiation of naïve T helper cells and shaping of the adaptive immune response. DCs express many different PRRs, among them the carbohydrate receptor DC-SIGN of the C type lectin family, recognizing high-mannose glycans. The authors also demonstrated that *Leptospira* binds in vitro and in vivo to DC-SIGN, through surface carbohydrates (Gaudart et al. 2008). However, DC-SIGN requires other PRRs to internalize the pathogens by endocytosis and trigger signaling. Which PRRs are involved in the cytokine production, and whether leptospires are killed by DCs and processed for antigen presentation remains to be elucidated.

2.4 NK and γ/δ T Cells

Natural killer (NK) cells are peculiar cytotoxic lymphocytes known to destroy parasite and virus infected cells. γ/δ T cells, named according to their particular T cell receptor, are other atypical large T lymphocytes involved in immediate epithelial defense against pathogens. Both NK and γ/δ T cells have been recently recognized as innate cells able to fight bacteria. Indeed, upon recognition of bacteria through PRRs of the TLR and NLR families, they produce IFN-γ, a potent macrophage activator (Reynolds and Dong 2013; Souza-Fonseca-Guimaraes et al. 2012). Moreover, these T subsets can develop immune memory, and can be quickly reactivated by the same antigen. A few studies addressed the recognition or activation of NK and γ/δ T cells by leptospires.

Firstly, Vinetz's group showed in vitro that peripheral blood monocyte cells (PBMC) derived from healthy volunteers and stimulated with high numbers of live or heat-killed *L. interrogans* serovar Copenhageni were able to produce Th1 cytokines such as IFN-γ, IL-12, and TNF. These cells also proliferated and preferentially expanded to γ/δ T cells and NK cells. Interestingly, at concentrations below 5×10^5 bacteria/ml, *Leptospira* did not induce these responses and the proliferation resulted in expansion of the classical α/β T cells (Klimpel et al. 2003). The expansion of atypical γ/δ T cells was observed in vivo in a case study of a woman with mild leptospirosis (Barry et al. 2006).

More recently, Zuerner et al. (2011) showed that compared to naïve animals, cattle vaccinated with a monovalent vaccine of *L. borgpeterseni* serovar Hardjo exhibited, upon challenge with the same serovar, increased amounts of NK and γ/δ T cells in the blood. These NK and γ/δ cells were prone to produce IFN-γ when exposed ex vivo to antigens from the same serovar, suggesting that they participate in the long-term protective role of the vaccine which decreased the renal colonization upon challenge one year after vaccination.

However, clinical studies comparing larger cohorts of hospitalized patients with leptospirosis to healthy ones gave conflicting results. Indeed, in a Brazilian cohort of 27 patients with leptospirosis, expansion of atypical γ/δ T cells was found in only 15 of the patients. Interestingly, the death rate was lower in this group, suggesting a protective role of the γ/δ T cells (Damasco et al. 2011). However, another study with 12 patients in La Reunion showed that acute leptospirosis was associated with a decrease in total T cells and in γ/δ T cell populations and that severe cases with the highest bacterial loads and liver dysfunction were associated with the lowest numbers of γ/δ T cells (Raffray et al. 2015). These discrepancies were not explained, but could be linked to the infecting serovar, the bacterial load, or the time of blood sampling; in the latter study, sampling was done upon admission of the patients, possibly before the onset of the expansion. Follow-up of case patients from admission would be very useful to clarify the condition of expansion or these atypical T cells and whether or not they have a protective role in human leptospirosis.

2.5 Antimicrobial Peptides

Antimicrobial peptides are small cationic peptides that specifically target and kill bacteria through destabilization of their negatively charged membranes. They are produced by almost all cells upon PRR stimulation, although in a species-specific manner. They also act as cytokines. For example, the human cathelicidin LL37 is involved in leukocyte recruitment (Selsted and Ouellette 2005).

Proteomic analysis of urine samples from patients with leptospirosis identified defensin α1 as a marker associated with kidney injury (Chagan-Yasutan et al. 2016), but the efficiency of this defensin to kill *Leptospira* has not been studied.

Five antimicrobial peptides from the cathelicidin family from different mammalian hosts have been tested in vitro against spirochetes, including *L. interrogans* (Sambri et al. 2002). Interestingly, the bactericidal effects of cathelicidins from cattle and sheep were more active against *L. interrogans* than the human LL37 or CRAMP, its equivalent in mice. Moreover, the bovine myeloid antimicrobial peptide 28 (BMAP28) known, like LL37, to bind lipopolysaccharide (LPS), was shown to efficiently block IL-6 and TNF production in bovine cells stimulated with LPS from *L. interrogans* serovar Hardjo (Guo et al. 2016). PG1 from pig was also very efficient in inactivating the serovars Hardjo and Pomona, but less active against Copenhageni and Icterohaemorrhagiae (Sambri et al. 2002).

Recently, Ko's group, in a very comprehensive study, performed whole blood transcriptomic analysis in Brazilian patients with acute leptospirosis and identified differential gene expression profiles that distinguished the survivor cohort versus the patients who died. They identified several risk factors, such as elevated pro-inflammatory genes, and confirmed some of these results by ELISA in serum. In particular, they found that elevated levels of IL-1ß and IL-18 receptors, and low levels of cathelicidin L37 and RANTES transcripts at the acute phase, were associated with a poor prognosis of severe leptospirosis and death (Lindow et al. 2016). Interestingly, administration of synthetic LL37 peptide could rescue hamsters from a lethal challenge with *L. interrogans* Copenhageni, suggesting that administration of synthetic cathelicidin could be of therapeutic use in humans (Lindow et al. 2016).

3 Recognition of Leptospiral MAMPs by TLRs and NLRs

3.1 CD14, TLR2, TLR4/Lipid A, and Lipopolysaccharide

Compared to most other spirochetes, *Leptospira* species are peculiar since their outer membrane contains lipopolysaccharide (LPS), anchored in the outer leaflet through the lipid A moiety, also called endotoxin. The O-antigen component differs according to serovars and constitutes the target of specific immunoglobulins (Ig) conferring humoral protection after infection (Challa et al. 2011; Chassin et al. 2009; Midwinter et al. 1990). Interestingly, the O-antigen part of the LPS, which is the target of neutralizing antibodies, has been shown to be decreased in liver during acute lethal infection in guinea pigs, although, when leptospires were retrieved from the kidneys, the O-antigen part was equivalent to those of leptospires grown in culture (Nally et al. 2005).

Early studies showed that LPS from pathogenic leptospires activates mouse macrophages and is mitogenic for B cells, but not for T and NK cells (Isogai et al. 1990a, b). Compared to LPS from Gram-negative bacteria, the endotoxic properties of the leptospiral LPS are reduced in mice and also in the limulus amebocyte lysate assay, based on the coagulation of the horseshoe crab hemolymph in the presence of endotoxin, an ancestral innate immune defense (Isogai et al. 1990a; Nahori et al. 2005; Werts et al. 2001).

The recognition of classical LPS from Gram-negative bacteria is mediated by soluble or membrane CD14 that delivers lipid A to TLR4 associated with the cofactor MD2. The structure of dimeric TLR4 in complex with MD2 and *E. coli* lipid A showed that lipid A binds both to TLR4 and MD2, itself bound to the inner face of the leucine reach repeat (LRR) domain of TLR4. One acyl chain of lipid A establishes contact with the LRR of the other TLR4/MD2/lipid A complex, which brings together two molecules of TLR4 and gathers the two intracellular (TIR) domains that initiate signaling through homotypic interactions with the TIR domains of cytosolic adaptors such as MyD88 or TRIF (Park et al. 2009).

Although leptospiral LPS classically binds to CD14, it is peculiar since, in human macrophages, it is not recognized by TLR4 but by TLR2, the receptor for lipoproteins (Werts et al. 2001). TLR2 recognition of leptospiral LPS has further been confirmed in bovine and pig cells infected with *L. interrogans* serovar Hardjo or Hebdomadis, respectively (Guo et al. 2015, 2016).

The structure of lipid A of *L. interrogans* has been elucidated by the Raetz's group (Que-Gewirth et al. 2004). Leptospiral lipids A from serovars Icterohaemorragiae, Lai (Que-Gewirth et al. 2004), and Manilae (Eshghi et al. 2015) are identical. Compared to classical endotoxin, leptospiral lipid A contains several modifications, including difference in acyl chain length, a methylation that blocks the reactivity of one of the two lateral phosphate groups, the lack of the second phosphate group, unsaturated secondary acyl chains, and only amide bonds to link the acyl chains to the disaccharide component. These structural differences may confer steric hindrance and explain the lack of recognition of leptospiral LPS through TLR4. The study of TLR2 and TLR4 sensing of purified leptospiral lipid A showed that, as expected, the leptospiral lipid A was not recognized by human TLR4, but neither was it recognized through TLR2 (Nahori et al. 2005). In addition, the leptospiral LPS was shown to signal through dimers of TLR2 and TLR1, suggesting that an unknown component, most probably a lipoprotein, tightly bound to the leptospiral LPS, confers the TLR2 specificity (Nahori et al. 2005). Unexpectedly, leptospiral lipid A was shown to be readily recognized by murine TLR4 (mTLR4) (Nahori et al. 2005). The structure of mTLR4/MD2 is not known, but functional studies showed that mTLR4/MD2 is more promiscuous than hTLR4/MD2, for example recognizing ligands like lipid IVA, an antagonist of hTLR4 (Saitoh et al. 2004). This unanticipated species specificity of sensing of leptospiral LPS by TLR4 could in part explain the difference in susceptibility between mice that are considered as asymptomatic reservoirs, and humans acutely sensitive to leptospirosis. Indeed, young mice naturally deficient for TLR4, such as C3H/HeJ and adult C57BL6/J TLR4 knockout (ko) mice, are very sensitive to the disease and constitute models, reproducing the features of severe, potentially fatal, human leptospirosis (Chassin et al. 2009; Pereira et al. 1998; Richer et al. 2015; Viriyakosol et al. 2006). These experiments demonstrated that TLR4 is a crucial

PRR that protects mice against leptospirosis. Surprisingly, TLR2 does not seem as important, since TLR2ko mice do not die from leptospirosis. However, double ko mice for both TLR2 and TLR4 died faster than TLR4ko mice, suggesting that TLR2 confers some additive protection (Chassin et al. 2009). It was also shown that, in mice at least, B cells are crucial for protection against leptospirosis (Adler and Faine 1976). Following infection of mice with *L. interrogans*, B cells produced specific IgM directed against the LPS through TLR4 recognition, as soon as 3 days post-infection, and later on, specific IgG dependent also on TLR2 (Chassin et al. 2009). The protection conferred by these Ig was demonstrated after passive transfer of serum in mice devoid of B cells. Intraperitoneal injection of 200 μl of serum collected from WT mice infected for 3 days (D3) or 20 days (D20), delayed death, or rescued them from infection with *L. interrogans* serovar Copenhageni strain Fiocruz L1-130, respectively (Chassin et al. 2009). Other roles of TLR2 and TLR4 have been found in kidney and liver, where their stimulation by *Leptospira* induced the secretion of protective mediators. For example, the TLR2- and TLR4-dependent IFN-γ activation of the phagocytic properties of macrophages was found to be dependent on the presence of T cells in the kidneys and on the presence of B cells in the liver. It was also shown that the expression of iNOS in the kidneys was dependent on both TLR2 and TLR4, expressed by T cells and epithelial cells (Chassin et al. 2009).

Several mutants in the biosynthetic pathway of LPS have been shown to attenuate the virulence of *Leptospira* (Eshghi et al. 2015; Marcsisin et al. 2013; Murray et al. 2010). This suggests that leptospires probably subvert the host recognition, through as yet unknown regulatory mechanisms or atypical features of LPS (Marcsisin et al. 2013; Que-Gewirth et al. 2004).

3.2 TLR2/LipL32

LipL32 is the major lipoprotein of pathogenic leptospires expressed during mammalian infection (Haake et al. 2000). LipL32 is very conserved among pathogenic leptospires and binds some components of the extracellular matrix (see chapter XXX). Bacterial lipoproteins are recognized by the innate immune system through TLR2, in association with TLR1 if the lipoprotein is triacylated or in association with TLR6 if the lipoprotein is diacylated. It was first shown that LipL32 activates TLR2 in human THP-1 monocytes (Werts et al. 2001). Then Yang's group showed that outer membrane preparations from pathogenic *L. santarosai* serovar Shermani induced in renal proximal tubule epithelial cells the expression of pro-inflammatory cytokines and KC and MIP2 chemokines, known to be important for neutrophil recruitment. Using transfected human HEK293 cells, they further showed that outer membrane proteins induced signaling through TLR2 and TLR1, but not TLR4 (Hung et al. 2006a, b). Stimulation of renal cells by outer membrane proteins or LipL32 also induced a TLR2-dependent iNOS production (Yang et al. 2006).

The direct interaction of LipL32 with TLR2 has been demonstrated by atomic force microscopy (Hsu et al. 2010). The presence of a calcium ion stabilizing LipL32 is required for efficient interaction with TLR2 (Lo et al. 2013).

However, the prominent inflammatory role of LipL32 through TLR2 has not been confirmed in vivo during infection in mammals, since TLR2ko mice infected with *L. interrogans* serovar Copenhageni strain Fiocruz L1-130 did not show higher bacterial loads, nor different inflammatory scores or reduced cytokine expression levels in kidneys 3 days post-infection (Chassin et al. 2009), and did not show differences in fibrosis 15 days post-infection (Fanton d'Andon et al. 2014), indirectly suggesting that LipL32 would not be a major player during acute leptospirosis in mice. This is in accordance with a recent study showing that LipL32 is not required for infection, since a *lipL32* mutant of *L. interrogans* serovar Manilae did not lose its virulence in hamsters or rats (Murray et al. 2009).

However, an elegant study by Yang's group recently showed that *L. santarosai* serovar Shermani caused nephritis in zebra fish embryos and expressed LipL32 in the first day post-infection. The renal injury, with malformations, swelling of the pronephric tubule and a complete perturbation of the potassium pump localization could be recapitulated by ectopic expression of LipL32 or injection of recombinant LipL32 in the embryos. The inflammatory effect of LipL32 was shown to be TLR2 dependent, suggesting that indeed LipL32 is responsible for kidney injury in zebra fish (Chang et al. 2016).

Haake's group recently showed that LipL32 is not localized at the leptospiral surface (Haake et al. 2000) but rather has a subcellular localization in the outer leaflet of the outer membrane (Pinne and Haake 2013). Importantly, this suggests that LipL32 may escape TLR2 recognition when *Leptospira* is alive, which is the case when leptospires are disseminating in blood and later on when they colonize the renal proximal tubules of the kidney (Ratet et al. 2014). In addition, in a breakthrough study, it was recently shown that LipL32 of *L. interrogans* serovar Copenhageni retrieved from rat urine contained post-translational modifications, including acetylation or tri-methylation of lysines that were not present when the bacteria were grown in EMJH (Witchell et al. 2014). Very interestingly, this modified LipL32 was less reactive with leptospirosis patient serum (Witchell et al. 2014). Of note, a major epitope of LipL32 recognized by the specific serum is in a region rich with lysine, suggesting that the modifications participate in the immune evasion from IgG, but also suggests that LipL32 signaling through TLR2 could also be attenuated when *Leptospira* is in the kidneys. Hopefully, future studies on the consequences of these methylations may give clues about the interaction of LipL32 and TLR2 in vivo and may explain some of the discrepant results, such as the apparent pro-inflammatory role of LipL32 activation of TLR2 in vitro not being associated with virulence of the bacteria.

Finally, the importance of TLR2 signaling to counteract pathogenic leptospires has recently been highlighted in two studies. First, the co-injection in hamsters with a synthetic agonist of TLR2 with *L. interrogans* serovar Automnalis alleviated the symptoms of leptospirosis and reduced lethality. This beneficial effect could be linked to an increased TLR2-dependent production of the anti-inflammatory IL-10 cytokine (Zhang et al. 2016). Second, Gomez's group showed that single nucleotide polymorphisms in *tlr2* and *tlr1* genes, known to alter their functions, have been associated with increased susceptibility risk of developing severe leptospirosis in an Argentinian population (Cedola et al. 2015).

3.3 TLR5/Flagellin

TLR5 is the flagellin receptor, expressed at the surface of epithelial cells and macrophages. Almost nothing is known about recognition of leptospires by this receptor. Stimulation of bone marrow-derived macrophages from WT and TLR5ko mice with live *L. interrogans* serovar Copenhageni strain Fiocruz L1-130 induced equivalent IL-1ß and IL-6 secretion, suggesting that *L. interrogans* is not sensed by TLR5 (Lacroix-Lamande et al. 2012). One study suggested that neutralizing anti-TLR5 antibodies lessened the cytokine secretion in human whole blood stimulated with *L. interrogans* serovar Bataviae. However, the lack of isotype controls puts the data into question (Goris et al. 2011).

Ongoing studies are focusing on TLR5 to understand whether the atypical periplasmic localization and/or flagellin composition of the leptospiral endoflagella does or does not impair the recognition by TLR5.

3.4 TRIF and MyD88 Responses

Myd88 and TRIF are adaptors of TLRs, responsible for the intracellular signaling after ligand binding on TLRs. MyD88 is the adaptor of all TLRs with the exception of TLR3, the receptor of double-stranded RNAs. TRIF is the adaptor of TLR3 but also TLR4, when this receptor is localized in endosomes. Myd88 signaling leads to activation of NF-κB and MAP kinases, mostly leading to pro-inflammatory cytokine secretion, whereas TRIF also stimulates the interferon responsive factor 3, leading to secretion of type I interferons. Although the TRIF response is better known as an anti-viral defense mediated by viral RNAs, TRIF also triggers type I interferons upon stimulation by LPS from bacteria (reviewed in Werts et al. 2006).

MyD88ko mice infected with sublethal doses of *L. interrogans* serovar Copenhageni recapitulated the highly sensitive phenotype of TLR2/TLR4 double ko mice, which indicated that TLR4 and TLR2 were the prominent receptors involved in defense against leptospirosis (Chassin et al. 2009), and by extension, that the other TLRs were not likely to play a significant role in the defense against leptospires.

As mentioned previously, expression of IFN-γ and iNOS mRNA after *L. interrogans* infection of mice is dependent on TLR4 and TLR2 activation, and these responses were also shown to be dependent on MyD88 (Chassin et al. 2009). In addition, both MyD88 and TRIF adaptors were required for the IL-1ß secretion induced by *Leptospira* (Lacroix-Lamande et al. 2012). Myd88, stimulated after TLR2 and TLR4 activation, has been shown to prime the expression of the pro-IL-1ß mRNA, although TRIF was required for the priming of the NLRP3 mRNA (Lacroix-Lamande et al. 2012).

However, the cytokine responses after leptospiral infection are not all dependent on MyD88 and TRIF. Interestingly, sublethal infection of MyD88ko with *L. interrogans* led to severe inflammation with IL-6, TNF, and RANTES secretion, suggesting that PRRs other than TLRs are involved in the deleterious inflammation occurring upon leptospirosis (Chassin et al. 2009).

Of note, a recent study using TRIFko mice showed that TRIF participates in the defense against leptospires and induces the expression of interferon-stimulated genes (Jayaraman et al. 2016). Additionally, it was shown that the TLR4-dependent IgM response, observed 3 days post-infection with *L. interrogans* (Chassin et al. 2009), was clearly decreased in TRIFko mice. This deficient TLR4/TRIF-dependent production of IgM most probably accounted for the higher bacterial loads in organs observed in kidneys of TRIFko mice compared to WT mice (Jayaraman et al. 2016). However, compared to WT, no difference in TNF or IL-6 responses was observed in infected TRIFko mice, leaving open the question of the PRRs involved in the *Leptospira*-induced inflammation.

3.5 NOD1 and NOD2/Peptidoglycan

Peptidoglycan (PG) is a single rigid macromolecule responsible for the shape of the bacterial cell. It is composed of chains of a motif of two sugars, N-acetyl glucosamine and N-acetyl muramic acid, the latter having short stem peptides that form cross-bridges. This structure is constantly remodeled to facilitate cell division and growth. Bacterial transglycosylases or host enzymes like lysozyme hydrolyze the peptidoglycan between the sugars and release muropeptides. These muropeptides are recognized by the NOD1 and NOD2 cytosolic receptors, which have different specificities according to the nature of the third amino acid of the stem peptide. NOD1 recognizes muropeptides with a meso-diaminopimelate residue (meso-DAP) in third position, usually found in peptidoglycans of Gram-negative bacteria, whereas NOD2 recognizes the muramyl dipeptide present in peptidoglycans of all bacteria (reviewed in Werts et al. 2006). In contrast to *Borrelia* and *Treponema* species, *Leptospira spp.* have a meso-dap peptidoglycan (Yanagihara et al. 1984). NOD receptors sense invasive intracellular bacteria, but also extracellular bacteria. Indeed, it has been shown that the extracellular human pathogen *H. pylori* triggers the secretion of IL-8 through NOD1 recognition of muropeptides injected by the Type 4 secretion system (Viala et al. 2004). *Leptospira* spp. are mostly considered as extracellular pathogens, but can also enter cells, and therefore should be recognized by NOD receptors.

Indeed, peptidoglycan purified from *L. interrogans* serovar Copenhageni strain Teramo was shown to induce the secretion of TNF in human PBMCs (Cinco et al.

1996), suggesting that leptospiral PG is inflammatory and therefore could be recognized by NOD proteins. However, infection of NOD1/NOD2dko mice with *L. interrogans* strain Fiocruz L1-130 led to renal colonization not different from that of WT mice 15 days post-infection, with the same bacterial loads, inflammatory score, and fibrosis, suggesting that NOD1 and NOD2 are not major players in the response of mice to leptospiral infection (Fanton d'Andon et al. 2014). Interestingly, a recent study determined the structure of the peptidoglycan of the saprophytic *L. biflexa* strain Patoc and showed that the leptospiral PG presented some chemical modifications (Slamti et al. 2011). Whether pathogenic leptospires are recognized or not through NOD receptors remains to be determined. If leptospires escape NOD1/NOD2 recognition, it will be interesting to study the consequences for the host, since NOD responses are important for mounting an effective adaptive response (Fritz et al. 2007).

3.6 Inflammasome NLRP3/GLP and LPS

NLRPs are other members of the NOD-like receptors family, involved in the "inflammasome" activation, leading to IL-1ß secretion and also to a peculiar cell death called pyroptosis. IL-1ß is a central cytokine in inflammation, with pleiotropic effects. Its production is tightly regulated and requires two signals. One signal emanates from TNF receptors or from TLR or NLR stimulation, leading to NF-κB activation, and is required to prime the expression of the pro-IL-1ß mRNA. The other signal activates the inflammasome, a platform of proteins gathering a NLRP protein, one or several adaptors, and caspase 1. The stimulation of the NLRP protein leads to the oligomerization of the NLRPs, the recruitment of their specific adaptors, and the activation of caspase 1 that cleaves the pro-IL-1ß to IL-1ß. A parallel pathway involves other caspases that trigger cell death. NLRPs sense microbial components such as toxins, but also react to danger signals such as urea and silicate crystals or UV that physically alter the cellular integrity (reviewed in Werts et al. 2006).

IL-1ß is triggered upon leptospiral infection in humans and animals (Lacroix-Lamande et al. 2012; Matsui et al. 2011; Wagenaar et al. 2009). Recently, the mechanism of IL-1ß secretion upon infection with *L. interrogans* serovar Copenhageni strain Fiocruz L1-130 was deciphered using BMMs from different TLR and NLRko mice (Lacroix-Lamande et al. 2012). *Leptospira* was known to inhibit the expression and function of the potassium pump (Na/K-ATPase), through the action of an outer membrane component, the glycolipoprotein (GLP) (Vinh et al. 1986). GLP is a toxin impairing water reabsorption in kidneys (Cesar et al. 2012). It was found that the potassium perturbation due to GLP constituted a signal to induce the NLRP3 inflammasome. It was also shown that leptospiral LPS and lipoproteins, through both TLR4 and TLR2, prime pro-IL-1ß and NLRP3 mRNA and act in synergy with the GLP to produce IL-1ß (Lacroix-Lamande et al. 2012).

Purified GLP induces different cytokines in PBMC (Diament et al. 2002; Dorigatti et al. 2005). Recently, it was also shown that GLP injection into murine lungs induces lung injury, with edema, inflammatory cytokines, and prostaglandin production, similar that observed with ouabain, a potent Na/K-ATPase inhibitor. These inflammatory effects, mostly independent of TLR4, were mediated through the blocking of the Na/K-ATPase triggering the activation the p38 MAP kinase (Goncalves-de-Albuquerque et al. 2014), a consequence of the formation of a signalosome involving the Src protein (Xie et al. 2013). However, in mouse cells, the leptospiral LPS, signaling through TLR4 and TLR2, stimulates both cytokines and MAP kinase activation (Nahori et al. 2005) and is likely to contaminate the GLP preparation. Therefore, it is difficult to be certain that the observed inflammatory effects are due solely to GLP.

3.7 Other PRRs

To the author's knowledge, TLR9 and TLR3 recognition of leptospires has been examined only in mouse studies (Fanton d'Andon et al. 2014; Lacroix-Lamande et al. 2012), suggesting that these responses are not major players in IL-1ß secretion or in renal colonization and fibrosis. The role in leptospiral infection of other nucleotide-sensing TLRs, such as TLR7 and TLR8, recognizing single-stranded RNA in endosomes, important for *Borrelia* spp. sensing, phagocytosis, and type I interferon production (see chapter XXX), remains to be elucidated.

4 Conclusions

It has been known for several decades that saprophytic *Leptospira* spp. are readily phagocytosed, whereas pathogenic leptospires largely escape phagocytosis, unless they are opsonized by specific Ig. Pathogenic leptospires trigger the expression and secretion of numerous cytokines and chemokines by phagocytes and epithelial cells, showing that leptospires are recognized by PRRs. However, the potential intracellular survival and deleterious effects of pathogenic leptospires on phagocytes are not clear. Hopefully, studies on innate recognition of leptospires will give clues about the potential mechanisms involved in avoiding phagocytosis and killing.

Although many questions are still unanswered, a few innate immunity studies have revealed important facets about leptospirosis (Fig. 1). Leptospires are mainly sensed by TLR2, through recognition of the two major components of the outer membrane, the LPS and the LipL32 lipoprotein. Importantly, the host specificity of TLR4 recognition likely explains the different susceptibility of humans versus mice

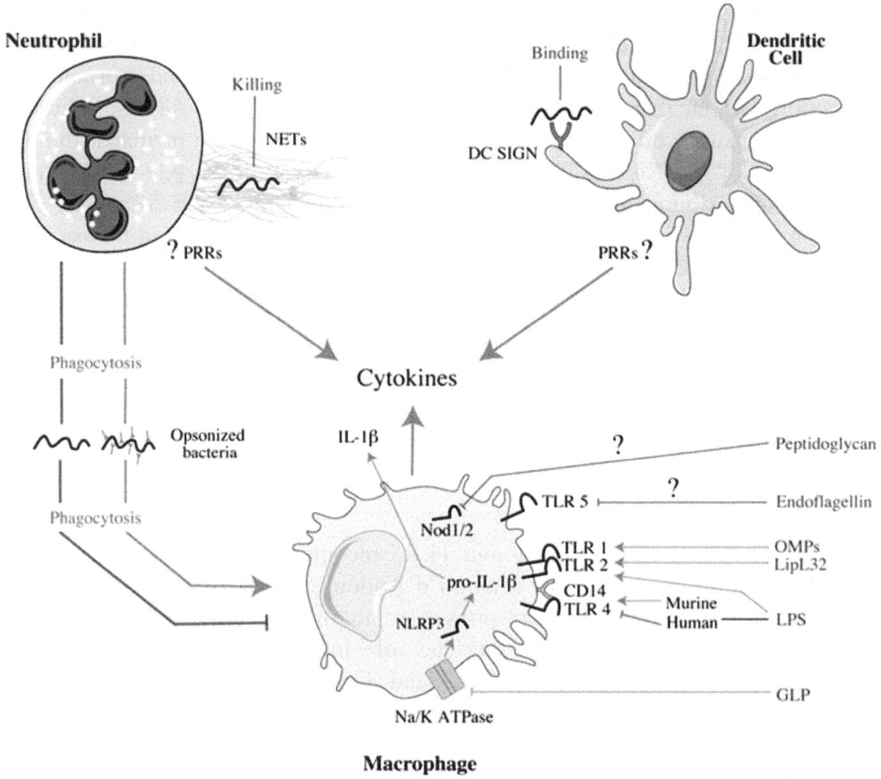

Fig. 1 Schematic representation of innate immune responses to pathogenic leptospires. Phagocytes are depicted with the identified PRRs (*in black*) involved in recognition of leptospiral MAMPs (*in blue*). Leptospires escape phagocytosis, unless opsonized by specific antibodies but are killed by neutrophil extracellular traps (NETs). Leptospires bind the DC-SIGN lectin on dendritic cells (DC). Leptospires stimulate phagocytes through different PRRs to produce cytokines. The LipL32 lipoprotein is a TLR2 agonist. LPS binds to CD14 and stimulates TLR4 in mouse macrophages, but not in human macrophages. LPS binds also to TLR2. GLP inhibits the Na/K-ATPase, which triggers the NLRP3 inflammasome, and in synergy with LPS and lipoproteins, stimulates the IL-1ß secretion by murine macrophages. Leptospires potentially escape the TLR5, TLR9, NOD1, and NOD2 responses in mouse macrophages, but this remains to be studied in human cells. Question marks indicate potential but unknown interactions

toward leptospirosis. Indeed, the humoral response directed against LPS, known to be critical to combat leptospirosis, the secretion of IFN-γ, priming the phagocytic functions, and IL-1ß secretion, essential to mount a robust inflammation, have been shown in mice to rely on TLR4, suggesting that humans, unable to sense the LPS through TLR4 may be impaired in their capacity to mount an adequate response to clear leptospires.

Interestingly, upon infection of their hosts, leptospires are able to modify LPS and LipL32, two major MAMPs. Compared to leptospires cultured in vitro, LPS of bacteria retrieved from guinea pigs' liver has decreased amount of O-antigen content, and LipL32 lipoprotein of bacteria retrieved from rat urine contains lysine methylations. Recognition of these modified LPS and methylated LipL32 by specific immune sera from leptospirosis patients is altered, suggesting that these modifications could play a role in immune evasion. Of note, compared to classical endotoxin, the leptospiral lipid A moiety presents different modifications, including a peculiar methylation, and is not recognized by human TLR4. Therefore, it would be very interesting to identify and test mutants of *Leptospira* unable to perform these methylations to confirm whether these genes could constitute virulence genes, able to subvert the TLR2 or TLR4 responses. In mouse kidneys, the decreased sensing of LipL32 or LPS could explain how leptospires succeed in establishing renal colonization. We also may speculate that it could give a clue about the lack of phenotype of the *lipL32* mutant in animal models. Indeed, if in vivo the methylated LipL32 is barely recognized by TLR2, its absence would not greatly affect the course of infection, and no phenotype would be expected. Importantly, this study about methylation of LipL32 in vivo should warn us about in vitro data that may not always reproduce the events occurring when *Leptospira* spp. infect their hosts.

The question about the PRRs involved in MyD88-independent inflammation observed in mice has recently found some potential answers. Indeed, the GLP may contribute to the deleterious inflammation through the blocking of Na/K-ATPase triggering inflammatory response and p38 MAP kinase.

It is now accepted that innate responses pave the way for adaptive immune responses. Sublethal leptospiral infection confers a strong and protective humoral antibody response, directed against the LPS, specific for each serovar. This immune response is generally short-lived, requiring immunization of animals or humans almost every year with killed whole *Leptospira* vaccines, when available. Therefore, a better knowledge of defective innate responses toward leptospires is instrumental in understanding why and how pathogenic leptospires escape the blood defense and succeed in colonizing the kidneys and fail to trigger a sustained memory response. This should also help to design rational therapeutic strategies against leptospires, aimed at restoring an efficient response, using adequate synthetic ligands of the PRRs, already largely used as adjuvants in the vaccine field.

Acknowledgements We thank Richard Wheeler for English editing and Ignacio Santecchia for the illustration, using resources from Servier Medical Art (www.servier.com).

References

Adler B (2014) Pathogenesis of leptospirosis: cellular and molecular aspects. Vet Microbiol 172:353–358

Adler B, Faine S (1976) Susceptibility of mice treated with cyclophosphamide to lethal infection with *Leptospira interrogans* Serovar pomona. Infect Immun 14:703–708

Banfi E, Cinco M, Bellini M, Soranzo MR (1982) The role of antibodies and serum complement in the interaction between macrophages and leptospires. J Gen Microbiol 128:813–816

Barry M, Wisnewski AV, Matthias MA, Inouye SK, Vinetz JM (2006) Suburban leptospirosis: atypical lymphocytosis and gamma-delta T cell response. Clin Infect Dis 43:1304–1307

Bens M, Vimont S, Ben Mkaddem S, Chassin C, Goujon JM, Balloy V, Chignard M, Werts C, Vandewalle A (2014) Flagellin/TLR5 signalling activates renal collecting duct cells and facilitates invasion and cellular translocation of uropathogenic *Escherichia coli*. Cell Microbiol 16:1503–1517

Blasi E, Ardizzoni A, Colombari B, Neglia R, Baschieri C, Peppoloni S, Cinco M (2007) NF-kB activation and p38 phosphorylation in microglial cells infected with *Leptospira* or exposed to partially purified leptospiral lipoproteins. Microb Pathog 42:80–87

Cedola M, Chiani Y, Pretre G, Alberdi L, Vanasco B, Gomez RM (2015) Association of Toll-like receptor 2 Arg753Gln and Toll-like receptor 1 Ile602Ser single-nucleotide polymorphisms with leptospirosis in an Argentine population. Acta Trop 146:73–80

Cesar KR, Romero EC, de Braganca AC, Blanco RM, Abreu PA, Magaldi AJ (2012) Renal involvement in leptospirosis: the effect of glycolipoprotein on renal water absorption. PLoS ONE 7:e37625

Chagan-Yasutan H, Chen Y, Lacuesta TL, Leano PS, Iwasaki H, Hanan F, Taurustiati D, Ohmoto Y, Ashino Y, Saitoh H, Kiyomoto H, Suzuki Y, Telan FO, Hattori T (2016). Urine levels of defensin α1 reflect kidney injury in leptospirosis patients. Int J Mol Sci 17(10):1637

Challa S, Nally JE, Jones C, Sheoran AS (2011) Passive immunization with *Leptospira* LPS-specific agglutinating but not non-agglutinating monoclonal antibodies protect guinea pigs from fatal pulmonary hemorrhages induced by serovar Copenhageni challenge. Vaccine 29:4431–4434

Chang MY, Cheng YC, Hsu SH, Ma TL, Chou LF, Hsu HH, Tian YC, Chen YC, Sun YJ, Hung CC, Pan RL, Yang CW (2016) Leptospiral outer membrane protein LipL32 induces inflammation and kidney injury in zebrafish larvae. Sci Rep 6:27838

Chassin C, Picardeau M, Goujon JM, Bourhy P, Quellard N, Darche S, Badell E, d'Andon MF, Winter N, Lacroix-Lamande S, Buzoni-Gatel D, Vandewalle A, Werts C (2009) TLR4- and TLR2-mediated B cell responses control the clearance of the bacterial pathogen, *Leptospira interrogans*. J Immunol 183:2669–2677

Cinco M, Banfi E, Soranzo MR (1981) Studies on the interaction between macrophages and leptospires. J Gen Microbiol 124:409–413

Cinco M, Vecile E, Murgia R, Dobrina P, Dobrina A (1996) *Leptospira interrogans* and *Leptospira* peptidoglycans induce the release of tumor necrosis factor alpha from human monocytes. FEMS Microbiol Lett 138:211–214

Craig SB, Collet TA, Wynwood SJ, Smythe LD, Weier SL, McKay DB (2013) Neutrophil counts in leptospirosis patients infected with different serovars. Trop Biomed 30:579–583

Damasco PV, Avila CA, Barbosa AT, Ribeiro Carvalho Mde M, Pereira GM, Lemos ER, Boia MN, Pereira MM (2011) Atypical lymphocytosis in leptospirosis: a cohort of hospitalized cases between 1996 and 2009 in State of Rio de Janeiro, Brazil. Rev Soc Bras Med Trop 44:611–615

Davis JM, Haake DA, Ramakrishnan L (2009) *Leptospira interrogans* stably infects zebrafish embryos, altering phagocyte behavior and homing to specific tissues. PLoS Negl Trop Dis 3: e463

De Silva NL, Niloofa M, Fernando N, Karunanayake L, Rodrigo C, De Silva HJ, Premawansa S, Handunnetti SM, Rajapakse S (2014) Changes in full blood count parameters in leptospirosis: a prospective study. Int Arch Med 7:31

Diament D, Brunialti MK, Romero EC, Kallas EG, Salomao R (2002) Peripheral blood mononuclear cell activation induced by *Leptospira interrogans* glycolipoprotein. Infect Immun 70:1677–1683

Dorigatti F, Brunialti MK, Romero EC, Kallas EG, Salomao R (2005) *Leptospira interrogans* activation of peripheral blood monocyte glycolipoprotein demonstrated in whole blood by the release of IL-6. Braz J Med Biol Res 38:909–914

Eshghi A, Henderson J, Trent MS, Picardeau M (2015) *Leptospira interrogans* lpxD homologue is required for thermal acclimatization and virulence. Infect Immun 83:4314–4321

Faine S, Shahar A, Aronson M (1964) Phagocytosis and its significance in leptospiral infection. Aust J Exp Biol Med Sci 42:579–588

Fanton d'Andon M, Quellard N, Fernandez B, Ratet G, Lacroix-Lamande S, Vandewalle A, Boneca IG, Goujon JM, Werts C (2014) *Leptospira interrogans* induces fibrosis in the mouse kidney through Inos-dependent, TLR- and NLR-independent signaling pathways. PLoS Negl Trop Dis 8:e2664

Fritz JH, Le Bourhis L, Sellge G, Magalhaes JG, Fsihi H, Kufer TA, Collins C, Viala J, Ferrero RL, Girardin SE, Philpott DJ (2007) Nod1-mediated innate immune recognition of peptidoglycan contributes to the onset of adaptive immunity. Immunity 26:445–459

Gaudart N, Ekpo P, Pattanapanyasat K, van Kooyk Y, Engering A (2008) *Leptospira interrogans* is recognized through DC-SIGN and induces maturation and cytokine production by human dendritic cells. FEMS Immunol Med Microbiol 53:359–367

Goncalves-de-Albuquerque CF, Burth P, Silva AR, de Moraes IM, Oliveira FM, Santelli RE, Freire AS, de Lima GS, da Silva ED, da Silva CI, Morandi V, Bozza PT, de Younes-Ibrahim M, Castro Faria Neto HC, de Castro Faria MV (2014) Murine lung injury caused by *Leptospira interrogans* glycolipoprotein, a specific Na/K-ATPase inhibitor. Respir Res 15:93

Gordon S (2016) Phagocytosis: an immunobiologic process. Immunity 44:463–475

Goris MG, Wagenaar JF, Hartskeerl RA, van Gorp EC, Schuller S, Monahan AM, Nally JE, van der Poll T, van't Veer C (2011) Potent innate immune response to pathogenic leptospira in human whole blood. PLoS ONE 6:e18279

Guo Y, Fukuda T, Nakamura S, Bai L, Xu J, Kuroda K, Tomioka R, Yoneyama H, Isogai E (2015) Interaction between leptospiral lipopolysaccharide and Toll-like receptor 2 in pig fibroblast cell line, and inhibitory effect of antibody against leptospiral lipopolysaccharide on interaction. Asian-Australas J Anim Sci 28:273–279

Guo Y, Ding C, Zhang B, Xu J, Xun M, Xu J (2016) Inhibitory effect of BMAP-28 on Leptospiral lipopolysaccharide-induced TLR2-dependent immune response in bovine cells. Jundishapur J Microbiol 9:e33926

Haake DA, Chao G, Zuerner RL, Barnett JK, Barnett D, Mazel M, Matsunaga J, Levett PN, Bolin CA (2000) The leptospiral major outer membrane protein LipL32 is a lipoprotein expressed during mammalian infection. Infect Immun 68:2276–2285

Hsu SH, Lo YY, Tung JY, Ko YC, Sun YJ, Hung CC, Yang CW, Tseng FG, Fu CC, Pan RL (2010) Leptospiral outer membrane lipoprotein LipL32 binding on Toll-like receptor 2 of renal cells as determined with an atomic force microscope. Biochemistry 49:5408–5417

Hu W, Ge Y, Ojcius DM, Sun D, Dong H, Yang XF, Yan J (2013) p53 signalling controls cell cycle arrest and caspase-independent apoptosis in macrophages infected with pathogenic *Leptospira* species. Cell Microbiol 15:1642–1659

Hung CC, Chang CT, Chen KH, Tian YC, Wu MS, Pan MJ, Vandewalle A, Yang CW (2006a) Upregulation of chemokine CXCL1/KC by leptospiral membrane lipoprotein preparation in renal tubule epithelial cells. Kidney Int 69:1814–1822

Hung CC, Chang CT, Tian YC, Wu MS, Yu CC, Pan MJ, Vandewalle A, Yang CW (2006b) Leptospiral membrane proteins stimulate pro-inflammatory chemokines secretion by renal tubule epithelial cells through Toll-like receptor 2 and p38 mitogen activated protein kinase. Nephrol Dial Transplant 21:898–910

Isogai E, Kitagawa H, Isogai H, Kurebayashi Y, Ito N (1986) Phagocytosis as a defense mechanism against infection with leptospiras. Zentralbl Bakteriol Mikrobiol Hyg A 261:65–74

Isogai E, Isogai H, Wakizaka H, Miura H, Kurebayashi Y (1989) Chemiluminescence and phagocytic responses of rat polymorphonuclear neutrophils to leptospires. Zentralbl Bakteriol 272:36–46

Isogai E, Isogai H, Fujii N, Oguma K (1990a) Biological effects of leptospiral lipopolysaccharide to mouse B, T and NK cells. Nihon Juigaku Zasshi 52:923–930

Isogai E, Isogai H, Fujii N, Oguma K (1990b) Macrophage activation by leptospiral lipopolysaccharide. Zentralbl Bakteriol 273:200–208

Jayaraman PA, Devlin AA, Miller JC, Scholle F (2016) The adaptor molecule TRIF contributes to murine host defense during leptospiral infection. Immunobiology 221:964–974

Jin D, Ojcius DM, Sun D, Dong H, Luo Y, Mao Y, Yan J (2009) *Leptospira interrogans* induces apoptosis in macrophages via caspase-8- and caspase-3-dependent pathways. Infect Immun 77:799–809

Klimpel GR, Matthias MA, Vinetz JM (2003) *Leptospira interrogans* activation of human peripheral blood mononuclear cells: preferential expansion of TCR gamma delta+ T cells vs TCR alpha beta+ T cells. J Immunol 171:1447–1455

Lacroix-Lamande S, d'Andon MF, Michel E, Ratet G, Philpott DJ, Girardin SE, Boneca IG, Vandewalle A, Werts C (2012) Downregulation of the Na/K-ATPase pump by leptospiral glycolipoprotein activates the NLRP3 inflammasome. J Immunol 188:2805–2814

Li L, Ojcius DM, Yan J (2007) Comparison of invasion of fibroblasts and macrophages by high- and low-virulence *Leptospira* strains: colonization of the host-cell nucleus and induction of necrosis by the virulent strain. Arch Microbiol 188:591–598

Li S, Ojcius DM, Liao S, Li L, Xue F, Dong H, Yan J (2010) Replication or death: distinct fates of pathogenic Leptospira strain Lai within macrophages of human or mouse origin. Innate Immun 16:80–92

Lindow JC, Wunder EA Jr, Popper SJ, Min JN, Mannam P, Srivastava A, Yao Y, Hacker KP, Raddassi K, Lee PJ, Montgomery RR, Shaw AC, Hagan JE, Araujo GC, Nery N Jr, Relman DA, Kim CC, Reis MG, Ko AI (2016) Cathelicidin insufficiency in patients with fatal leptospirosis. PLoS Pathog 12:e1005943

Liu B, Wang Y, Guo X, Zhu W, Zhang Y, He P (2014) Carboxyfluorescein diacetate succinimidyl ester labeling method to study the interaction between *Leptospira* and macrophages. J Microbiol Methods 107:205–213

Lo YY, Hsu SH, Ko YC, Hung CC, Chang MY, Hsu HH, Pan MJ, Chen YW, Lee CH, Tseng FG, Sun YJ, Yang CW, Pan RL (2013) Essential calcium-binding cluster of *Leptospira* LipL32 protein for inflammatory responses through the Toll-like receptor 2 pathway. J Biol Chem 288:12335–12344

Marcsisin RA, Bartpho T, Bulach DM, Srikram A, Sermswan RW, Adler B, Murray GL (2013) Use of a high-throughput screen to identify *Leptospira* mutants unable to colonize the carrier host or cause disease in the acute model of infection. J Med Microbiol 62:1601–1608

Matsui M, Rouleau V, Bruyere-Ostells L, Goarant C (2011) Gene expression profiles of immune mediators and histopathological findings in animal models of leptospirosis: comparison between susceptible hamsters and resistant mice. Infect Immun 79:4480–4492

McGrath H, Adler B, Vinh T, Faine S (1984) Phagocytosis of virulent and avirulent leptospires by guinea-pig and human polymorphonuclear leukocytes in vitro. Pathology 16:243–249

Merien F, Baranton G, Perolat P (1997) Invasion of Vero cells and induction of apoptosis in macrophages by pathogenic *Leptospira interrogans* are correlated with virulence. Infect Immun 65:729–738

Midwinter A, Faine S, Adler B (1990) Vaccination of mice with lipopolysaccharide (LPS) and LPS-derived immuno-conjugates from *Leptospira interrogans*. J Med Microbiol 33:199–204

Murray GL, Srikram A, Hoke DE, Wunder EA Jr, Henry R, Lo M, Zhang K, Sermswan RW, Ko AI, Adler B (2009) Major surface protein LipL32 is not required for either acute or chronic infection with *Leptospira interrogans*. Infect Immun 77:952–958

Murray GL, Srikram A, Henry R, Hartskeerl RA, Sermswan RW, Adler B (2010) Mutations affecting *Leptospira interrogans* lipopolysaccharide attenuate virulence. Mol Microbiol 78:701–709

Nahori MA, Fournie-Amazouz E, Que-Gewirth NS, Balloy V, Chignard M, Raetz CR, Saint Girons I, Werts C (2005) Differential TLR recognition of leptospiral lipid A and lipopolysaccharide in murine and human cells. J Immunol 175:6022–6031

Nally JE, Chow E, Fishbein MC, Blanco DR, Lovett MA (2005) Changes in lipopolysaccharide O antigen distinguish acute versus chronic *Leptospira interrogans* infections. Infect Immun 73:3251–3260

Papa A, Kotrotsiou T (2015) Cytokines in human leptospirosis. Trans R Soc Trop Med Hyg 109:749–754

Park BS, Song DH, Kim HM, Choi BS, Lee H, Lee JO (2009) The structural basis of lipopolysaccharide recognition by the TLR4-MD-2 complex. Nature 458:1191–1195
Pereira MM, Andrade J, Marchevsky RS, Ribeiro dos Santos R (1998) Morphological characterization of lung and kidney lesions in C3H/HeJ mice infected with *Leptospira interrogans* serovar icterohaemorrhagiae: defect of CD4+ and CD8+ T-cells are prognosticators of the disease progression. Exp Toxicol Pathol 50:191–198
Pinne M, Haake DA (2013) LipL32 is a subsurface lipoprotein of *Leptospira interrogans*: presentation of new data and reevaluation of previous studies. PLoS ONE 8:e51025
Que-Gewirth NL, Ribeiro AA, Kalb SR, Cotter RJ, Bulach DM, Adler B, Girons IS, Werts C, Raetz CR (2004) A methylated phosphate group and four amide-linked acyl chains in *Leptospira interrogans* lipid A. The membrane anchor of an unusual lipopolysaccharide that activates TLR2. J Biol Chem 279:25420–25429
Raffray L, Giry C, Thirapathi Y, Binois F, Moiton MP, Lagrange-Xelot M, Ferrandiz D, Jaffar-Bandjee MC, Gasque P (2015) High leptospiremia is associated with low gamma-delta T cell counts. Microbes Infect 17:451–455
Raffray L, Giry C, Vandroux D, Kuli B, Randrianjohany A, Pequin AM, Renou F, Jaffar-Bandjee MC, Gasque P (2016) Major neutrophilia observed in acute phase of human leptospirosis is not associated with increased expression of granulocyte cell activation markers. PLoS ONE 11: e0165716
Ratet G, Veyrier FJ, Fanton d'Andon M, Kammerscheit X, Nicola MA, Picardeau M, Boneca IG, Werts C (2014) Live imaging of bioluminescent *Leptospira interrogans* in mice reveals renal colonization as a stealth escape from the blood defenses and antibiotics. PLoS Negl Trop Dis 8: e3359
Reynolds JM, Dong C (2013) Toll-like receptor regulation of effector T lymphocyte function. Trends Immunol 34:511–519
Richer L, Potula HH, Melo R, Vieira A, Gomes-Solecki M (2015) Mouse model for sublethal *Leptospira interrogans* infection. Infect Immun 83:4693–4700
Rittig MG, Jagoda JC, Wilske B, Murgia R, Cinco M, Repp R, Burmester GR, Krause A (1998) Coiling phagocytosis discriminates between different spirochetes and is enhanced by phorbol myristate acetate and granulocyte-macrophage colony-stimulating factor. Infect Immun 66:627–635
Saitoh S, Akashi S, Yamada T, Tanimura N, Kobayashi M, Konno K, Matsumoto F, Fukase K, Kusumoto S, Nagai Y, Kusumoto Y, Kosugi A, Miyake K (2004) Lipid A antagonist, lipid IVa, is distinct from lipid A in interaction with Toll-like receptor 4 (TLR4)-MD-2 and ligand-induced TLR4 oligomerization. Int Immunol 16:961–969
Sambri V, Marangoni A, Giacani L, Gennaro R, Murgia R, Cevenini R, Cinco M (2002) Comparative in vitro activity of five cathelicidin-derived synthetic peptides against *Leptospira*, *Borrelia* and *Treponema pallidum*. J Antimicrob Chemother 50:895–902
Scharrig E, Carestia A, Ferrer MF, Cedola M, Pretre G, Drut R, Picardeau M, Schattner M, Gomez RM (2015) Neutrophil extracellular traps are involved in the innate immune response to infection with *Leptospira*. PLoS Negl Trop Dis 9:e0003927
Segal AW (2005) How neutrophils kill microbes. Annu Rev Immunol 23:197–223
Selsted ME, Ouellette AJ (2005) Mammalian defensins in the antimicrobial immune response. Nat Immunol 6:551–557
Slamti L, de Pedro MA, Guichet E, Picardeau M (2011) Deciphering morphological determinants of the helix-shaped *Leptospira*. J Bacteriol 193:6266–6275
Souza-Fonseca-Guimaraes F, Adib-Conquy M, Cavaillon JM (2012) Natural killer (NK) cells in antibacterial innate immunity: angels or devils? Mol Med 18:270–285
Toma C, Okura N, Takayama C, Suzuki T (2011) Characteristic features of intracellular pathogenic *Leptospira* in infected murine macrophages. Cell Microbiol 13:1783–1792
Toma C, Murray GL, Nohara T, Mizuyama M, Koizumi N, Adler B, Suzuki T (2014) Leptospiral outer membrane protein LMB216 is involved in enhancement of phagocytic uptake by macrophages. Cell Microbiol 16:1366–1377

Tourneur E, Ben Mkaddem S, Chassin C, Bens M, Goujon JM, Charles N, Pellefigues C, Aloulou M, Hertig A, Monteiro RC, Girardin SE, Philpott DJ, Rondeau E, Elbim C, Werts C, Vandewalle A (2013) Cyclosporine A impairs nucleotide binding oligomerization domain (Nod1)-mediated innate antibacterial renal defenses in mice and human transplant recipients. PLoS Pathog 9:e1003152

Truong KN, Coburn J (2011) The emergence of severe pulmonary hemorrhagic leptospirosis: questions to consider. Front Cell Infect Microbiol 1:24

Veiga E, Cossart P (2006) The role of clathrin-dependent endocytosis in bacterial internalization. Trends Cell Biol 16:499–504

Viala J, Chaput C, Boneca IG, Cardona A, Girardin SE, Moran AP, Athman R, Memet S, Huerre MR, Coyle AJ, DiStefano PS, Sansonetti PJ, Labigne A, Bertin J, Philpott DJ, Ferrero RL (2004) Nod1 responds to peptidoglycan delivered by the *Helicobacter pylori* cag pathogenicity island. Nat Immunol 5:1166–1174

Vinh T, Adler B, Faine S (1986) Glycolipoprotein cytotoxin from *Leptospira interrogans* serovar copenhageni. J Gen Microbiol 132:111–123

Viriyakosol S, Matthias MA, Swancutt MA, Kirkland TN, Vinetz JM (2006) Toll-like receptor 4 protects against lethal *Leptospira interrogans* serovar icterohaemorrhagiae infection and contributes to in vivo control of leptospiral burden. Infect Immun 74:887–895

Wagenaar JF, Gasem MH, Goris MG, Leeflang M, Hartskeerl RA, van der Poll T, van't Veer C, van Gorp EC (2009) Soluble ST2 levels are associated with bleeding in patients with severe leptospirosis. PLoS Negl Trop Dis 3:e453

Wang B, Sullivan J, Sullivan GW, Mandell GL (1984a) Interaction of leptospires with human polymorphonuclear neutrophils. Infect Immun 44:459–464

Wang B, Sullivan JA, Sullivan GW, Mandell GL (1984b) Role of specific antibody in interaction of leptospires with human monocytes and monocyte-derived macrophages. Infect Immun 46:809–813

Werts C (2010) Leptospirosis: a Toll road from B lymphocytes. Chang Gung Med J 33:591–601

Werts C, Tapping RI, Mathison JC, Chuang TH, Kravchenko V, Saint Girons I, Haake DA, Godowski PJ, Hayashi F, Ozinsky A, Underhill DM, Kirschning CJ, Wagner H, Aderem A, Tobias PS, Ulevitch RJ (2001) Leptospiral lipopolysaccharide activates cells through a TLR2-dependent mechanism. Nat Immunol 2:346–352

Werts C, Girardin SE, Philpott DJ (2006) TIR, CARD and PYRIN: three domains for an antimicrobial triad. Cell Death Differ 13:798–815

Wilson-Welder JH, Frank AT, Hornsby RL, Olsen SC, Alt DP (2016) Interaction of bovine peripheral blood polymorphonuclear cells and *Leptospira* species; Innate responses in the natural bovine reservoir host. Front Microbiol 7:1110

Witchell TD, Eshghi A, Nally JE, Hof R, Boulanger MJ, Wunder EA Jr, Ko AI, Haake DA, Cameron CE (2014) Post-translational modification of LipL32 during *Leptospira interrogans* infection. PLoS Negl Trop Dis 8:e3280

Xie JX, Li X, Xie Z (2013) Regulation of renal function and structure by the signaling Na/K-ATPase. IUBMB Life 65:991–998

Xue F, Zhao X, Yang Y, Zhao J, Yang Y, Cao Y, Hong C, Liu Y, Sun L, Huang M, Gu J (2013) Responses of murine and human macrophages to leptospiral infection: a study using comparative array analysis. PLoS Negl Trop Dis 7:e2477

Yanagihara Y, Kamisango K, Yasuda S, Kobayashi S, Mifuchi I, Azuma I, Yamamura Y, Johnson RC (1984) Chemical compositions of cell walls and polysaccharide fractions of spirochetes. Microbiol Immunol 28:535–544

Yang CW (2007) Leptospirosis renal disease: understanding the initiation by Toll-like receptors. Kidney Int 72:918–925

Yang CW, Hung CC, Wu MS, Tian YC, Chang CT, Pan MJ, Vandewalle A (2006) Toll-like receptor 2 mediates early inflammation by leptospiral outer membrane proteins in proximal tubule cells. Kidney Int 69:815–822

Zhang L, Zhang C, Ojcius DM, Sun D, Zhao J, Lin X, Li L, Li L, Yan J (2012) The mammalian cell entry (Mce) protein of pathogenic *Leptospira* species is responsible for RGD motif-dependent infection of cells and animals. Mol Microbiol 83:1006–1023

Zhang W, Zhang N, Xie X, Guo J, Jin X, Xue F, Ding Z, Cao Y (2016). TLR2 agonist Pam3CSK4 alleviates the pathology of leptospirosis in hamster. Infect Immun 84(12):3350–3357

Zuerner RL, Alt DP, Palmer MV, Thacker TC, Olsen SC. (2011) A Leptospira borgpetersenii serovar Hardjo vaccine induces a Th1 response, activates NK cells, and reduces renal colonization. Clin Vaccine Immunol 18(4):684–691. doi:10.1128/CVI.00288-10. Epub 2011 Feb 2. PMID:21288995

Leptospiral Genomics and Pathogenesis

Dieter Bulach and Ben Adler

Abstract Until about 15 years ago, the molecular and cellular basis for pathogenesis in leptospirosis was virtually unknown. The determination of the first full genome sequence in 2003 was followed rapidly by other whole genome sequences, whose availability facilitated the development of transposon mutagenesis and then directed mutagenesis of pathogenic *Leptospira* spp. The combination of genomics, transcriptomics and mutant construction and characterisation has resulted in major progress in our understanding of the taxonomy and biology of *Leptospira*. The most recent advances are analysed and discussed in this chapter.

Contents

Electronic supplementary material The online version of this chapter (https://doi.org/10.1007/82_2018_87) contains supplementary material, which is available to authorized users.

D. Bulach
Melbourne Bioinformatics, The University of Melbourne, Carlton, VIC, Australia
e-mail: dieter.bulach@unimelb.edu.au

B. Adler (✉)
Department of Microbiology, Monash University, Clayton, VIC, Australia
e-mail: ben.adler@monash.edu

Current Topics in Microbiology and Immunology (2018) 415:189–214
DOI 10.1007/82_2018_87

Published Online: 26 April 2018

1 Introduction

Spirochetes constitute a distinct group of eubacteria that diverged early in evolution. Among the spirochetes, *Leptospira* forms the most deeply branching group (Paster et al. 1991). The genus possesses some characteristics not found in other genera, most notably *Treponema* and *Borrelia*, such as the presence of lipopolysaccharide (LPS) in its outer membrane (Adler and de la Pena Moctezuma 2010), a feature that it shares with *Brachyspira*. Also in contrast to most other pathogenic spirochetes, leptospires can survive free-living in the environment, as well as in the kidneys of their carrier hosts. The genus *Leptospira* also contains saprophytic species that never cause disease and intermediate species that may infect animals and humans, but usually result in a milder disease (Fouts et al. 2016; Levett 2015). It is therefore not unexpected that the mechanisms of pathogenesis in acute infection and those of colonisation of the renal tubules differ from those found in other bacterial species.

2 Bacterial Genomics and Isolate Sequencing

Isolate sequencing is now broadly used in public health microbiology. While isolate sequencing might not be routinely applied to the management of *Leptospira*-associated outbreaks, the large-scale comparison of the genome sequence of isolates has the potential to provide overviews that will enable the development of useful diagnostic tools and potentially a broader understanding of the genetics of this genus through investigation of isolate to isolate genomic differences.

In the 14 years since the first leptospiral genome sequence was published (Ren et al. 2003), a significant evolution of sequencing technology has occurred. In practical terms, bacterial genome sequencing has gone from being a multi-year research project with a million-dollar budget, to a routine procedure that might take about one week and cost as little as US$20 to produce read data that can be used to assemble a high-quality draft genome sequence of a bacterial isolate.

Large-scale comparison of isolate genomes requires reliable and reproducible primary data in a standardised format and uniform analysis methods. One typical example in *Leptospira* genomics of different analysis methods hindering comparative genomics was the initial comparison of the *L. interrogans* serovar Lai and serovar Copenhageni genome sequences with the *L. borgpetersenii* serovar Hardjo genome sequences, where there was a substantial difference in the number of annotated genes due to differences in genome annotation methods (Bulach et al. 2006). Large-scale bacterial comparative genomics methods have developed rapidly over the past four years, driven primarily by demand from public health laboratories in the USA and UK in particular, where isolate sequencing is replacing isolate typing methods based on PCR or serology and comparative genomics is used to investigate outbreaks. The changing scale at which isolate sequencing is now being undertaken is illustrated by the number of *Salmonella enterica* isolates for which read data have been published at the Sequence Read Archive (SRA; https://www.ncbi.nlm.nih.gov/sra) and at the European Nucleotide Archive (ENA; https://www.ebi.ac.uk/ena) and summarised at the Pathogen Portal (https://www.ncbi.nlm.nih.gov/pathogens). Just over 200 *S. enterica* read sets were available at the end of 2014; at the end of 2017, there were over 90,000 *S. enterica* isolate read sets. Predominantly, these read sets are produced using the Illumina sequencing technology. Consequently, the comparative genomic analysis methods often require the input data to be Illumina read sets. Moreover, there is a minimum standard for these read sets at about 50× read coverage of the genome sequence. At this level of read coverage, the read set is useful both for performing de novo assembly for the production of a draft genome sequence and for performing read mapping, where this level of coverage is sufficient for the reliable calling of small sequence differences between isolates and the reference genome sequence. These small sequence differences are collectively used in the comparative genomics methods.

Relatively simple summary metrics about the read set for an isolate and the output produced by the read set from de novo assembly provide a means to evaluate the quality of a read set, including the quality of the input genomic DNA, sequencing library preparation and appropriate operation of the sequencing instrument. For *Leptospira*, a good quality 50× paired-end read set should produce a draft genome sequence an average of 4.5 Mb, broken into about 200 contigs (based on averages determined from the de novo assembly of currently available *Leptospira* read sets).

2.1 The Genus **Leptospira***: Read Data and Assemblies*

A combination of read data and genome sequences, draft or closed, is available on the public sequence databases. Collectively, there are genomic data available for 591 isolates (Table 1). For the purposes of the comparative analysis presented in this chapter, 129 isolates had genomic data that did not meet minimum criteria (read sets: mono-isolate, >25× Illumina read coverage; assemblies: <400 contigs). Some 395 isolates had suitable Illumina read set data available, while 67 isolates comprising 28 isolates for which there was a closed genome sequence (Table 2) and 39 isolates for which there was a published assembly, but the read data were either not published or the read data were not from the Illumina sequencing platform. In total, 462 isolates have been included in the analysis. For the 67 isolates for which there were no Illumina read data, the genome sequences, draft or closed, have been

Table 1 Isolates sequenced from the genus *Leptospira* (November 2017)

Status	Number
Isolate sequence data not used in this study[a]	129
Suitable assembly available	68
Suitable Illumina read set available	394
Total	591

[a]Isolates with sequence data (assembly or read sets) that did not meet minimum criteria were excluded:
Read sets Illumina, paired end, 25× read coverage. *Note* Illumina read sets from Yale School of Public Health have been omitted because published reads are arranged in an unusual order
Assemblies included closed genomes for which no read data had been submitted and draft genomes which had been produced from non-Illumina read data. Note: systematic errors introduced by the particular sequencing technology (e.g. Roche-454 and long homopolymeric runs of bases) may not have been corrected in the assemblies by the respective sequence authors

Table 2 Complete *Leptospira* genome sequences

Species	Number
Leptospira alstonii	1
Leptospira biflexa	2
Leptospira borgpetersenii	9
Leptospira interrogans	13
Leptospira kmetyi	1
Leptospira santarosai	2
Total	28

converted into a synthetic Illumina 150 base, paired-end read sets with 50× read coverage using ART (Huang et al. 2012).

2.2 *Genetic Diversity in the Genus* Leptospira

There are currently 22 recognised species classified in the genus *Leptospira*, although at least 12 new species have recently been isolated from soil or water in New Caledonia (Thibeaux et al. 2017). These 22 current species can be divided into three main groups based on the relationship between 16S rRNA gene of the type strain for each of the species. These groups are nominally the saprophyte (S), intermediate (I) and pathogen (P) groups (Table 3).

Table 3 Species in the genus *Leptospira*

Species	GenBank accession[a]	Type strain (including synonyms)	Genome[b]	Type strain SRA read set[c]	Clade[d]
L. alexanderi	AY631880	L 60; ATCC 700520	6	SRR403925	P
L. alstonii	AY631881	79601; ATCC BAA-2439	6	SRR611260	P
L. biflexa	AY631876	Patoc I; ATCC 23582	2	GCA_000017685.1	S
L. borgpetersenii	AY887899	Veldrat Bataviae 46; ATCC 43292	39		P
L. broomii	AY796065	5399; ATCC BAA-1107; KIT 5399	1	SRR403924	I
L. fainei	AY631885	BUT 6	1	SRR507740	I
L. idonii	AB721966	Eri-1; DSM 26084; JCM 18486	0		S
L. inadai	Z21634	10; ATCC 43289	2	SRR403922	I
L. interrogans	Z12817	RGA	413	SRR715768, SRR4235351	P
L. kirschneri	AY631895	3522C; ATCC 49945	31	SRR403923	P
L. kmetyi	AB279549	Bejo-Iso9; KIT Bejo-Iso9; WHO LT1101	1	SRR507728	P
L. licerasiae	EF612284	VAR 010; ATCC BAA-1110; KIT VAR 010	6	SRR1801751, SRR397970	I
L. mayottensis	KJ847187	200901116; CIP 110703; DSM 28999	2	SRR554115	P
L. meyeri	AY631878	Iowa City Frog; ATCC 43287	2		S
L. noguchii	AY631886	CZ 214; ATCC 43288	15	SRR507741	P
L. santarosai	AY631883	LT 821; ATCC 43286	41	GCA_000313175.2	P
L. terpstrae	AY631888	LT 11-33; ATCC 700639	1	SRR710513	S
L. vanthielii	AY631897	WaZ Holland; ATCC 700522	1	SRR710255	S
L. weilii	AY631877	Celledoni; ATCC 43285	18		P
L. wolbachii	AY631879	CDC; ATCC 43284	1	SRR649354	S

(continued)

Table 3 (continued)

Species	GenBank accession[a]	Type strain (including synonyms)	Genome[b]	Type strain SRA read set[c]	Clade[d]
L. wolffi	EF025496	Khorat-H2; KIT Khorat-H2; WHO LT1686	1	SRR507743	I
L. yanagawae	AY631882	Sao Paulo; ATCC 700523	1	GCA_000332475.2	S

[a]16s rRNA gene sequence of the type strain
[b]Number of isolates for which there is a draft genome sequence and/or Whole Genome Shotgun read set
[c]Assembly accession number where no read set is available
[d]Clade: saprophyte (S), intermediate (I), pathogen (P)

2.3 *Comparison of the Genome Sequences of Isolates*

The 16S rRNA gene, which is present in all bacteria, has been used as the basis for inferring the relationship between bacterial genomes. The relationship between isolates is inferred from differences in the conserved 16S rRNA gene sequence between isolates. Likewise, in comparative genomics, regions conserved in the genomes of all the isolates being compared are selected and sequence differences within these conserved regions are used to build a relationship which is presented as a phylogeny. The conserved region of the genome is commonly referred to as the 'core genome' and will necessarily vary depending on the group of isolates being compared. At a technical level, the identification of the core genome is achieved by mapping reads (the Illumina read set) for each isolate to a related reference genome sequence. The core genome is the region of the reference genome to which reads from all of the isolates have mapped.

2.3.1 The Core Genome for *Leptospira*

The available genomic data for the type strains provide a useful representation of the genetic diversity found in the genus *Leptospira*. Genomic data for 18 of the 22 species are represented (Table 3). Using the genome sequence of *L. interrogans* strain Fiocruz L1-130 as the reference genome onto which reads are mapped, we find that there is an extremely small core genome using the read mapping approach. A summary of the isolate comparative analysis underpinning the trees presented in figures is available via a figshare repository (Bulach, 2018). Figure 1 shows the relationship inferred from the core genome where the core genome is less than 500 bases and has 82 variable sites. Despite the small number of variable sites, the strains are distributed into three clades consistent with the saprophyte (S), intermediate (I) and pathogen (P) groups. While genome-wide comparison of *Leptospira*

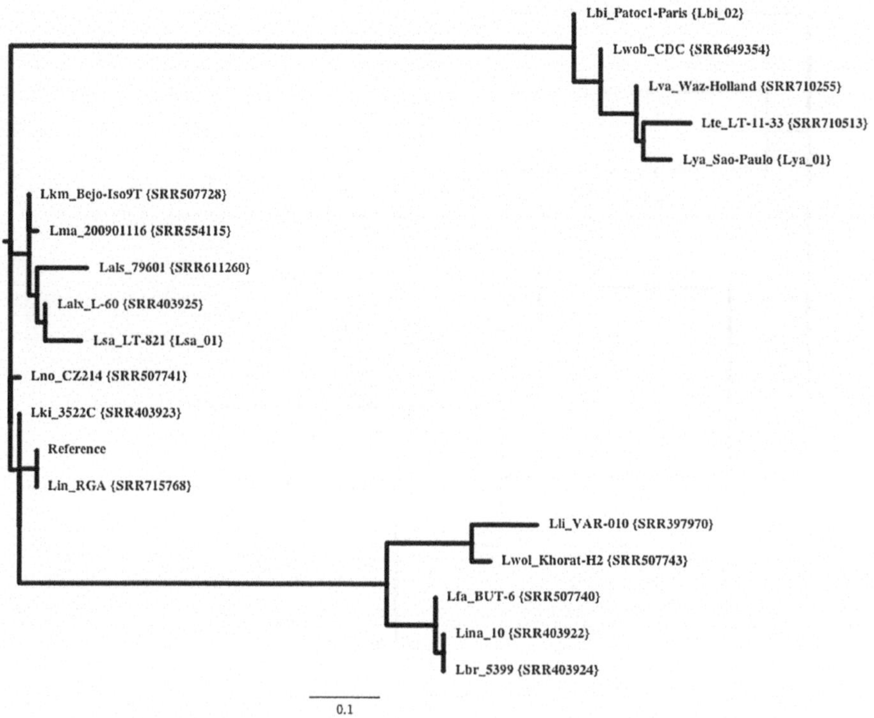

Fig. 1 Unrooted tree showing the inferred relationship between the core genomes of the set of 18 species type strains (isolates) for which genomic data are available. *L. interrogans* serovar Copenhageni strain Fiocruz L1-130 was used as the reference genome sequence. The tree is based on 82 variable sites in a core genome of less than 500 bases. Taxon labels include an abbreviated species name, the isolate name and the read set in brackets. Synthetic read set identifiers start with 'L', while published read sets start with 'ERR' 'SRR' or 'DRR'. Abbreviated species names are listed as part of Supplementary Table 1

at a protein level from the three groups has been put to considerable use to predict virulence proteins, genome-wide comparison across the whole genus at a nucleotide level is not useful because of the small core genome.

2.3.2 The Saprophyte Group

The first leptospiral saprophyte to be sequenced was *L. biflexa* serovar Patoc (Picardeau et al. 2008). Genomic data are now available for eight isolates from the saprophyte group; these isolates represent six species. Approximately 18% of the genome is conserved across this group, and 522 proteins were identified as conserved using Roary (Page et al. 2015). The relationship between the isolates is shown in Fig. 2. The average genome size is 4.06 Mb (range: 3.95–4.18 Mb).

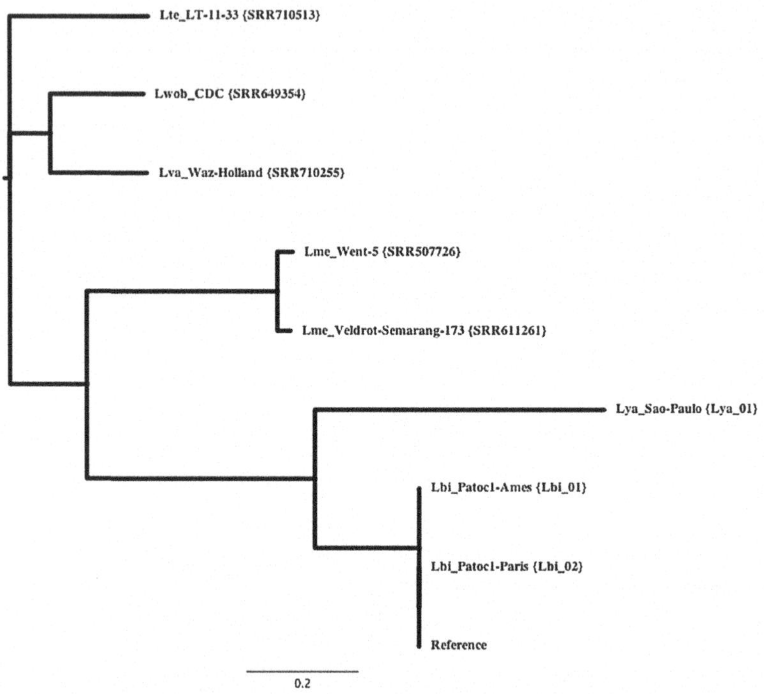

Fig. 2 Unrooted tree showing the inferred relationship between the core genomes of the set of eight saprophyte group isolates for which genomic data are available. *L. biflexa* serovar Patoc strain 'Patoc 1 (Ames)' was used as the reference. The core genome comprises ~18% of the reference genome and includes 55,126 bp where sequence differs in one or more isolates. The greatest pairwise SNP distance was between isolates Lya_Sao-Paulo and Lwob_CDC with 33,700 SNP differences and the smallest was between the two Patoc 1 isolates (8 SNPs). The species abbreviations used are as follows: Lya: Lbi: *L. biflexa*; Lme: *L. meyeri*; Lwob: *L. wolbachii*; Lya: *L. yanagawae*; Lva: *L. vanthielii*; Lte: *L. terpstrae*. Tree inferred using FastTree

2.3.3 The Intermediate Group

Eleven isolates from the intermediate group have genomic data available; these isolates represent the five recognised species in this group. The relationship between the isolates is shown in Fig. 3. The Roary analysis revealed 83 proteins conserved across the group. Using the read mapping approach with *L. licerasiae* strain MMD4847 as a reference, only approximately 2.5% of the genome is included in the core genome. This substantially smaller core genome compared to the saprophyte group indicates greater diversity within this group. Despite the significant genetic diversity, the range of genome sizes (4.19–4.46 Mb; avg = 4.30 Mb) is small. No closed genome is available for an intermediate group isolate, so the precise chromosomal organisation is not known.

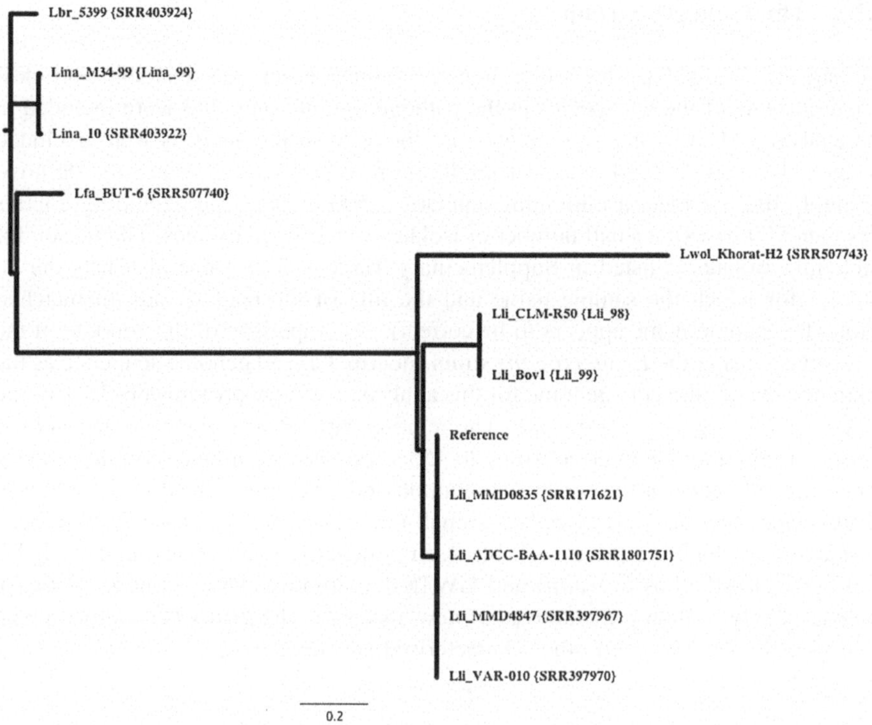

Fig. 3 Unrooted tree showing the inferred relationship between the core genomes of the set of 11 intermediate group isolates for which genomic data are available. *L. licerasiae* strain MMD4847 was used as the reference genome sequence

2.3.4 The Purpose of Taxonomy

The considerable genomic distance between the three genomic groups of the genus *Leptospira* brings into question if it should remain as a single genus. The purpose of classification is to collect together items with similar characteristics; one of the key characteristics that has become integral to classification has been genetic similarity. Given that this overview genetic similarity is not a unifying feature, the single unifying feature for the genus *Leptospira* is cell morphology (genomic relationship at nucleotide sequence level might be a minimum requirement for genus-level grouping of taxa).

Certainly, there are features that are unique to the saprophyte and pathogen groups that would warrant their classification into separate taxonomic entities. Given the diversity of the intermediate group, it may be difficult to identify features common to this group.

2.3.5 The Pathogen Group

Among the 572 isolates for which genome sequence data are available and classified into one of the ten species in the pathogen group, only 443 were included in the analysis. Most of the 129 isolates not included in the analysis were excluded because the read data did not meet minimum read coverage (25×) or the de novo assembly did not meet a minimum standard (>500 contigs and excluding contigs less than 500 bases); a small number of isolates were mixed isolates. The reason for excluding isolates is listed in Supplementary Table 1. The Table also lists the 40 isolates for which the sample name and the title of the read set are mismatched (Note: the sample name appears to be correct). A comparison of the genome of the 443 isolates using the *L. interrogans* strain Fiocruz L1-130 genome sequence as the reference reveals the core genome for this analysis covers approximately 10% of the Fiocruz L1-130 genome sequence. The relationship between the core genomes is shown in Fig. 4 and is inferred using 20,455 sites that vary in one or more isolates. Predominantly each of the species is monophyletic, i.e. represented in the tree as a single clade. This analysis identifies a small number of isolates that may have been misclassified which are listed in Table 4. Significantly, two isolates, namely ICFT (currently classified as *L. weilii*) and GWTS-1 (currently classified as *L. alstonii*), are each likely to be representatives of new species in the genus *Leptospira*, given their genetic distance from other characterised species.

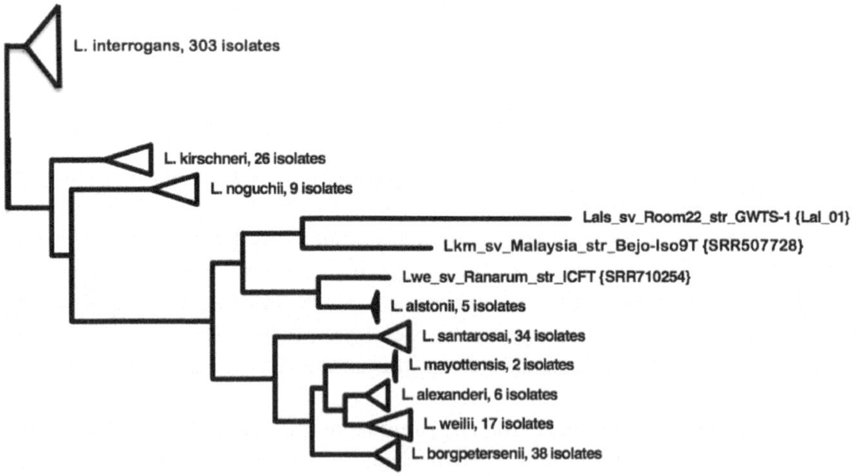

Fig. 4 Unrooted tree showing the inferred relationship between the core genomes of the set of 443 pathogen group isolates for which genomic data are available. *L. interrogans* serovar Copenhageni strain Fiocruz L1-130 was used as the reference genome sequence

Table 4 Misclassified isolates based on core genome comparison

Published classification	Isolate name	Comment
Leptospira alstonii	GWTS #1	Not grouped, may be an example of a new species
Leptospira weilii	ICFT	Not grouped, may be an example of a new species
Leptospira interrogans	M05	Probably *L. borgpetersenii*
Leptospira interrogans	HAI1536	Probably *L. noguchii*
Leptospira santarosai	CBC613	Probably *L. kirschneri*
Leptospira kirschneri	200801925	Probably *L. borgpetersenii*

2.3.6 Subspecies Classification

Subspecies classification has been an essential component of the classification of *Leptospira* over the past century. Serology and the determination of serovar have been central to this process. In connecting the past to the future, it is important to examine the relationship between genomics and serovar classification. We know from past work that genes contributing to lipopolysaccharide (LPS) biosynthesis have moved horizontally within the genus *Leptospira* with the Hardjo serovar being characterised from both *L. interrogans* and *L. borgpetersenii*. Presented here is a survey of the genomic relationship between isolates classified into the species *L. interrogans* and *L. borgpetersenii*. Examination of the isolates exclusively from the *L. borgpetersenii* core genome clade provides greater detail as compared to where all 443 pathogen group isolates were compared. The 38 sequenced isolates in the *L. borgpetersenii* core genome clade were compared using the *L. borgpetersenii* serovar Hardjo strain L550 as the reference genome. More than 85% of the L550 genome sequence was included in the core genome for this comparison. The Roary analysis identified 2113 proteins out of approximately 3300 proteins predicted to be encoded on the genomes of each of the 38 isolates being compared. The inferred tree (Fig. 5) shows that these isolates are divided into three clades, with the predominant serovars in each of the clades being either Ballum, Hardjobovis or Javanica. Among the serovars for which there are multiple isolates, the Hardjobovis isolates form an exclusive monophyletic clade, perhaps indicating there is a stable vertical inheritance of genes involved in LPS biosynthesis. However, analysis of a larger group of isolates for which serovar has been determined needs to be undertaken before broad ranging overviews can be made. The limitation of this comparison is highlighted by the fact that nine of the 38 isolates did not have serovar-type information included in the metadata and 11 of the 38 isolates were the only representative of a serovar in the analysis.

Nearly half of the 303 isolates with genomic data suitable for analysis and that are part of the *L. interrogans* clade have been classified as either serovars Autumnalis (n = 70) or Copenhageni (n = 66), and most of these isolates arise from leptospirosis outbreaks in Thailand/Laos and Brazil, respectively. Despite the over-representation of these two serovars, in total 49 serovars are represented in the

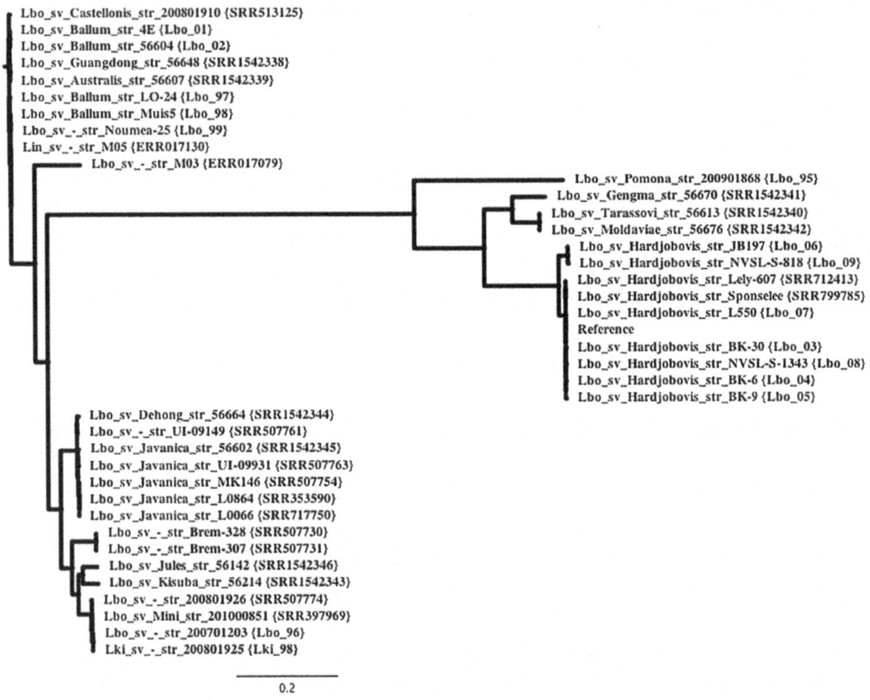

Fig. 5 Unrooted tree showing the inferred relationship between the core genomes of the set of 38 *L. borgpetersenii* clade isolates for which genomic data are available. *L. borgpetersenii* serovar Hardjo strain L550 was used as the reference genome to derive the core genome alignment used to infer the relationship shown. The alignment contained 76,018 variable sites. The label for each taxon shows the abbreviated species name, the serovar, isolate name and a reference to the genomic data set being used in the comparative analysis

genomic comparison, although 25 serovars have just one representative isolate (Table 5). Forty isolates have no serovar information. The *L. interrogans* isolates have been collected from a range of geographic locations and over more than a century, so the set of isolates provides a reasonable overview of the genetic diversity for the species. Approximately 87% of the *L. interrogans* strain Fiocruz L1-130 reference genome occurs as the core genome for the 303 isolates. The relationship between the core genomes (Supplementary Fig. 1) is based on 137,506 sites that vary in one or more of the isolates. Using Roary, 1801 proteins were identified as conserved across all isolates; this is about the half protein coding regions for each isolate. A further 467 protein coding regions were present in up to 95% of the isolates, in conjunction with the 2333 predicted coding regions that are present in between 15 and 95% of isolates. The diversity in genomic content provides an opportunity to use informatics approaches to infer gene characteristics, for example, through association with adaption to particular hosts. Importantly, these approaches depend on comprehensive information about isolates. However,

Table 5 Serovar count of sequenced *L. interrogans* isolates

Serovar	Count[a]
Autumnalis	70
Copenhageni	66
Pomona	12
Manilae	10
Pyrogenes	10
Bataviae	9
Canicola	9
Grippotyphosa	7
Lai	7
Icterohaemorrhagiae	6
Hebdomadis	4
Hardjoprajitno	3
Medanensis	3
Abramis	2
Australis	2
Bangkinang	2
Bratislava	2
Jalna	2
Linhai	2
Muenchen	2
Naam	2
Valbuzzi	2
Wolffi	2
Zanoni	2

[a]Serovars represented once: Bindjei, Bulgarica, Djasiman, Fort-Bragg, Guidae, Haemolytica, Honghe, Jonsis, Kunming, Liangguang, Lora, Mankarso, Mengma, Mengrum, Monymusk, Mooris, Paidjan, Perameles, Qunjian, Rachmati, Saxkoebing, Sentot, Sumatrana, Szwajizak, Trinidad. A total of 40 isolates did not have serovar information listed. Total Count: 303

the dearth of information currently available is notable: what should be essential basic information is not available for more than half of the 303 isolates. Such information should include year and location of isolation, isolation host and perhaps serovar and any associated publications about the isolate. The collection and curation of isolate information could be performed using the existing Bacterial Isolate Genome Sequence Database (BIGSdb) which is part of PubMLST (currently hosts the allele database for the **leptospira** and **leptospira3** Multi-locus sequence type (MLST) schemes, https://pubmlst.org/leptospira/). As well as collecting isolate information, BIGSdb can be used to organise data derived from genomic read data sets, for example MLST. Currently, BIGSdb is not up to date with nearly 30 of the 303 sequenced isolates having no MLST classification despite using the current, online allele database.

The tree in image (Supplementary Fig. 1) showing the relationship between the 303 *L. interrogans* isolates has a clade that contains nearly all the serovar Copenhageni and serovar Icterohaemorrhagiae isolates along with the Monymusk isolate, a Canicola isolate and isolates that have no serovar information. Among the 78 isolates in this group, only a small number of core SNP differences are observed (less than 300 SNPs across a core genome that comprises about 87% of the *L. interrogans* strain Fiocruz L1-130 reference genome sequence). The isolates arising from Brazil have less than 30 pairwise SNP differences. However, the most remarkable feature of this group of isolates is that M20, a strain isolated in 1938 in Denmark and passaged in vitro for more than half a century, is part of this group and has as few as 60 pairwise SNP differences from the Brazilian isolates. The high level of conservation is confirmed by similarly small pairwise SNP differences between the isolates in the Copenhageni/Icterohaemorrhagiae group that have been collected from a range of geographic locations over time.

The close, exclusive genetic relationship between Copenhageni/Icterohaemorrhagiae/Monymusk isolates (so far) highlights two isolates as being unusual, namely the Tande strain (classified as serovar Canicola) and strain 56619 (classified as Copenhageni). These strains may represent a misclassification or an example of a horizontal genetic exchange event leading to a change in the LPS phenotype. In general, serovar is associated with single stable genetic lineages; that is if isolates that have no serovar information are disregarded. Apparently, monophyletic serovars (based on this limited analysis of 303 isolates) include serovars: Pomona, Canicola, Medanensis, Autumnalis and Lai/Naam/Mengma/Mengrum/Qunjian. Serovars that have multiple lineages (in this analysis a lineage contains at least two isolates) include serovars Pyrogenes and Grippotyphosa. Moreover, these monophyletic 'serovar' clades are often associated with a particular MLST. Thus, MLST could be a useful independant cross check for serovar classification.

2.4 Overview

Genomics is unlikely to replace serology as the primary means of diagnosis of individual infections, nor for that matter is MLST; both have a cost and technical barrier that preclude global use, especially in lower per capita income settings where the prevalence of leptospirosis is high. Observations from this review highlight some of the remarkable features we already suspected about *Leptospira*, namely that there are stable lineages, but the example of the Copenhageni/Icterohaemorrhagiae/Monymusk lineage provides evidence that this is a long-standing, globally distributed lineage that is virtually clonal. Taking this into account, specifically the diversity and variation in the gene content between groups of related isolates observed for the *L. interrogans* isolates, the genetic differences between groups of isolates probably evolved through long, independent co-evolution with an animal host. The observations made for *L. interrogans* may not apply to other species from the genus *Leptospira.*

The observation that isolates of the same serovars generally form monophyletic groups provides a useful cross-check for serologically based serovar classification. Several isolates based on their aberrant grouping could be retested. Moreover, where the serological classification is confirmed, and aberrant, such isolates are likely to have arisen from recombination events. Such isolates will be useful to develop an understanding of genetic exchange within and between leptospiral species.

We are now at a stage where a small amount of effort can enable the incremental establishment of a genomic resource for the genus *Leptospira*. The genomic resource should enable access to genomic data for isolates and information about isolates. GenBank/ENA provides a means to permanently archive WGS read sets of isolates, with the prospect that such data sets will be managed by the databases to derive draft genome sequences and annotations of these draft genome sequences (ASM Pathogen Genomics Conference, Washington Sept 2017). However, those submitting the data should endeavour to meet minimum data set standards. The deficiency of the public databases is that it is difficult to update entries. Taking this into consideration, it is essential that some minimum information about isolates is included with the database submission, i.e. isolate name, date of isolation, isolation host, country/region of isolation. Expanded and curated isolate information might be collected at BIGSdb where updating and organising isolate information can be managed more easily.

The utility of the genomic data we are now accumulating will be useful only if we take the time to organise it and make it easily accessible.

3 Whole Genome Transcriptomics

3.1 In Vitro Comparisons

There have been several studies that have investigated genome-wide transcriptional responses under in vitro conditions that mimicked a condition likely to be encountered in vivo. The most obvious condition for *Leptospira* was temperature, given that pathogenic species must make the transition from environmental to physiological temperature in order to survive and multiply. Indeed, such temperature shifts were instrumental in identifying virulence genes, such as the major virulence regulon in *Shigella* spp. (Adler et al. 1989). Similar studies have identified genes required for host adaptation in other spirochetes (Revel et al. 2002; Yang et al. 2000).

Even before the advent of whole genome transcriptome methods, it was apparent that in vivo-like conditions could affect the expression of a range of leptospiral proteins, such as LigA, LigB, LipL36 and LruB (Cullen et al. 2002; Matsunaga et al. 2005; Nally et al. 2001, 2007). Indeed, the expression of lipopolysaccharide (LPS), the major outer membrane component, can also vary under different in vivo conditions (Nally et al. 2005). More recently, a study examined the expression of a limited number of 'virulence-associated' genes under conditions of temperature upshift or oxidative stress (Fraser and Brown 2017); a rather complex interplay emerged, depending upon the conditions and the strain used. Notably, most of the genes have not been shown to be essential for virulence and in fact some have been shown to be not required.

A major drawback with the use of such simulated in vivo studies is that usually only a single condition has been examined. Nonetheless, genome-wide transcription profiles have been investigated in leptospires grown at different temperatures or shifted from environmental to physiological temperature (Lo et al. 2006; Qin et al. 2006), following upshift to physiological osmolarity (Matsunaga et al. 2007), in the presence of guinea pig serum (Patarakul et al. 2010) or under iron-depleted conditions (Lo et al. 2010). These studies produced large gene data sets, the significance of which can be difficult to define. In most cases, the proportion of genes of known, or predicted, function was low. Of the 25 most highly upregulated under physiological upshift, 22 encoded proteins of unknown function (Matsunaga et al. 2007). Likewise, the majority of genes differentially regulated upon overnight upshift to 37 °C after in vitro growth at 30 °C either had no defined function, or had a general function prediction only, such as (not surprisingly) involvement in signal transduction mechanisms or cell wall biosynthesis (Lo et al. 2006). Of 55 genes upregulated upon exposure to serum, 30 were involved in cellular metabolism, processing or signalling (Patarakul et al. 2010); the remainder had no known function. A similar situation was observed under iron-limiting conditions, with the great majority of differentially regulated genes having no known, or generally predicted, function only (Lo et al. 2010); the exception was, unsurprisingly, genes involved in inorganic ion transport and metabolism.

It is difficult to compare individual studies where different methods were used. Accordingly, Adler et al. (2011) compared transcriptome data from some of the above studies which used similar methods and analyses. Of the several hundred genes in those data sets, only 14 were differentially expressed under at least three conditions; of these, only six were absent from the saprophyte *L. biflexa*. Two genes (also present in *L. biflexa*) encode a heme oxygenase and permease; heme acquisition is a known virulence attribute (Murray et al. 2008, 2009b). A third gene encodes the lipoprotein LigB, although the role of *ligB* in virulence is uncertain (Croda et al. 2008; Pappas and Picardeau 2015). The other pathogen-specific genes encode proteins with either unknown, or broadly defined, function such as DNA repair, sugar phosphatase. For a list of the larger number of genes that were differentially expressed under only two conditions, the reader is referred to Adler et al. (2011).

Using a different approach, Zhukova et al. (2017) mapped the genome-wide transcriptional start sites in pathogenic *Leptospira* grown at 30 or 37 °C. Most of the sites were the same at either temperature.

A further drawback of transcriptome studies is the quite wide discrepancy that may occur between the expression profiles observed at gene and protein level (Lo et al. 2009). Taken together, the insights into pathogenic mechanisms provided by in vitro gene expression studies have been limited. Differentially expressed genes usually encoded proteins of unknown or general function only or reflected the necessary change in nutrition and metabolism in transition among different environments.

3.2 *In Vivo Transcriptomics*

In an attempt to overcome some of these limitations, Caimano et al. (2014) took a different approach. Leptospires were grown in dialysis membrane chambers implanted in rat peritoneal cavity. While the conditions would still be different from those encountered in animal tissues (the bacteria were grown in EMJH medium), the aim was to approximate more closely the host environment. RNA-seq was then used to compare the transcript profile with that of leptospires grown in vitro at 30 ° C, identifying 110 genes upregulated in vivo. However, as with the previous studies, the great majority of genes encoded proteins with no known function (64 genes) or proteins with metabolic or regulatory function (32 genes). Seven genes involved in motility were consistent with the essential role that motility plays in virulence (Lambert et al. 2012; Wunder et al. 2016). The known virulence gene encoding heme oxygenase was also identified. Only eight genes were annotated as involved in pathogenesis, but only one of these, *colA*, has been confirmed by mutagenesis (see 4.2), which showed a minor non-essential role (Kassegne et al. 2014). Interestingly, almost none of the small number of known virulence genes (see 4.2) was upregulated in this rat chamber system.

So even with a transcriptome study that more closely resembled the host environment, no new virulence genes were identified. Overall, the transcriptomics approach has yielded disappointing results which are consistent with the notion that *Leptospira* utilises virulence mechanisms that differ from those in other bacterial species. It is probable that the set of genes of unknown function upregulated under some of the conditions studied hides a treasure trove of unique leptospiral virulence factors.

4 Construction of Mutants. Transposon (Tn) and Directed Mutagenesis

Understanding of leptospiral virulence factors at the molecular level has lagged some way behind what has been elucidated in many other bacterial species, due largely to the lack, until relatively recently, of methods for genetically manipulating pathogenic *Leptospira* spp. The development of a transposon mutagenesis system using *Himar* (Bourhy et al. 2005; Murray et al. 2009a) and of methods for targeted gene inactivation (Croda et al. 2008) has finally allowed a classical molecular genetics approach to defining leptospiral virulence genes. A high-throughput screen also distinguished for the first time genes involved in acute hamster infection or mouse renal colonisation (Marcsisin et al. 2013). Here again, however, most genes were of unknown function. The reader is referred to the chapter by M. Picardeau in this volume for a detailed description of genetic methods for *Leptospira* spp.

4.1 Mutants that Retain Virulence. Functional Redundancy

Perhaps the greatest initial surprise arising from mutagenesis studies was the number of mutants that retained virulence despite the inactivation of genes encoding proteins that possessed key hallmarks associated with virulence. Such attributes include predicted surface exposure, absence of saprophytic *Leptospira* spp., similarity to known virulence proteins, role in chemotaxis, interaction with host proteins or tissues, and known in vivo expression.

Arguably, the most notable surprise was the LipL32 lipoprotein. LipL32 is the most abundant protein in the leptospiral cell, is highly conserved, is restricted to pathogenic *Leptospira* spp., and interacts with a range of host extracellular matrix proteins (Murray, 2013). However, a *lipL32* mutant retained both virulence for hamsters and the ability to colonise rat kidneys (Murray et al. 2009c). Likewise, the third most abundant leptospiral protein LipL41 was shown to not be required for acute infection of hamsters (King et al. 2013). Other putative virulence factors shown to not be required, at least for acute hamster infection, include chemotaxis proteins CheB and CheX, endostatin-like proteins LenB and LenE, metalloprotease, the lipoprotein LigC, the Fur family regulator PerR and the lipoprotein LigB (Croda et al. 2008; Murray et al. 2009a). Indeed, subsequent inactivation of the *cheAWDBY* operon resulted in a mutant with reduced chemotactic responses, but full retention of virulence (Lambert et al. 2015). These results indicate a high level of functional redundancy and also are consistent with the concept that '*Leptospira* does it differently'.

4.2 Virulence Genes with Defined Functions

4.2.1 LPS Biosynthesis

It should come as no surprise that leptospiral LPS is essential for virulence. Nevertheless, unequivocal proof only became available when two Tn mutants with truncated LPS were shown to be attenuated for the ability to cause acute leptospirosis in hamsters and for renal colonisation in mice (Marcsisin et al. 2013; Murray et al. 2010). The structure of leptospiral LPS remains unknown, but appears to be very complex. The LPS biosynthesis locus varies in size among the species of *Leptospira*, but may comprise as many as 100 genes (Fouts et al. 2016).

Accordingly, the precise nature of the alteration in these mutants could not be determined. Interestingly, out of a total of over 600 random Tn mutants in serovar Manilae, only three were found in the LPS biosynthesis locus (Murray et al. 2009a), consistent with the essential role of LPS in leptospiral viability.

The lipid A component of bacterial LPS anchors the LPS molecule in the outer membrane and is essential for viability of most, but not all, gram-negative bacteria (Moffatt et al. 2010). The structure of leptospiral lipid A shows some unusual features (Que-Gewirth et al. 2004), but in general the lipid A biosynthesis pathway resembles that of other bacteria, with the LpxD acyltransferase catalysing the third step in the pathway. Surprisingly, *Leptospira* was found to have two *lpxD* homologs and Tn mutants could be generated in both genes, designated *lpxD1* and *lpxD2*, without affecting viability (Eshghi et al. 2015). However, only the *lpxD1* mutant was attenuated for virulence in gerbils, possibly because of sensitivity to physiological temperature of 37 °C and increased outer membrane permeability. Interestingly, inactivation of either gene resulted in only very minor changes to lipid A and LPS, raising questions about the precise biochemical activity of the LpxD enzymes (Eshghi et al. 2015).

4.2.2 Motility

Motility is a key virulence factor in many bacterial species (Chaban et al. 2015), and the same is true for *Leptospira*. Even before the advent of modern molecular methods, spontaneous non-motile mutants were recognised as avirulent (Faine and van der Hoeden, 1964). Recent studies in which different genes involved in flagellar biosynthesis, structure or activity were inactivated have confirmed these findings. Mutation of the flagellar switch protein, FliY, abrogated virulence in guinea pigs (Liao et al. 2009) as did inactivation of the flagellar motor switch protein FliM in hamsters (Fontana et al. 2016). Similarly, inactivation of the flagellar structural protein FlaA2, but not FlaA1, produced a non-motile phenotype which was attenuated in the hamster model. A possible reason for the retention of virulence in the *flaA1* mutant was that it retained some translational motility (Lambert et al. 2012). Interestingly, the FlaA proteins were previously considered to comprise the flagellar sheath (Kelson et al. 1988), but both *flaA* mutants produced sheathed flagella. Subsequently, FcpA was identified as a flagellar sheath protein (Wunder et al. 2016); an *fcpA* mutant was attenuated in the hamster acute infection model. Clearly, leptospiral motility is essential for virulence.

4.2.3 Iron Acquisition

As with almost all bacterial species, the ability to acquire iron is essential for both leptospiral growth and survival; in the mammalian host, the primary source of iron is heme. The leptospiral heme oxygenase enzyme, encoded by *hemO*, mediates the acquisition of iron from heme. Inactivation of *hemO* resulted in attenuation of acute disease in hamsters, but the mutant retained the ability to colonise hamster kidneys (Murray et al. 2008; 2009b).

4.2.4 Oxidative Stress Response

Pathogenic leptospires must be able to survive in different ecological niches. In the carrier state, they must colonise the proximal renal tubules. Then, they must be able to survive in a range of environments (soil, water, etc.) before colonising another carrier host or causing acute infection in a susceptible host. During the transition from the environment to the mammalian host, several stress response genes are upregulated. Accordingly, inactivation of some of these genes has resulted in attenuation of virulence. Leptospiral catalase mediates resistance to H_2O_2. When the corresponding *katE* gene was inactivated, the resulting mutant was completely attenuated in hamsters, consistent with KatE mediating resistance to reactive oxygen species produced by the host innate immune response. A similar phenotype resulted from inactivation of the molecular chaperone ClpB. The mutant was more susceptible to oxidative stress and was attenuated in gerbils (Lourdault et al. 2011). Mutation of the leptospiral Hsp90 homolog, HtpG, yielded similar results; an *htpG* mutant was unable to cause disease in hamsters (King et al. 2014). Interestingly, in this case, the mutant was not more susceptible to oxidative stress. Thus, although the *htpG* gene has a defined function, the mechanism of virulence attenuation remains undefined.

4.2.5 Cell Entry

Leptospira is generally not considered to be an intracellular pathogen (Adler 2014), although leptospires may be observed transiently inside both phagocytic and non-phagocytic cells. However, in the tissues of patients and experimentally infected animals, the majority of leptospires are located in the extracellular milieu. It was therefore perhaps surprising that *Leptospira* was found to have a homolog of a cell entry protein, Mce, in *Mycobacterium*, a classical intracellular pathogen (Zhang et al. 2012). Although recombinant leptospiral Mce bound mammalian integrins, an *mce* mutant showed only a slight reduction in virulence for hamsters, consistent with a minor role in virulence and again highlighting the degree of functional redundancy in pathogenic *Leptospira*. A further confounding issue was the very low virulence of the parent strain (LD_{50} 10^7). Thus, Mce may have a virulence-associated function, but is clearly not essential.

4.2.6 Collagenase

Leptospiral collagenase, ColA, could hydrolyse several host collagens in vitro. However, inactivation of *colA* resulted in only a minor (less than 2 logs) reduction in LD_{50} from the very high LD_{50} of 10^7 of the parent strain (Kassegne et al. 2014). These results suggest a minor, non-essential role in pathogenesis for collagenase.

4.2.7 Sphingomyelinases/Phospholipases

Lysis of mammalian erythrocytes by leptospiral culture supernatants has been known for decades, and sphingomyelinases have been identified (Segers et al. 1990, 1992). Leptospiral genomes may contain several genes encoding sphingomyelinase-like enzymes (Fouts et al. 2016), although the enzyme active site may not be universally present (Narayanavari et al. 2012). Despite the fact that these genes are absent in the saprophyte *L. biflexa* (Picardeau et al. 2008), suggesting a role in pathogenesis rather than just in nutrient acquisition in the environment, proof of an unequivocal role in virulence, or even in haemolysis, has been elusive. Recent genetic studies identified for the first time the Sph2 sphingomyelinase as a major haemolysin that was expressed much more strongly under physiological osmolarity (Narayanavari et al. 2015). However, the *sph2* Tn mutant was not assessed for virulence.

The leptospiral LB361 protein was identified as phospholipase C and was shown to elevate intracellular Ca^{2+} levels resulting in macrophage death (Zhao et al. 2014). However, the claim of a novel virulence factor was not supported by testing of the *lb361* mutant for virulence in an animal model.

4.3 *Virulence Genes of Unknown or Undefined Function*

Several genes have been demonstrated unequivocally to be essential for virulence but whose function is either unknown or poorly understood. Ironically, the very first virulence-essential gene to be identified genetically was *loa22*, encoding a putative lipoprotein with a predicted OmpA domain (Ristow et al. 2007). The precise function of Loa22 has not been determined, but it is hypothesised to mediate linkage between the peptidoglycan and outer membrane. The mechanism by which inactivation of *loa22* mediates attenuation is likewise unclear, but the presence of a Loa22 homolog in the saprophyte *L. biflexa* would suggest an indirect role in pathogenesis.

The LruA surface protein is a key factor in provoking leptospiral uveitis in horses via an autoimmune mechanism, but the function of LruA remains unclear (Verma et al. 2013). Why LruA should also bind mammalian Apolipoprotein A1 (ApoA1) is likewise uncertain (Zhang et al. 2013). Interestingly, a deletion mutant with no detectable LruA was attenuated in hamsters and bound more ApoA1 than wild type or than a mutant that expressed truncated LruA, prompting the suggestion, purely speculative, that LruA may be shed from the leptospiral surface as a molecular decoy.

A genetic locus comprising four genes was identified by random Tn mutagenesis (Eshghi et al. 2014). Inactivation of the first gene, *lb139*, encoding a predicted sensor phosphatase resulted in complete attenuation of virulence in hamsters. The *lb139* mutant showed reduced expression of all three downstream genes, encoding an anti-sigma factor antagonist and proteins of unknown function. However, RNA-seq analysis identified a total of 143 genes that were differentially expressed in the mutant. Many of these genes have no known function, but the data set included genes involved in motility and signal transduction; the mutant showed partial decrease in motility. Notably, downregulated genes included some that are known to be upregulated at physiological osmolarity, such as *sph2* and *ligB*. Of course, it is not possible from such a complex data set to identify the specific mechanisms responsible for attenuation. Nevertheless, this study described the first potential virulence regulon in *Leptospira* and clearly warrants further study.

In an extension of the high-throughput approach for identifying attenuated mutants (Marcsisin et al. 2013), Lourdault et al. (2016) used Tn-seq to screen a pool of 42 Tn mutants for in vivo fitness in hamsters, followed by virulence testing for any mutants with reduced fitness. A mutant with an inactivated putative adenylate/guanylate cyclase gene was partially, but significantly attenuated in hamsters; the precise mechanism of attenuation is unknown. Interestingly, none of the over 20 mutants in predicted signal transduction genes showed reduced in vivo fitness, again consistent with the high degree of functional redundancy in pathogenic *Leptospira* spp.

5 Conclusions

The availability of whole genome sequences has expanded our understanding of leptospiral taxonomy, with the recognition of multiple species clustered into three groups, saprophytic, intermediate and pathogenic. Recent studies have identified several new species in all three groups. Interestingly, genomic analysis has suggested the possibility of two clades within the pathogen group, containing species of high and low pathogenicity (C. Goarant, personal communication). Clearly, much remains to be discovered in this area.

Mutagenesis of *Leptospira*, although still relatively inefficient, has identified a substantial number of virulence genes and factors. Some of them are now well characterised, while the basis for attenuation of others remains obscure. Many mutants in genes with hallmarks of virulence genes have retained virulence, indicating that leptospires display a high level of functional redundancy. Genes of unknown function are over-represented in pathogenic *Leptospira* spp., suggesting that leptospires possess unique, as-yet uncharacterised virulence mechanisms.

References

Adler B, Sasakawa C, Tobe T, Makino S, Komatsu K, Yoshikawa M (1989) A dual transcriptional activation system for the 230 kb plasmid genes coding for virulence-associated antigens of *Shigella flexneri*. Mol Microbiol 3:627–635

Adler B, de la Pena Moctezuma A (2010) *Leptospira* and leptospirosis. Vet Microbiol 140:287–296

Adler B, Lo M, Seemann T, Murray GL (2011) Pathogenesis of leptospirosis: the influence of genomics. Vet Microbiol 153:73–81

Adler B (2014) Pathogenesis of leptospirosis: cellular and molecular aspects. Vet Microbiol 172:353–358

Bourhy P, Louvel H, Saint Girons I, Picardeau M (2005) Random insertional mutagenesis of *Leptospira interrogans*, the agent of leptospirosis, using a mariner transposon. J Bacteriol 187:3255–3258

Bulach DM (2018) Adler book: chapter on leptospiral genomics: archive containing nullarbor reports, excel spreadsheet version of tables. figshare. https://doi.org/10.4225/49/5a8670974d186

Bulach DM, Zuerner RL, Wilson P, Seemann T, McGrath A, Cullen PA, Davis J, Johnson M, Kuczek E, Alt DP, Peterson-Burch B, Coppel RL, Rood JI, Davies JK, Adler B (2006) Genome reduction in *Leptospira borgpetersenii* reflects limited transmission potential. Proc Nat Acad Sci U S A 103:14560–14565

Caimano MJ, Sivasankaran SK, Allard A, Hurley D, Hokamp K, Grassmann AA, Hinton JCD, Nally JE (2014) A model system for studying the transcriptomic and physiological changes associated with mammalian host-adaptation by *Leptospira interrogans* serovar Copenhageni. PLoS Pathog 10:e10004004

Chaban B, Hughes HV, Beeby M (2015) The flagellum in bacterial pathogens: for motility and a whole lot more. Semin Cell Dev Biol 46:91–103

Croda J, Figueira CP, Wunder EA Jr, Santos CS, Reis MG, Ko AI, Picardeau M (2008) Targeted mutagenesis in pathogenic *Leptospira* species: disruption of the LigB gene does not affect virulence in animal models of leptospirosis. Infect Immun 76:5826–5833

Cullen PA, Cordwell SJ, Bulach DM, Haake DA, Adler B (2002). Global analysis of outer membrane proteins from *Leptospira interrogans* serovar Lai. Infect Immun 70:2311–2270

Eshghi A, Becam J, Lambert A, Sismeiro O, Dillies M-A, Jagla B, Wunder EA, Ko AI, Coppee J-Y, Goarant C, Picardeau M (2014) A putative regulatory genetic locus modulates virulence in the pathogen *Leptospira interrogans*. Infect Immun 82:2542–2552

Eshghi A, Henderson J, Trent MS, Picardeau M (2015) *Leptospira interrogans lpxD* homologue is required for thermal acclimatization and virulence. Infect Immun 83:4314–4321

Faine S, Hoeden JVD (1964) Virulence-linked colonial and morphological variation in *Leptospira*. J Bacteriol 88:1493–1496

Fontana C, Lambert A, Benaroudj N, Gasparini D, Gorgette O, Cachet N, Bomchil N, Picardeau M (2016) Analysis of a spontaneous non-motile and avirulent mutant shows that FliM is required for full endoflagella assembly in *Leptospira interrogans*. PLoS ONE 11:e0152916

Fouts DE, Matthias MA, Adhikarla H, Adler B, Amorim-Santos L, Berg DE, Bulach D, Buschiazzo A, Chang YF, Galloway RL, Haake DA, Haft DH, Hartskeerl R, Ko AI, Levett PN, Matsunaga J, Mechaly AE, Monk JM, Nascimento AL, Nelson KE, Palsson B, Peacock SJ, Picardeau M, Ricaldi JN, Thaipandungpanit J, Wunder EA Jr, Yang XF, Zhang JJ, Vinetz JM (2016) What makes a bacterial species pathogenic?: comparative genomic analysis of the genus Leptospira. PLoS Negl Trop Dis 10:e0004403

Fraser T, Brown PD (2017) Temperature and oxidative stress as triggers for virulence gene expression in pathogenic *Leptospira* spp. Front Microbiol 8

Huang W, Li L, Myers JR, Marth GT (2012) ART: a next-generation sequencing read simulator. Bioinformatics 28:593–594

Kassegne K, Hu W, Ojcius DM, Sun D, Ge Y, Zhao J, Yang XF, Li L, Yan J (2014) Identification of collagenase as a critical virulence factor for invasiveness and transmission of pathogenic *Leptospira* species. J Infect Dis 209:1105–1115

Kelson JS, Adler B, Chapman AJ, Faine S (1988) Identification of leptospiral flagellar antigens by gel electrophoresis and immunoblotting. J Med Microbiol 26:47–53

King AM, Bartpho T, Sermswan RW, Bulach DM, Eshghi A, Picardeau M, Adler B, Murray GL (2013) Leptospiral outer membrane protein LipL41 is not essential for acute leptospirosis but requires a small chaperone protein, Lep, for stable expression. Infect Immun 81:2768–2776

King AM, Pretre G, Bartpho T, Sermswan RW, Toma C, Suzuki T, Eshghi A, Picardeau M, Adler B, Murray GL (2014) High-temperature protein G is an essential virulence factor of *Leptospira interrogans*. Infect Immun 82:1123–1131

Lambert A, Picardeau M, Haake DA, Sermswan RW, Srikram A, Adler B, Murray GA (2012) FlaA proteins in *Leptospira interrogans* are essential for motility and virulence but are not required for formation of the flagellum sheath. Infect Immun 80:2019–2025

Lambert A, Wong Ng J, Picardeau M (2015) Gene inactivation of a chemotaxis operon in the pathogen *Leptospira interrogans*. FEMS Microbiol Lett 362:1–8

Levett PN (2015) Systematics of *Leptospiraceae*. Curr Top Microbiol Immunol 387:11–20

Liao S, Sun A, Ojcius D, Wu S, Zhao J, Yan J (2009) Inactivation of the *fliY* gene encoding a flagellar motor switch protein attenuates mobility and virulence of *Leptospira interrogans* strain Lai. BMC Microbiol 9:253

Lo M, Bulach DM, Powell DR, Haake DA, Matsunaga J, Paustian ML, Zuerner RL, Adler B (2006) Effects of temperature on gene expression patterns in *Leptospira interrogans* serovar Lai as assessed by whole-genome microarrays. Infect Immun 74:5848–5859

Lo M, Cordwell SJ, Bulach DM, Adler B (2009) Comparative transcriptional and translational analysis of leptospiral outer membrane protein expression in response to temperature. PLoS Negl Trop Dis 3:e560

Lo M, Murray GL, Khoo CA, Haake DA, Zuerner RL, Adler B (2010) Transcriptional response of *Leptospira interrogans* to iron limitation and characterization of a PerR homolog. Infect Immun 78:4850–4859

Lourdault K, Cerqueira GM, Wunder EA, Picardeau M (2011) Inactivation of *clpB* in the pathogen *Leptospira interrogans* reduces virulence and resistance to stress conditions. Infect Immun 79:3711–3717

Lourdault K, Matsunaga J, Haake DA (2016) High-throughput parallel sequencing to measure fitness of Leptospira interrogans transposon insertion mutants during acute infection. PLoS Negl Trop Dis 10:e0005117

Marcsisin RA, Bartpho T, Bulach DM, Srikram A, Sermswan RW, Adler B, Murray GL (2013) Use of a high-throughput screen to identify Leptospira mutants unable to colonize the carrier host or cause disease in the acute model of infection. J Med Microbiol 62:1601–1608

Matsunaga J, Sanchez Y, Xu X, Haake DA (2005) Osmolarity, a key environmental signal controlling expression of leptospiral proteins LigA and LigB and the extracellular release of LigA. Infect Immun 73:70–78

Matsunaga J, Lo M, Bulach DM, Zuerner RL, Adler B, Haake DA (2007) Response of *Leptospira interrogans* to physiologic osmolarity: Relevance in signaling the environment-to-host transition. Infect Immun 75:2864–2874

Moffatt JH, Harper M, Harrison P, Hale JDF, Vinogradov E, Seemann T, Henry R, Crane B, St. Michael F, Cox AD, Adler B, Nation RL, Li J, Boyce JD (2010) Colistin resistance in *Acinetobacter baumannii* is mediated by complete loss of lipopolysaccharide production. Antimicrob Ag Chemother 54:4971–4977

Murray GL, Ellis KM, Lo M, Adler B (2008) *Leptospira interrogans* requires a functional heme oxygenase to scavenge iron from hemoglobin. Microb Infect 10:791–797

Murray GL, Morel V, Cerqueira GM, Croda J, Srikram A, Henry R, Ko AI, Dellagostin OA, Bulach DM, Sermswan RW, Adler B, Picardeau M (2009a) Genome-wide transposon mutagenesis in pathogenic *Leptospira* species. Infect Immun 77:810–816

Murray GL, Srikram A, Henry R, Puapairoj A, Sermswan RW, Adler B (2009b) *Leptospira interrogans* requires heme oxygenase for disease pathogenesis. Microb Infect 11:311–314

Murray GL, Srikram A, Hoke DE, Wunder EA Jr, Henry R, Lo M, Zhang K, Sermswan RW, Ko AI, Adler B (2009c) Major surface protein LipL32 is not required for either aute or cronic infection with *Leptospira interrogans*. Infect Immun 77:952–958

Murray GL, Srikram A, Henry R, Hartskeerl RA, Sermswan RW, Adler B (2010) Mutations affecting *Leptospira interrogans* lipopolysaccharide attenuate virulence. Mol Microbiol 78:701–709

Murray GL (2013) The lipoprotein LipL32, an enigma of leptospiral biology. Vet Microbiol 162:305–314

Nally JE, Artiushin S, Timoney JF (2001) Molecular characterization of thermoinduced immunogenic proteins Q1p42 and Hsp15 of *Leptospira interrogans*. Infect Immun 69:7616–7303

Nally JE, Chow E, Fishbein MC, Blanco DR, Lovett MA (2005) Changes in lipopolysaccharide O antigen distinguish acute versus chronic *Leptospira interrogans* infections. Infect Immun 73:3251–3260

Nally JE, Whitelegge JP, Bassilian S, Blanco DR, Lovett MA (2007) Characterization of the outer membrane proteome of *Leptospira interrogans* expressed during acute lethal infection. Infect Immun 75:766–773

Narayanavari SA, Sritharan M, Haake DA, Matsunaga J (2012) Multiple leptospiral sphingomyelinases (or are there?). Microbiol 158:1137–1146

Narayanavari SA, Lourdault K, Sritharan M, Haake DA, Matsunaga J (2015) Role of *sph2* gene regulation in hemolytic and sphingomyelinase activities produced by *Leptospira interrogans*. PLoS Negl Trop Dis 9:e0003952

Page AJ, Cummins CA, Hunt M, Wong VK, Reuter S, Holden MT, Fookes M, Falush D, Keane JA, Parkhill J (2015) Roary: rapid large-scale prokaryote pan genome analysis. Bioinformatics 31:36913693

Pappas CJ, Picardeau M (2015) Control of gene expression in *Leptospira* spp. by transcription activator-like effectors demonstrates a potential role for LigA and LigB in *Leptospira interrogans* virulence. Appl Environ Microbiol 81:7888–7892

Paster BJ, Dewhirst FE, Weisburg WG, Tordoff LA, Fraser GJ, Hespell RB, Stanton TB, Zablen L, Mandelco L, Woese CR (1991) Phylogenetic analysis of the spirochetes. J Bacteriol 173:6101–6109

Patarakul K, Lo M, Adler B (2010) Global transcriptomic response of *Leptospira interrogans* serovar Copenhageni upon exposure to serum. BMC Microbiol 10:31

Picardeau M, Bulach DM, Bouchier C, Zuerner RL, Zidane N, Wilson PJ, Creno S, Kuczek ES, Bommezzadri S, Davis JC, McGrath A, Johnson MJ, Boursaux-Eude C, Seemann T, Rouy Z, Coppel RL, Rood JI, Lajus A, Davies JK, Medigue C, Adler B (2008) Genome sequence of the saprophyte *Leptospira biflexa* provides insights into the evolution of *Leptospira* and the pathogenesis of leptospirosis. PLoS ONE 3:e1607

Qin J-H, Sheng Y-Y, Zhang Z-M, Shi Y-Z, He P, Hu B-Y, Yang Y, Liu S-G, Zhao G-P, Guo X-K (2006) Genome-wide transcriptional analysis of temperature shift in *Leptospira interrogans* serovar lai strain 56601. BMC Microbiol 6:51

Que-Gewirth NLS, Ribeiro AA, Kalb SR, Cotter RJ, Bulach DM, Adler B, Girons IS, Werts C, Raetz CRH (2004) A methylated phosphate group and four amide-linked acyl chains in *Leptospira interrogans lipid* A: the membrane anchor of an unusual lipopolysaccharide that activates TLR2. J Biol Chem 279:25420–25429

Ren SX, Fu G, Jiang XG, Zeng R, Miao YG, Xu H, Zhang YX, Xiong H, Lu G, Lu LF, Jiang HQ, Jia J, Tu YF, Jiang JX, Gu WY, Zhang YQ, Cai Z, Sheng HH, Yin HF, Zhang Y, Zhu GF, Wan M, Huang HL, Qian Z, Wang SY, Ma W, Yao ZJ, Shen Y, Qiang BQ, Xia QC, Guo XK, Danchin A, Saint Girons I, Somerville RL, Wen YM, Shi MH, Chen Z, Xu JG, Zhao GP (2003) Unique physiological and pathogenic features of *Leptospira interrogans* revealed by whole-genome sequencing. Nature 422:888–893

Revel AT, Talaat AM, Norgard MV (2002) DNA microarray analysis of differential gene expression in *Borrelia burgdorferi*, the Lyme disease spirochete. Proc Nat Acad Sci U S A 99:1562–1567

Ristow P, Bourhy P, McBride FW, Figueira CP, Huerre M, Ave P, Girons IS, Ko AI, Picardeau M (2007) The OmpA-like protein Loa22 is essential for leptospiral virulence. PLoS Pathog 3:e97

Segers RP, van der Drift A, de Nijs A, Corcione P, van der Zeijst BA, Gaastra W (1990) Molecular analysis of a sphingomyelinase C gene from *Leptospira interrogans* serovar hardjo. Infect Immun 58:2177–2185

Segers RP, van Gestel JA, van Eys GJ, van der Zeijst BA, Gaastra W (1992) Presence of putative sphingomyelinase genes among members of the family *Leptospiraceae*. Infect Immun 60:1707–1710

Thibeaux R, Iraola G, Ferres I, Bierque E, Girault D, Soupe-Gilbert M, Picardeau M, Goarant C (2017) Of soils, *Leptospira* and humans. In: Tenth international leptospirosis society conference. Palmerston North, New Zealand

Verma A, Stevenson B, Adler B (2013) Leptospirosis in horses. Vet Microbiol 167:61–66

Wunder EA, Figueira CP, Benaroudj N, Hu B, Tong BA, Trajtenberg F, Liu J, Reis MG, Charon NW, Buschiazzo A, Picardeau M, Ko AI (2016) A novel flagellar sheath protein, FcpA, determines filament coiling, translational motility and virulence for the *Leptospira* spirochete. Mol Microbiol 101:457–470

Yang X, Goldberg MS, Popova TG, Schoeler GB, Wikel SK, Hagman KE, Norgard MV (2000) Interdependence of environmental factors influencing reciprocal patterns of gene expression in virulent *Borrelia burgdorferi*. Mol Microbiol 37:1470–1479

Zhang K, Murray GL, Seemann T, Srikram A, Bartpho T, Sermswan RW, Adler B, Hoke DE (2013) Leptospiral LruA is required for virulence and modulates an interaction with mammalian Apolipoprotein AI. Infect Immun 81:3872–3879

Zhang L, Zhang C, Ojcius DM, Sun D, Zhao J, Lin X, Li L, Li L, Yan J (2012) The mammalian cell entry (Mce) protein of pathogenic *Leptospira* species is responsible for RGD motif-dependent infection of cells and animals. Mol Microbiol 83:1006–1023

Zhao JF, Chen HH, Ojcius DM, Zhao X, Sun D, Ge YM, Zheng LL, Lin X, Li LJ, Yan J (2014) Identification of *Leptospira interrogans* phospholipase C as a novel virulence factor responsible for intracellular free calcium ion elevation during macrophage death. PLoS ONE 8:e75652

Zhukova A, Fernandes LG, Hugon P, Pappas CJ, Sismeiro O, Coppée J-Y, Becavin C, Malabat C, Eshghi A, Zhang J-J, Yang FX, Picardeau M (2017) Genome-wide transcriptional start site mapping and sRNA identification in the pathogen *Leptospira interrogans*. Front Cell Infect Microbiol 7

Complement Immune Evasion by Spirochetes

Angela S. Barbosa and Lourdes Isaac

Abstract The complement system plays an important role in the innate and acquired immune response against pathogens. A sophisticated network of activating and regulating proteins allows the distinction between intact and damaged host and non-host surfaces such as bacteria and other parasites. Non-host structures trigger the alternative pathway which may lead to their elimination by phagocytosis or cell lysis. In addition, complement proteins such as C1q, mannose binding lectin (MBL), and ficolins act as pathogen pattern-recognition molecules. Biological functions such as opsonization, activation of B lymphocytes and production of antibodies, degranulation of mast cells and basophils, and cell lysis that are important for elimination of microorganisms are dependent on complement activation. However, several pathogens including spirochetes have developed several specialized mechanisms to evade the complement system, thereby contributing to survival in the host. In this review, we give a brief overview of complement activation and regulation, and discuss in detail the strategies used by spirochetes from the genera *Borrelia*, *Leptospira*, and *Treponema* to overcome complement activation.

Contents

A.S. Barbosa (✉)
Laboratory of Bacteriology, Butantan Institute, São Paulo, Brazil
e-mail: angela.barbosa@butantan.gov.br

L. Isaac (✉)
Department of Immunology, Institute of Biomedical Sciences,
University of São Paulo, São Paulo, Brazil
e-mail: louisaac@icb.usp.br

Current Topics in Microbiology and Immunology (2018) 415:215–238
DOI 10.1007/82_2017_47

Published Online: 23 September 2017

1 The Complement System

Spirochetes are Gram-negative bacteria that are generally considered to be extracellular. Innate immunity against these pathogens includes activation of the complement system (alternative and lectin pathways), phagocytosis, and inflammatory response. Adaptive immunity includes humoral immunity, and is mediated by specific antibodies and other soluble molecules. Antibodies may control the infection by neutralizing toxins, blocking cell invasion, facilitating phagocytosis of bacteria, and complement system activation through the classical pathway.

The importance of this system becomes evident in complement-deficient patients that are more susceptible to recurrent and severe infections, including those caused by Gram-negative bacteria (Macedo and Isaac 2016). Once activated, the complement system generates several important biological functions such as: (i) opsonization by complement activated fragments (C3b; iC3b) facilitating bacterial phagocytosis by leukocytes mediated by complement receptors (CR) CR1, CR3, and CR4; (ii) formation of the membrane complement attack complex (MAC; $C5b6789_n$) which may lead to bacterial lysis; (iii) chemo-attraction of leukocytes to the inflammatory site (fragments C3a and C5a, and C3aR, C5aR1 and C5aR2); (iv) B lymphocyte activation and production of specific antibodies (fragment C3d; CR2; Carroll and Isenman 2012); and, (v) release of several inflammatory mediators (anaphylatoxins C3a and C5), which amplifies the inflammatory response, among other important functions (reviewed in Walport 2001; Ricklin et al. 2010).

The complement system is composed of more than 50 proteins present either bound to cell surfaces or soluble in plasma. This system may be activated by at least three different pathways: alternative pathway (innate immunity), lectin (innate immunity), and classical (acquired immunity) pathways. All three converge to the central component C3 with formation of C3 convertases and C5 convertases which cleave C3 and C5, respectively. Subsequently, activation proceeds via a common terminal pathway (Fig. 1), culminating in the formation of the MAC.

The alternative pathway is continuously activated by the spontaneous hydrolysis of the thioester bond present in the C3 α-chain in the native protein (Pangburn and Müller-Eberhard 1983). Once the thiolactone ring formed by Cys^{988} and Gln^{991} is disrupted, the native C3 molecule adopts a different structural conformation, designated $C3H_2O$ (Isenman 1983; Straight and McKee 1983). The inactive protease Factor B is then able to bind to $C3H_2O$ ($C3H_2OB$; Pangburn and Müller-Eberhard 1983) and is further cleaved by the enzyme Factor D into two fragments (Ba and Bb). The larger fragment Bb is an active serine protease and remains attached to the

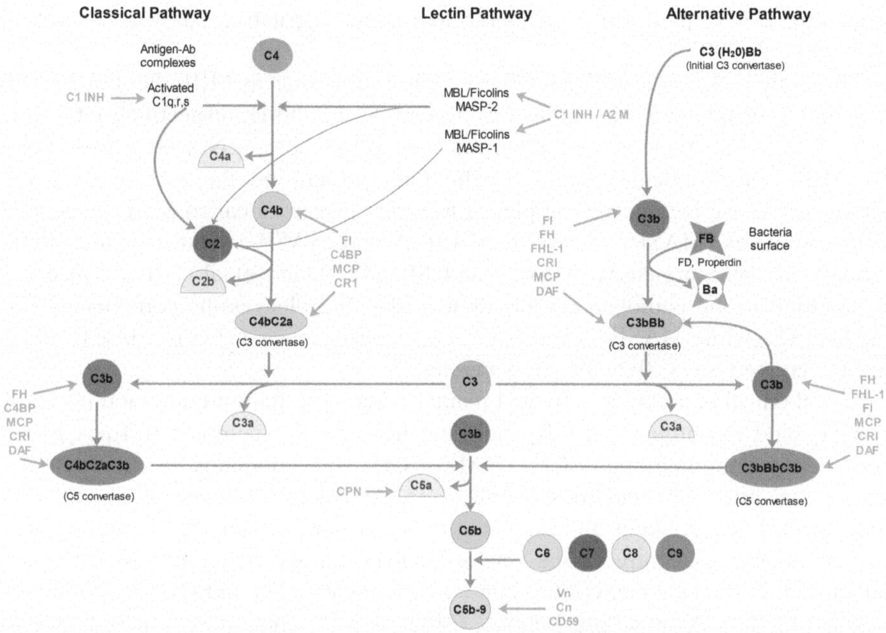

Fig. 1 Overview of the activation of the complement system. Activation can be triggered by at least three distinct pathways: alternative, lectin, and classical pathways. These pathways all lead to the generation of distinct C3 convertases that cleave C3 into fragments C3a and C3b. The association of C3b with C3 convertases generates C5 convertases that cleave C5 into C5a and C5b fragments. All three pathways converge to a common terminal pathway which leads to the formation of the membrane attack complex (C5b-9). Several complement regulatory proteins are responsible in regulating positively (properdin) or negatively (indicated in green font) the activation of this system. Factor H: FH; Factor D: FD; Factor B: FB; Factor I: FI; Factor H-like-1: FHL-1; CR1 complement receptor1: CR1; membrane cofactor protein: MCP; decay-accelerating factor: DAF; mannose binding lectin: MBL; MBL-associated serine protease: MASP; MBL; C1 inhibitor: C1INH; alpha2 macroglobulin; carboxypeptidase N: CPN; A2M; vitronectin: Vn; clusterin: Cn; membrane attack complex: C5b-9

complex ($C3H_2OBb$), known as the initial C3 convertase of the alternative pathway. The C3 convertase cleaves C3 into two fragments, C3a and C3b. C3b exhibits a structural protein conformation similar to that of $C3H_2O$ and is also able to bind to Factor B. The complex C3bB is cleaved by Factor D resulting in the second and more stable C3 convertase (C3bBb), which cleaves many C3 molecules, generating more C3 convertases, thereby amplifying the alternative pathway. This loop amplification does not discriminate whether fragment C3b is generated during activation of the alternative, lectin or classical pathways (Lachmann 2009). The positive regulatory protein properdin binds to the C3 convertases $C3H_2OBb$ and C3bBb which increases their half-lives. Properdin can also bind directly to the surface of some pathogens. Once the C3b fragment binds to this surface-associated properdin, a new platform is available to initiate alternative pathway activation (Spitzer et al. 2007).

The lectin pathway is initiated when certain lectins such as MBL, ficolins (ficolin 1, ficolin 2, or ficolin 3, also known as ficolins M, L, and H, respectively), CL-K1, or CL-L1 bind to abundant carbohydrates such as mannose or N-acetylglucosamine (by MBL) or acetylated sugars (by ficolins) present on the surface of many pathogens. These lectins are complexed to serine proteases called MBL-associated serine proteases (MASP) (MASP-1, MASP-2, and MASP-3) and to non-catalytic variants of these proteases, MAp19 and MAp44 (Dobó et al. 2016). Once the lectins bind to the pathogen carbohydrates, MASP-2 cleaves the components C4, and later C2, while MASP-1 cleaves C2. As a consequence, C3 convertase (C4b2a) and C5 convertase (C4b2a3b) are generated.

The classical pathway is activated mainly when specific immunoglobulins (IgM or IgG) bind to antigens. A C1q molecule from the C1 complex ($C1qr_2s_2$) then binds to the immunoglobulin CH_4 (IgM) or CH_3 (IgG) domains located in the Fc region, activating C1r and subsequently C1s. Activated C1s cleaves C4 into C4a and C4b. C4 also contains an intra-chain thioester bond which allows the fragment C4b to attach covalently to the microorganism surface. C2 binds to C4b, and activated C1s enzyme cleaves C2 into two fragments: C2a and C2b. C2a remains attached to C4b. As mentioned above, C4b2a is a C3 convertase which is able to cleave intact C3 molecules.

Furthermore, proteins such as C1q, pentraxins, and C-reactive protein (CRP) can also bind directly to pathogens and initiate the activation of the classical pathway even in the absence of specific antibodies (Albertí et al. 1993). As mentioned above, C1q is also considered a pattern-recognition molecule because it recognizes molecules present in pathogens such as viruses and Gram-negative bacteria (reviewed in Meri 2016).

All C5 convertases of the alternative, lectin, or classical pathways cleave C5 into two fragments: C5a and C5b. C5b binds non-covalently first to the component C6 and then to C7 (DiScipio et al. 1983, Preissner et al. 1985). The complex C5b67 inserts into the cell membrane and recruits C8, followed by several molecules of the component C9 to form the MAC ($C5b6789_n$). MAC generates pores in the cell surface through which ions, small molecules, and water (Podack et al. 1980; Müller-Eberhard 1985) enter culminating in cell lysis. Similar pore formation is observed on cell surfaces targeted by $CD8^+$ cytotoxic T lymphocytes (Tschopp et al. 1986).

2 Regulation of Complement System Activation

The complement system requires tight regulation in order to avoid tissue damage, excessive inflammatory response, and unnecessary consumption of complement proteins (Meri 2016; Medjeral-Thomas and Pickering 2016). Figure 1 illustrates the means by which regulatory proteins inhibit the complement system activation pathways. Several pathogens adopt the strategy of binding or expressing proteins similar to host regulatory proteins to avoid complement activation and in this way may escape from killing.

The alternative pathway must be tightly regulated, since C3b is part of the C3 convertase (C3bBb) that cleaves intact C3 molecules, generating more C3b, in this way creating a positive feedback amplification loop. One of the main regulators of the alternative pathway is Factor H, a 155 kDa plasma protein (443 ± 106 μg/mL; de Paula et al. 2003) composed of 20 complement control protein (CCP) domains. Factor H regulates activation by: (i) binding to C3b and inhibiting the formation of new C3 convertases (Pangburn 1986); (ii) accelerating the decay of C3 convertase (C3bBb) (Pangburn 1986); and (iii) acting as a cofactor for the cleavage of C3b by the enzyme Factor I (Whaley and Ruddy 1976; Weiler et al. 1976; Pangburn et al. 1977). Factor H inhibitory activity is enhanced by the presence of polyanions such as sialic acid, phospholipids, and glycosaminoglycans (such as chondroitin sulfate, dermatan, and heparin sulfate) on host cell surfaces or proteoglycans (reviewed in Meri 2016). In the case of complement resistant pathogens, other ligands that favor Factor H binding have been described, as discussed later. In this way, distinguishing host and non-host in the alternative pathway is an important recognition mechanism in innate immunity. Despite this, several microorganisms exploit host Factor H and are able to cause infection (Fraga et al. 2016). The first four Factor H CCP domains (CCP1-CCP4) located on the N-terminus are considered the most important C3b binding region (Kühn et al. 1995). The remaining 16 CCPs present several binding sites for different targets. CCPs 7, 19, and 20 are important to prevent complement activation on host cells and are considered binding sites for heparin (Pangburn et al. 1991; Blackmore et al. 1996), glycosaminoglycan and extracellular matrix molecules (Hellwage et al. 2002; Jokiranta et al. 2005).

Factor H is the most abundant protein of the Factor H family, which also includes Factor H-like (FHL)-1 protein (generated by an alternative splicing of the *fH* gene), and five homologs called Factor H-related (FHR) proteins: FHR-1, FHR-2, FHR-3, FHR-4, and FHR-5. The physiological role of FHR proteins is not yet completely understood. All FHRs lack CCP domains similar to CCP1-4 present on the Factor H molecule, exactly where most of the C3b binding occurs, which strongly suggests that FHR proteins do not play an important role in the regulation of the alternative pathway. In this way, binding of FHR to the cell surface could compete with Factor H and consequently diminish complement regulation. FHR proteins are present in relatively low concentrations, and their role in pathogen defense has been considered (Skerka et al. 2013; Medjeral-Thomas and Pickering 2016).

The regulation of the alternative pathway is dependent on regulatory proteins that are part of the cell surface. CR1 (CD35) and membrane cofactor protein (MCP, CD46) bind to deposited C3b and replace Factor H in cofactor function in the cleavage of C3b mediated by Factor I (Fig. 1). Decay-accelerating factor (DAF/CD55) is another regulatory protein present on the cell surface, able to bind to C3b and accelerate C3 convertase decay. However, DAF does not exhibit cofactor activity for Factor I.

C1 inhibitor (C1INH) regulates the classical and lectin pathways. This inhibitor is a member of the serine protease inhibitor (serpin) family and interacts reversibly with C1 proteins (C1q, C1r, and C1s) to favor C1 complex assembly, preventing its auto-activation (Ziccardi 1982, 1985). However, after C1 complex activation, C1INH binds irreversibly to the activated serine proteases C1r and C1s, generating the tetrameric C1INH-C1s-C1r-C1INH complex (Salvesen et al. 1985; Sim et al. 1979). A few studies have demonstrated that pathogens can bind to C1INH. Specifically, *Bordetella pertussis* (Marr et al. 2007, 2011), *Borrelia recurrentis* (Grosskinsky et al. 2010), and *Plasmodium falciparum* (Mejia et al. 2016) have been shown to interact with C1INH.

C4b binding protein (C4BP) is a relatively abundant plasma glycoprotein (200–500 μg/mL) (de Paula et al. 2003). The C4BP molecule comprises seven identical C4BP α-chains (75 kDa) and a unique C4BP β (45 kDa) polypeptide chain. The C4BP α-chain contains eight CCPs and C4BP β-chain contains three CCPs. C4BP regulates the activation of the classical and lectin pathways because it is a cofactor for the cleavage of C4b by Factor I, and accelerates the decay of the C3 convertase (C4bC2a) of both pathways (Fujita et al. 1978; Daha and Van Es 1980; Blom et al. 2003). The CCP2 and CCP3 domains in the C4BP α-chain are considered important for the interaction with C4b (Blom et al. 1999, 2001), whereas α-chain CCPs 1-3 are necessary for the binding to heparin (Hessing et al. 1990). C4BP α-chain CCP domains are also involved in interactions with several pathogens, including spirochetes, as will be discussed below.

The MAC formation is regulated mainly by the membrane protein CD59 which is present on the majority of cells, inhibiting host cell lysis (Morgan and Harris 1999). CD59 is considered the most important MAC regulator. It binds to C8 in the C5b-8 complex and also binds to C9. Consequently, it inhibits the complete formation of the pore in the membrane. Other plasma regulatory proteins such as vitronectin and clusterin regulate MAC formation by binding to C7 in C5b-7 forming a non-lytic complex. Both regulatory proteins also bind to C9 molecules, thereby preventing polymerization.

3 Complement Evasion Mechanisms by Spirochetes

Spirochetes are known to cause important human diseases. In order to survive and disseminate in the host, these highly motile bacteria must evade the host's immune responses. Like many other pathogens, spirochetes have evolved sophisticated strategies to avoid killing by the complement system, among which we highlight: (i) the acquisition of negative regulators of complement on their surfaces, (ii) the cleavage of complement components by secreted proteases, (iii) the inhibition of complement pathways by direct interaction of bacterial proteins with complement components, (iv) the mimicking of the host's complement regulatory proteins, and (v) the acquisition of host proteases that cleave complement components (Fig. 2). In the following sections, the specific mechanisms used by *Borrelia*, *Leptospira*, and *Treponema* to overcome complement activation will be discussed in detail.

3.1 Borrelia *spp.*

3.1.1 Lyme Disease Spirochetes

Lyme disease spirochetes are heterogeneous in their capacity to resist complement-mediated bacteriolysis, being classified as serum-resistant, partially serum-resistant or serum-sensitive (Brade et al. 1992; Breitner-Ruddock et al. 1997; van Dam et al. 1997). As a result of complement activation, the MAC is formed and inserted into the bacterial membrane of serum-sensitive *Borrelia*, leading to disruption and death of the target organisms. With the exception of *B. garinii*, the serum susceptibility profile of Lyme disease-causing spirochetes correlates with pathogenesis in humans (Kraiczy 2016).

Diverse molecular mechanisms contribute to serum resistance of *Borrelia* spp. These highly flexible and adaptable spirochetes are equipped with strategies to face host challenges and survive in the complex tick-mammal life cycle (Radolf et al. 2012). Serum-resistant isolates hijack both human and tick complement regulators to protect themselves against clearance by the innate immune system. Control of both the alternative and the classical pathways is achieved by the recruitment of proteins of the Factor H family as well as C4BP (Kraiczy et al. 2001a; Herzberger et al. 2007; Pietikäinen et al. 2010).

Factor H interacts with surface lipoproteins of *Borrelia* called complement regulator-acquiring surface proteins (CRASPs). Five CRASPs have been described: CRASP-1 (also referred to as CspA, BbCRASP-1, BBA68, FHBP), CRASP-2 (CspZ, BbCRASP-2, BBH06), CRASP-3 (ErpP, BbCRASP-3, BBN38), CRASP-4 (ErpC, BbCRASP-4), and CRASP-5 (ErpA, BbCRASP-5, ErpI, ErpN, BBP38, BBL39, OspE) (Kraiczy and Stevenson 2013). While Factor H binds to all five CRASPs, FHL-1 interacts with CRASP-1 and -2, and FHR proteins with CRASP-3, -4, and -5 (Kraiczy and Stevenson 2013; Kraiczy et al. 2001b).

CRASP-1 plays a major role in mediating complement resistance of *B. burgdorferi* s.s. Inactivation of the single copy *CRASP-1* gene of a serum-resistant *Borrelia* isolate rendered the bacterium sensitive to human serum, and complementation of the mutant with *CRASP-1* restored its ability to resist complement-mediated killing (Brooks et al. 2005). In vitro, studies have demonstrated that CRASP-1 controls complement activation at multiple steps. Firstly, by recruiting the soluble regulators Factor H/FHL-1, C3b is inactivated by Factor I on the bacterial surface (Haupt et al. 2007; Kraiczy et al. 2004). Secondly, by interacting with components of the terminal pathway such as C7, C8, and C9, MAC assembly at the borrelial surface is impaired (Hallström et al. 2013). CRASP-1 is

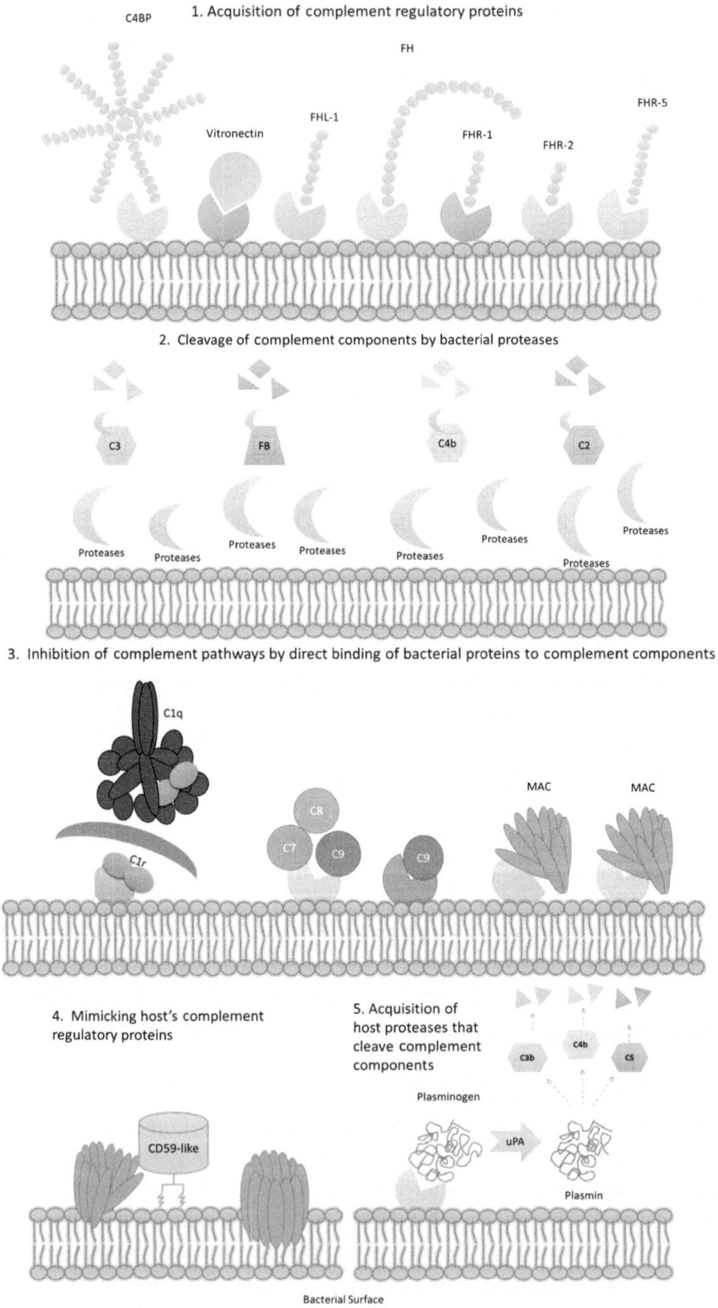
1. Acquisition of complement regulatory proteins
C4BP
FH
Vitronectin
FHL-1
FHR-1
FHR-2
FHR-5
2. Cleavage of complement components by bacterial proteases
C3
FB
C4b
C2
Proteases
Proteases
Proteases
Proteases
Proteases
Proteases
Proteases
Proteases
3. Inhibition of complement pathways by direct binding of bacterial proteins to complement components
C1q
C1r
C8
C7
C9
C9
MAC
MAC
4. Mimicking host's complement regulatory proteins
CD59-like
5. Acquisition of host proteases that cleave complement components
C3b
C4b
C5
Plasminogen
uPA
Plasmin
Bacterial Surface

◂**Fig. 2** Complement evasion mechanisms of spirochetes. *1* Acquisition of complement regulatory proteins. Serum-resistant *Borrelia*, *Leptospira*, and *Treponema* recruit FH, an important soluble regulator of the alternative pathway. Binding of other members of the FH family such as FHL-1 and FHR-1 has also been reported for *Borrelia* and *Leptospira*, and Lyme disease *borreliae* also bind FHR-2 and FHR-5. Inhibition of the classical pathway is achieved through acquisition of C4BP by *Borrelia* and *Leptospira*. Vitronectin binding by pathogenic *Leptospira* allows control of the terminal pathway by reducing MAC deposition. *2* Cleavage of complement components by bacterial proteases. Pathogenic *Leptospira* strains have a protease-dependent mechanism to control complement activation. Secreted proteases cleave proteins of all three pathways including C3, FB, C2, and C4b. *3* Inhibition of complement pathways by direct binding of bacterial proteins to complement components. Interaction of *Borrelia* with C1r leads to downregulation of the classical pathway and binding to the late complement components C7, C8, and C9, or to the lytic MAC allows control of the terminal pathway. *Leptospira* has also been shown to interact with C9. *4* Mimicking host complement regulatory proteins. *Borrelia burgdorferi* expresses a CD59-like protein that inhibits MAC formation. *5* Acquisition of host proteases that cleave complement components. *Borrelia* and *Leptospira* produce plasminogen-binding proteins. In the presence of specific activators, plasmin is generated and cleaves C3b, and/or C4b, and/or C5, contributing to complement evasion.

not produced during established mammalian infection, but the protein is readily detectable during both tick-to-mammal and mammal-to-tick transmission. It is plausible to assume that CRASP-1 protects *Borrelia* from complement attack within the tick and during the very early stages of dissemination within the mammalian host (Bykowski et al. 2007; von Lackum et al. 2005).

Conversely, CRASP-2, which also binds to Factor H/FHL-1, is abundant during established infection, but is only barely detectable during transmission stages (Bykowsky et al. 2007). Despite CRASP-2 being produced at high levels during infection, *Borrelia* CRASP-2 mutants are resistant to complement-mediated killing in vitro and are capable of infecting the murine host and inducing arthritis (Coleman et al. 2008), leading to the hypothesis that this protein may not have an essential role in borrelial persistence in the host.

CRASP-3 (ErpP), -4 (ErpC), and -5 (ErpA) are 17–20 kDa OspE/F-related surface proteins with binding specificities for Factor H and FHR-1, -2, and -5 (Hellwage et al. 2001; Haupt et al. 2007; Alitalo et al. 2002; Hammerschmidt et al. 2012; Hovis et al. 2006a; McDowell et al. 2003; Metts et al. 2003; Siegel et al. 2010). Despite the tenfold higher concentration of Factor H in human plasma, CRASP-3, -4, and -5 bind preferentially to FHR proteins in vitro with greatest affinity for FHR-2 (Haupt et al. 2007; Siegel et al. 2010; Hammerschmidt et al. 2012). Live *borreliae* expressing these CRASPs are unable to bind Factor H, but do interact with FHR proteins (Hammerschmidt et al. 2012). FHR-1 was shown to inhibit C5 convertase activity and MAC formation, whereas FHR-5 inhibits C3 convertase activity and has cofactor activity (Heinen et al. 2009; McRae et al. 2005). Nevertheless, C3 activation products, C6 and MAC, are deposited on *B. garinii* expressing CRASP-4, which indicates that acquisition of FHR-1, FHR-2, and FHR-5 on the borrelial surface is insufficient to confer complement resistance (Hammerschmidt et al. 2012). Due to overlapping binding sites, FHR proteins can compete with Factor H and reduce complement control at the surface of the pathogen, which could explain the susceptible phenotype mentioned above. Erp proteins are produced both during tick feeding and after transmission to the mammalian host (Bykowski et al. 2007), but a definite role for these particular proteins in immune evasion remains to be clarified.

Another known ligand for all five CRASPs is human plasminogen (Hallström et al. 2010). Once bound to these borrelial proteins, plasminogen can be converted to its active form, plasmin. CRASP-1-bound plasmin was shown to degrade C3b (Hammerschmidt et al. 2014), but given the low proteolytic activity observed, it is still premature to assume that cleavage of host complement molecules by host-acquired proteases such as plasminogen is a bona fide *borreliae* complement evasion strategy.

Lyme disease-causing spirochetes also interact with the complement classical and lectin pathways inhibitor C4BP, through a not yet identified 43 kDa outer membrane protein designated P43. When bound to the borrelial surface, C4BP continues to display cofactor activity for Factor I-mediated cleavage of C4b (Pietikäinen et al. 2010). This attribute could contribute to bacterial persistence during chronic infections in the presence of specific antibodies, but additional studies are required to further evaluate the role of C4BP in borrelial complement resistance.

A novel mechanism employed by *B. burgdorferi* s.s. to down-regulate the classical pathway of complement has been recently described. The surface protein BBK32 was shown to block activation of the "bouquet-like" C1 complex composed of C1q, C1r, and C1s (Garcia et al. 2016). BBK32 interacts directly with C1r, and inhibits C1r autocatalysis within the C1 complex. As a consequence, C1s is not activated. Serine proteases of the lectin and alternative pathways are not targeted by BBK32 (Garcia et al. 2016).

Recently, it was demonstrated that *B. bavariensis* strains resist complement-mediated killing, but fail to bind complement regulators such as FH or FHL-1 and apparently do not possess surface or secreted proteases with complement regulatory activity. In these strains, serum resistance seems to be mediated by two surface proteins named BGA66 and BGA71 that belong to the PFam 54 protein family (Hammerschmidt et al. 2016). Except for CRASP-1 (BBA68), none of the other paralogs of this protein family binds Factor H or FHL-1. BGA66 and BGA71 inhibit complement by interfering with the late steps of the cascade. They interact with C6, C7, and C9, and probably affect MAC assembly by hindering proper insertion of the C5b-8 complex into the target cell membrane bilayer. They also inhibit the subsequent binding of C9 to the preformed complex, and C9 polymerization (Hammerschmidt et al. 2016).

Serum-resistant strains of *B. burgdorferi* also produce a not yet identified 80-kDa surface protein with antigenic and functional similarities to human CD59. This CD59-like molecule binds to C9 and weakly to the C8 β subunit. Treatment of serum-resistant strains of *Borrelia* with $F(ab')_2$ anti-CD59 renders these bacteria sensitive to bacteriolysis, pointing to an inhibitory role of this molecule in MAC assembly (Pausa et al. 2003).

Borrelia also benefits from tick salivary proteins with anti-complement activity to successfully infect hosts (de Taeye et al. 2013). Levels of *Ixodes scapularis* salivary protein 15 (Salp15), known to inhibit T cell activation, are specifically increased in *B. burgdorferi*-infected ticks. Salp15 binds to the outer surface protein OspC from *B. burgdorferi*, and contributes to borrelial survival in naïve mice (Ramamoorthi et al. 2005). Salp15 inhibits deposition of C5b-9 complexes on serum-sensitive *B. burgdorferi* "sensu lato" isolates, thus conferring protection against complement-mediated killing (Schuijt et al. 2008). *Ixodes scapularis* salivary protein 20 (Salp20) is another anti-complement tick protein that inhibits the alternative pathway of complement. This protein has been shown to partially protect the serum-sensitive strain *B. garinii* from complement-mediated killing (Tyson et al. 2007). Salp20 binds to and displaces properdin, a positive regulator of the alternative pathway, from the C3 convertase. As a consequence, an accelerated decay of the C3 convertase occurs (Tyson et al. 2008).

Borrelia also co-opts a tick salivary lectin pathway inhibitor (TSLPI). It has been demonstrated that TSLPI does not bind to *Borrelia*, but levels of *TSLPI* expression are increased after tick attachment (Schuijt et al. 2011). Pre-incubation of human serum with recombinant TSLPI inhibits C4 deposition upon complement activation via the lectin pathway. Besides inhibiting MBL, TSLPI also interferes with ficolin 2 activity, since C4 deposition mediated specifically by ficolin 2 is impaired in the presence of this tick salivary protein. Inhibition of *TSLPI* in *Borrelia*-infected nymphs by RNA interference resulted in reduced bacterial loads, pointing to a crucial role of TSLPI for *Borrelia* survival during feeding in the tick midgut (Schuijt et al. 2011). In addition, transmission of *Borrelia* to mice from *TSLPI*-silenced nymphs was compromised, given the lower bacterial burden at distinct sites of infection. Moreover, TSLPI-immunized mice presented lower *Borrelia* loads, further confirming a role for TSLPI in the tick-mammal infectious cycle (Schuijt et al. 2011).

3.1.2 Relapsing Fever Spirochetes

Relapsing fever spirochetes are also recognized as bacterial pathogens of medical relevance. The human-adapted species *B. recurrentis* and *B. duttonii* are responsible for the most serious cases of the disease (Cutler 2015) and have been shown to control complement activation in vitro (Meri et al. 2006). *B. hermsii* is also equipped with immune evasion strategies to persist in human serum and cause disease (Hovis et al. 2004).

Relapsing fever *borreliae* have also been shown to interact with host negative complement regulators leading to downregulation of the complement cascade (Meri et al. 2006). *B. hermsii* binds Factor H and FHL-1 through a protein named Factor H-binding protein A (FhbA) (Hovis et al. 2004, 2006b). Factor H binding mediates C3b cleavage on the borrelial surface (McDowell et al. 2003). However, a *fhbA* deletion mutant was recently generated in *B. hermsii* and retained virulence in a mouse model. Moreover, the strain remained serum-resistant, thus indicating that

Factor H binding is not essential for *B. hermsii* to evade complement attack (Fine et al. 2014). Other yet unknown mechanisms may confer protection against complement activation in these relapsing fever spirochetes.

Additional Factor H-binding proteins have been identified in relapsing fever *borreliae. B. hermsii* produces a protein from the CRASP family designated BhCRASP-1, and *B. parkeri* and *B. recurrentis* produce surface lipoproteins termed BpcA, and HcpA, respectively. All of them bind FHR-1, plasminogen, and Factor H (Rossmann et al. 2007; Schott et al. 2010; Grosskinsky et al. 2009). Ectopic expression of BhCRASP- 1 and BpcA in the serum-sensitive *B. burgdorferi* B313 strain enhanced protection against complement-mediated killing (Rossmann et al. 2007; Schott et al. 2010). *B. recurrentis*, *B. duttonii*, and *B. hermsii* are potentially able to inhibit the classical and lectin pathways as well. The outer membrane lipoprotein CihC from *B. recurrentis* and *B. duttonii* interacts with C4BP and C1INH, and BHA007 from *B. hermsii* binds C4BP (Grosskinsky et al. 2010; Lewis et al. 2014). Serum-sensitive *B. burgdorferi* ectopically expressing CihC becomes more resistant to complement-mediating killing (Grosskinsky et al. 2010).

3.2 Leptospira *spp.*

Leptospira has different strategies to escape the host´s innate immune response. Pioneering studies in the mid-1960s reported a correlation between serum resistance and virulence (Johnson and Muschel 1965; Johnson and Harris 1967). Pathogenic *Leptospira* spp. circumvent complement activation by hijacking complement regulatory proteins through surface-exposed proteins, by targeting key complement molecules with secreted bacterial proteases or acquired host proteases (reviewed in Fraga et al. 2016). Conversely, saprophyte *Leptospira* spp. lack complement evasion strategies and are consequently highly sensitive to serum killing (Meri et al. 2005; Barbosa et al. 2009).

As with *Borrelia*, control of both the alternative and the classical pathways on the leptospiral membrane is achieved by the recruitment of proteins of the Factor H family as well as C4BP (Meri et al. 2005; Castiblanco-Valencia et al. 2012; Barbosa et al. 2009). *Leptospira* has also been shown to bind vitronectin, and acquisition of this terminal pathway regulatory protein may additionally help these spirochetes overcome complement activation (da Silva et al. 2015). A number of leptospiral outer membrane proteins have been reported to bind complement regulatory proteins, but only a few have been characterized at a functional level (reviewed in Fraga et al. 2016).

L. interrogans endostatin-like outer membrane proteins A and B (LenA/LenB) were the first Factor H-binding proteins described in this spirochete. In addition to Factor H, LenA also binds FHR-1 (Verma et al. 2006; Stevenson et al. 2007). A few years later, *Leptospira* immunoglobulin-like proteins A and B (LigA and LigB), some of the best studied surface proteins from pathogenic *Leptospira*, were reported to bind complement regulators. Multiple roles have been attributed to them including adhesion, invasion, and immune evasion properties (Choy et al. 2007; Castiblanco-Valencia et al. 2012). LigA and LigB are large proteins composed of 13 and 12 bacterial immunoglobulin-like (Big) domains, respectively (Matsunaga et al. 2003), and were shown to bind Factor H and C4BP in vitro through distinct sites. Since complement regulatory activities of Factor H and C4BP bound to Lig proteins are maintained, these interactions potentially contribute to the regulation of more than one complement activation pathway simultaneously (Castiblanco-Valencia et al. 2012). Ectopically expressed LigB on the saprophyte *L. biflexa* enhances bacterial survival in normal human serum. This acquired resistance can be attributed to binding of C3b and Factor H by these bacteria (Choy 2012). In addition, reduced MAC deposition was observed on *ligA*- and *ligB*-transformed *L. biflexa* compared to the wild type strain (Castiblanco-Valencia et al. 2016a). Lig proteins also bind other members of the Factor H family such as FHL-1 and FHR-1, but the functional consequences of these interactions have not yet been determined (Castiblanco-Valencia et al. 2012). More recently, both LigA and LigB have been characterized as plasminogen-binding proteins in vitro (Castiblanco-Valencia et al. 2016b). The active serine protease plasmin, generated from its zymogen plasminogen in the presence of specific activators, degrades fibrin meshes, being a key enzyme of the fibrinolytic system. In addition, plasmin targets an array of other substrates, including complement components (Barthel et al. 2012). Besides cleaving fibrinogen, plasmin bound to Lig proteins was shown to cleave C3b and C5. While degradation of host fibrinogen may contribute to leptospiral invasion, cleavage of C3b and C5 may lead to complement downregulation, thus favoring bacterial immune evasion. Besides LigA and LigB, other leptospiral proteins have been described as plasminogen ligands, but only a few were shown to interfere with complement activation in vitro, namely the elongation factor Tu (EF-Tu; Wolff et al. 2013) and *Leptospira* 23 kDa surface adhesion (Lsa23; Siqueira et al. 2016). The relevance of these interactions on the leptospiral surface still needs further investigation.

Leptospiral complement regulator-acquiring protein A (LcpA) is another surface-exposed *Leptospira* protein that simultaneously binds multiple human complement molecules. By interacting with Factor H, C4BP, vitronectin, and the terminal pathway component C9, LcpA may control early and late steps of the complement cascade (Barbosa et al. 2009; da Silva et al. 2015). Factor H and C4BP remain functional once bound to LcpA, and erythrocyte lysis caused by MAC deposition is partially impaired in vitro in the presence of this leptospiral lipoprotein (Barbosa et al. 2010; da Silva et al. 2015).

Based only on binding affinities, a few other leptospiral proteins were shown to interact with complement regulators. These include EF-Tu (Wolff et al. 2013), Lsa30, and Lsa23 (Souza et al. 2012; Siqueira et al. 2016). Determining the biological relevance of the acquisition of negative complement regulators by *Leptospira* proteins remains a challenge. The generation of site-directed mutant strains would be a promising way to address this question, but the simultaneous expression of multiple immune evasion proteins to overcome complement activation may hamper assessing the role of individual proteins involved in serum resistance.

Pathogenic *Leptospira* also has a protease-dependent mechanism to evade complement. All three activation pathways are targeted by leptospiral secreted proteins, which have been shown to cleave C3, C3b, iC3b, FB, C2, and C4b. Both purified and in-serum complement proteins are degraded, suggesting that leptospiral proteases are active in a physiological environment and may contribute to virulence (Fraga et al. 2014). Cleavage assays with protease inhibitors strongly suggest the involvement of metalloproteases in the proteolytic activities observed. Genome analyses revealed the presence of four genes coding for thermolysins, which are zinc metalloproteases from the peptidase M4 family. One of them, encoded by the gene *LIC13322* of *L. interrogans* serovar Copenhageni, was shown to cleave the key complement component C3. It is worth mentioning that leptospiral thermolysin catalytic domains are approximately 30% similar to those present in metalloproteases of other pathogens known to cleave complement proteins (Fraga et al. 2014). A wide variety of bacteria produce proteases that proteolytically inactivate complement components as a mechanism to protect themselves from lytic complement attack, and pathogenic *Leptospira* spp. are no exception to this universal immune evasion strategy.

3.3 Treponema *spp.*

Most of the studies that assessed the interaction between complement and *Treponema* were conducted with *T. denticola*, a periodontal disease-associated spirochete. As periodontitis progresses, the treponemal population expands vertiginously, and may account for 40% of the total bacterial community (Ellen and Galimanas 2005; Dewhirst et al. 2000). *T. denticola* is one of the bacteria that compose the "red complex," and has been associated with disease severity (Socransky et al. 1998). The crevicular fluid of subgingival contains high levels of complement proteins (Schenkein and Genco 1977), and in order to survive in this hostile microenvironment, *T. denticola* as well as other periodontal pathogens have evolved mechanisms to circumvent host immune defenses.

Like *Borrelia* and *Leptospira*, oral *Treponema* also varies in its capacity to resist complement-mediated killing. Isolates of *T. socranskii*, *T. medium*, *T. pectinovorum*, *T. maltophilum*, and a panel of *T. denticola* strains have been shown to be serum resistant, whereas less than 5% of *T. vincentii* and *T. amylovorum* survive in normal human serum (Miller et al. 2014). While acquisition of Factor H is crucial for *T. denticola* complement evasion, the molecular mechanisms underlying serum resistance of all other oral *Treponema* species mentioned above remain unknown. As they are unable to bind complement regulatory proteins such as Factor H, C4BP, and vitronectin, further studies are necessary to elucidate the basis for complement resistance of these treponemal species (Miller et al. 2014).

The interaction of *T. denticola* with Factor H has been well characterized over the last ten years and has some peculiar features. *T. denticola* binds Factor H via the small surface-exposed 11.4-kDa FhbB lipoprotein (McDowell et al. 2005, 2007).

Once bound to *T. denticola*, Factor H is cleaved by the serine protease dentilisin, an important virulence factor that contributes to tissue destruction over the course of periodontal disease, and has been shown to hydrolyze other host proteins including complement C3 (Yamazaki et al. 2006), fibrinogen (Bamford et al. 2007), type IV collagen and fibronectin (Ishihara et al. 1996). Factor H is fully degraded by dentilisin over time, and Factor H cleavage products lose cofactor activity due to complete degradation of CCP1-4 domains (Miller et al. 2016).

Phylogenetic analyses allowed classification of FhbB into three clusters called FhbB1, FhbB2, and FhbB3, which display similar affinities for Factor H, despite sequence variations (Miller et al. 2013, 2014). However, only FhbB1 strains present dentilisin activity, and, as a consequence, have a greater ability to cleave Factor H. FhbB2 and FhbB3 strains, which lack detectable dentilisin activity, are unable, or have an attenuated capacity, to degrade Factor H. It is assumed that FhbB1 isolates producing high levels of dentilisin are likely to impact on disease severity, but additional studies are needed to further validate this issue (Miller et al. 2016).

Cleaving a negative complement regulatory protein may seem rather contradictory in the context of bacterial persistence within the host, and one may wonder how this particular mechanism would benefit a bacterium known to proliferate during disease progression. In periodontally healthy subjects, treponemes account for approximately 1% of the total bacteria (Paster et al. 2001), but the population of anaerobic spirochetes increases dramatically in the subgingival crevice of diseased subjects (Ellen and Galimanas 2005; Dewhirst et al. 2000). It has been hypothesized that evading complement-mediated killing loses relevance as periodontitis progresses. In this scenario, physical protection would be conferred by the developed oral biofilm (Miller et al. 2013). Cleavage of Factor H by dentilisin would cause depletion of this complement regulator in the gingival crevicular fluid leading to deregulation of complement activation locally. As a consequence, decay of the alternative pathway C3 convertase (C3bBb) and Factor I-mediated cleavage and inactivation of C3b would be impaired with undesirable quantities of C3b accumulating on host cells. Complement-mediated destruction of the periodontium may, therefore, propitiate a favorable environment for the proliferation of anaerobic periopathogens.

Factor H has also been shown to bind to and negatively regulate CRP, whose levels are elevated in patients with chronic periodontitis (Gomes-Filho et al. 2011). On the surface of apoptotic cells, CRP binds C1q and activates the classical pathway of complement, resulting in C3b deposition and bacterial opsonization (Volanakis 1982). Further progression of the cascade can be controlled by the interaction of CRP with the soluble complement regulators Factor H or C4BP, that inhibit complement at the level of C3 convertase (Gershov et al. 2000; Sjöberg et al. 2006). The scarcity of Factor H in the periodontal pocket would then prevent proper regulation of complement activation leading to exacerbated inflammation and tissue destruction. Interestingly, FhbB and CRP bind to the Factor H molecule at CCP7, but it remains to be evaluated if they compete for binding to Factor H (Miller et al. 2012).

An overview of *Borrelia*, *Leptospira*, and *Treponema* proteins that interact with the human complement system is presented in Table 1.

Table 1 Overview of *Borrelia*, *Leptospira*, and *Treponema* proteins that interact with human complement components and plasminogen

	Genospecies/species	Host proteins	Complement pathway affected
Borrelia			
Lyme disease *borreliae*			
CRASP-1	*B. burgdorferi* *B. afzelii* *B. spielmanii*	FH/FHL-1 C7/C8/C9/MAC Plasminogen	Alternative/terminal pathways
CRASP-2	*B. burgdorferi* *B. afzelii* *B. spielmanii*	FH/FHL-1 Plasminogen	Alternative pathway
CRASP-3	*B. burgdorferi* *B. afzelii* *B. spielmanii*	FHR-1/FHR-2/FHR-5 Plasminogen	–
CRASP-4	*B. burgdorferi* *B. afzelii*	FHR-1/FHR-2 Plasminogen	–
CRASP-5	*B. burgdorferi* *B. afzelii* *B. garinii* *B. lusitaniae*	FHR-1/FHR-2/FHR-5 Plasminogen	–
P43	*B. burgdorferi* *B. garinii*	C4BP	Classical/Lectin pathways (?)
BBK32	*B. burgdorferi*	C1r	Classical pathway
BGA66	*B. bavariensis*	C7/C8/C9	Terminal pathway
BGA71	*B. bavariensis*	C7/C8/C9	Terminal pathway
CD59-like protein	*B. burgdorferi*	C8/C9	Terminal pathway
Relapsing fever *borreliae*			
FhbA	*B. hermsii*	FH/FHL-1	Alternative pathway(?)
BhCRASP-1	*B. hermsii*	FH/FHR-1 Plasminogen	Alternative pathway
BpcA	*B. parkeri*	FH/FHR-1 Plasminogen	Alternative pathway
HcpA	*B. recurrentis*	FH/FHR-1 Plasminogen	Alternative pathway
CihC	*B. recurrentis* *B. duttonii*	C4PB/C1-Inh	Classical/Lectin pathways
BHA007	*B. hermsii*	C4BP	Classical/Lectin pathways
Leptospira			
LigA/LigB	*L. interrogans*	FH/FHL-1/FHR-1 C4BP Plasminogen	Alternative/Classical pathways
LcpA	*L. interrogans*	FH C4BP Vitronectin C9	Alternative/Classical/Terminal pathways

(continued)

Table 1 (continued)

	Genospecies/species	Host proteins	Complement pathway affected
LenA/LenB	*L. interrogans*	FH/FHR-1	Alternative pathway (?)
EF-Tu	*L. interrogans*	FH	Alternative pathway
Lsa23	*L. interrogans*	FH C4BP	Alternative/Classical pathways (?)
Lsa30	*L. interrogans*	C4BP	Classical pathway (?)
Treponema			
FhbB	*T. denticola*	FH	Alternative pathway

References are mentioned in the text

4 Concluding Remarks

Spirochetes cause important human diseases, and there is a search for long-lasting and efficacious vaccines. The complement system plays an important role in limiting infections caused by *Borrelia*, *Leptospira*, and *Treponema*. In response, pathogenic spirochetes have evolved different strategies to evade complement-mediated killing. Identifying and understanding the virulence factors and bacterial ligands involved in these immune evasion mechanisms will allow us to develop new and better immunotherapies that will bring long-lasting protection in the future.

Acknowledgements Original work from the authors' laboratories was financed by Fundação de Amparo à Pesquisa do Estado de São Paulo (FAPESP) and Conselho Nacional de Pesquisa e Desenvolvimento Científico (CNPq). We thank Sara Isaac and Kamila Oliveira Nunes for helping to prepare the figures. The authors also thank Shaker Chuck Farah and Lazara Elena Santiesteban Lores for their critical review of the text, useful comments, and suggestions.

References

Albertí S, Marqués G, Camprubí S, Merino S, Tomás JM, Vivanco F, Benedí VJ (1993) C1q binding and activation of the complement classical pathway by *Klebsiella pneumoniae* outer membrane proteins. Infect Immun 61:852–860

Alitalo A, Meri T, Lankinen H, Seppälä I, Lahdenne P, Hefty PS, Akins D, Meri S (2002) Complement inhibitor factor H binding to Lyme disease spirochetes is mediated by inducible expression of multiple plasmid-encoded outer surface protein E paralogs. J Immunol 169:3847–3853

Bamford CV, Fenno JC, Jenkinson HF, Dymock D (2007) The chymotrypsin-like protease complex of *Treponema denticola* ATCC 35405 mediates fibrinogen adherence and degradation. Infect Immun 75:4364–4372

Barbosa AS, Abreu PA, Vasconcellos SA, Morais ZM, Gonçales AP, Silva AS, Daha MR, Isaac L (2009) Immune evasion of *Leptospira* species by acquisition of human complement regulator C4BP. Infect Immun 77:1137–1143. doi:10.1128/IAI.01310-08

Barbosa AS, Monaris D, Silva LB, Morais ZM, Vasconcellos SA, Cianciarullo AM, Isaac L, Abreu PA (2010) Functional characterization of LcpA, a surface-exposed protein of *Leptospira* spp. that binds the human complement regulator C4BP. Infect Immun 78:3207–3216. doi:10.1128/IAI.00279-10

Barthel D, Schindler S, Zipfel PF (2012) Plasminogen is a complement inhibitor. J Biol Chem 287:18831–18842. doi:10.1074/jbc.M111.323287

Blackmore TK, Sadlon TA, Ward HM, Lublin DM, Gordon DL (1996) Identification of a heparin binding domain in the seventh short consensus repeat of complement factor H. J Immunol 157:5422–5427

Blom AM, Webb J, Villoutreix B, Dahlbäck B (1999) A cluster of positively charged amino acids in the N-terminal modules of C4BP α-chain is crucial for the C4b binding and factor I cofactor function. J Biol Chem 274:19237–19245

Blom AM, Kask L, Dahlbäck B (2001) Structural requirements for the complement regulatory activities of C4b-binding protein. J Biol Chem 276:27136–27144

Blom AM, Kask L, Dahlbäck B (2003) CCP1-4 of the C4b-binding protein alpha-chain are required for factor I mediated cleavage of C3b. Mol Immunol 39:547–556. doi:10.1016/S0161-5890(02)00213-4

Brade V, Kleber I, Acker G (1992) Differences of two *Borrelia burgdorferi* strains in complement activation and serum resistance. Immunobiology 185:453–465

Breitner-Ruddock S, Würzner R, Schulze J, Brade V (1997) Heterogeneity in the complement-dependent bacteriolysis within the species of *Borrelia burgdorferi*. Med Microbiol Immunol 185:253–260

Brooks CS, Vuppala SR, Jett AM, Alitalo A, Meri S, Akins DR (2005) Complement regulator-acquiring surface protein 1 imparts resistance to human serum in *Borrelia burgdorferi*. J Immunol 175:3299–3308

Bykowski T, Woodman ME, Cooley AE, Brissette CA, Brade V, Wallich R, Kraiczy P, Stevenson B (2007) Coordinated expression of *Borrelia burgdorferi* complement regulator-acquiring surface proteins during the Lyme disease spirochete's mammal-tick infection cycle. Infect Immun 75:4227–4236

Carroll MC, Isenman DE (2012) Regulation of humoral immunity by complement. Immunity 37:199–207. doi:10.1016/j.immuni.2012.08.002

Castiblanco-Valencia MM, Fraga TR, Silva LB, Monaris D, Abreu PA, Strobel S, Józsi M, Isaac L, Barbosa AS (2012) Leptospiral immunoglobulin-like proteins interact with human complement regulators factor H, FHL-1, FHR-1, and C4BP. J Infect Dis 205:995–1004. doi:10.1093/infdis/jir875

Castiblanco-Valencia MM, Fraga TR, Breda LC, Vasconcellos SA, Figueira CP, Picardeau M, Wunder E, Ko AI, Barbosa AS, Isaac L (2016a) Acquisition of negative complement regulators by the saprophyte *Leptospira biflexa* expressing LigA or LigB confers enhanced survival in human serum. Immunol Lett 173:61–68. doi:10.1016/j.imlet.2016.03.005

Castiblanco-Valencia MM, Fraga TR, Pagotto AH, Serrano SM, Abreu PA, Barbosa AS, Isaac L (2016b) Plasmin cleaves fibrinogen and the human complement proteins C3b and C5 in the presence of *Leptospira interrogans* proteins: a new role of LigA and LigB in invasion and complement immune evasion. Immunobiology 221:679–689. doi:10.1016/j.imbio.2016.01.001

Choy HA (2012) Multiple activities of LigB potentiate virulence of *Leptospira interrogans*: inhibition of alternative and classical pathways of complement. PLoS ONE 7:e41566. doi:10.1371/journal.pone.0041566

Choy HA, Kelley MM, Chen TL, Møller AK, Matsunaga J, Haake DA (2007) Physiological osmotic induction of *Leptospira interrogans* adhesion: LigA and LigB bind extracellular matrix proteins and fibrinogen. Infect Immun 75:2441–2450

Coleman AS, Yang X, Kumar M, Zhang X, Promnares K, Shroder D, Kenedy MR, Anderson JF, Akins DR, Pal U (2008) *Borrelia burgdorferi* complement regulator-acquiring surface protein 2 does not contribute to complement resistance or host infectivity. PLoS ONE 3:3010e. doi:10.1371/journal.pone.0003010

Cutler SJ (2015) Relapsing fever borreliae: a global review. Clin Lab Med 35:847–865. doi:10.1016/j.cll.2015.07.001

Daha MR, Van Es LA (1980) Relative resistance of the F-42-stabilized classical pathway C3 convertase to inactivation by C4-binding protein. J Immunol 125:2051–2054

da Silva LB, Miragaia Ldos S, Breda LC, Abe CM, Schmidt MC, Moro AM, Monaris D, Conde JN, Józsi M, Isaac L, Abreu PA, Barbosa AS (2015) Pathogenic *Leptospira* species acquire factor H and vitronectin via the surface protein LcpA. Infect Immun 83:888–897. doi:10.1128/IAI.02844-14

de Paula PF, Barbosa JE, Junior PR, Ferriani VP, Latorre MR, Nudelman V, Isaac L (2003) Ontogeny of complement regulatory proteins—concentrations of factor h, factor I, c4b-binding protein, properdin and vitronectin in healthy children of different ages and in adults. Scand J Immunol 58:572–577

de Taeye SW, Kreuk L, van Dam AP, Hovius JW, Schuijt TJ (2013) Complement evasion by *Borrelia burgdorferi*: it takes three to tango. Trends Parasitol 29:119–128. doi:10.1016/j.pt.2012.12.001

Dewhirst FE, Tamer MA, Ericson RE, Lau CN, Levanos VA, Boches SK, Galvin JL, Paster BJ (2000) The diversity of periodontal spirochetes by 16S rRNA analysis. Oral Microbiol Immunol 15:196–202

DiScipio RG, Smith CA, Müller-Eberhard HJ, Hugli TE (1983) The activation of human Complement component C5 by a fluid phase C5 convertase. J Biol Chem 258:10629–10636

Dobó J, Pál G, Cervenak L, Gál P (2016) The emerging roles of mannose-binding lectin-associated serine proteases (MASPs) in the lectin pathway of complement and beyond. Immunol Rev 274:98–111. doi:10.1111/imr.12460

Ellen RP, Galimanas VB (2005) Spirochetes at the forefront of periodontal infections. Periodontol 2000 38:13–32

Fine LM, Miller DP, Mallory KL, Tegels BK, Earnhart CG, Marconi RT (2014) The *Borrelia hermsii* factor H binding protein FhbA is not required for infectivity in mice or for resistance to human complement in vitro. Infect Immun 82:3324–3332. doi:10.1128/IAI.01892-14

Fraga TR, dos Santos Courrol D, Castiblanco-Valencia MM, Hirata IY, Vasconcellos SA, Juliano L, Barbosa AS, Isaac L (2014) Immune evasion by pathogenic *Leptospira* strains: the secretion of proteases that directly cleave complement proteins. J Infect Dis 209:876–886. doi:10.1093/infdis/jit569

Fraga TR, Isaac L, Barbosa AS (2016) Complement evasion by pathogenic *Leptospira*. Front Immunol 7:623. doi:10.3389/fimmu.2016.00623

Fujita T, Gigli I, Nussenzweig V (1978) Human C4-binding protein. II. Role in proteolysis of C4b by C3b-inactivator. J Exp Med 148:1044–1051

Garcia BL, Zhi H, Wager B, Höök M, Skare JT (2016) *Borrelia burgdorferi* BBK32 inhibits the classical pathway by blocking activation of the C1 complement complex. PLoS Pathog 12: e1005404. doi:10.1371/journal.ppat.1005404

Gershov D, Kim S, Brot N, Elkon KB (2000) C-Reactive protein binds to apoptotic cells, protects the cells from assembly of the terminal complement components, and sustains an antiinflammatory innate immune response: implications for systemic autoimmunity. J Exp Med 192:1353–1364 (Erratum in: J Exp Med 2001 193(12):1439)

Gomes-Filho IS, Freitas Coelho JM, da Cruz SS, Passos JS, Teixeira de Freitas CO, Aragão Farias NS, Amorim da Silva R, Silva Pereira MN, Lima TL, Barreto ML (2011) Chronic periodontitis and C-reactive protein levels. J Periodontol 82:969–978. doi:10.1902/jop.2010.100511

Grosskinsky S, Schott M, Brenner C, Cutler SJ, Kraiczy P, Zipfel PF, Simon MM, Wallich R (2009) *Borrelia recurrentis* employs a novel multifunctional surface protein with anti-complement, anti-opsonic and invasive potential to escape innate immunity. PLoS ONE 4:e4858. doi:10.1371/journal.pone.0004858

Grosskinsky S, Schott M, Brenner C, Cutler SJ, Simon MM, Wallich R (2010) Human complement regulators C4b-binding protein and C1 esterase inhibitor interact with a novel outer surface protein of *Borrelia recurrentis*. PLoS Negl Trop Dis 4:e698. doi:10.1371/journal.pntd.0000698

Hallström T, Haupt K, Kraiczy P, Hortschansky P, Wallich R, Skerka C, Zipfel PF (2010) Complement regulator-acquiring surface protein 1 of *Borrelia burgdorferi* binds to human bone morphogenic protein 2, several extracellular matrix proteins, and plasminogen. J Infect Dis 202:490–498. doi:10.1086/653825

Hallström T, Siegel C, Mörgelin M, Kraiczy P, Skerka C, Zipfel PF (2013) CspA from *Borrelia burgdorferi* inhibits the terminal complement pathway. MBio 4 pii: e00481-13. doi:10.1128/mBio.00481-13

Hammerschmidt C, Hallström T, Skerka C, Wallich R, Stevenson B, Zipfel PF, Kraiczy P (2012) Contribution of the infection-associated complement regulator-acquiring surface protein 4 (ErpC) to complement resistance of *Borrelia burgdorferi*. Clin Dev Immunol 2012:349657. doi:10.1155/2012/349657

Hammerschmidt C, Koenigs A, Siegel C, Hallström T, Skerka C, Wallich R, Zipfel PF, Kraiczy P (2014) Versatile roles of CspA orthologs in complement inactivation of serum-resistant Lyme disease spirochetes. Infect Immun 82:380–392. doi:10.1128/IAI.01094-13

Hammerschmidt C, Klevenhaus Y, Koenigs A, Hallström T, Fingerle V, Skerka C, Pos KM, Zipfel PF, Wallich R, Kraiczy P (2016) BGA66 and BGA71 facilitate complement resistance of *Borrelia bavariensis* by inhibiting assembly of the membrane attack complex. Mol Microbiol 99:407–424. doi:10.1111/mmi.13239

Haupt K, Kraiczy P, Wallich R, Brade V, Skerka C, Zipfel PF (2007) Binding of human factor H-related protein 1 to serum-resistant *Borrelia burgdorferi* is mediated by borrelial complement regulator-acquiring surface proteins. J Infect Dis 196:124–133

Heinen S, Hartmann A, Lauer N, Wiehl U, Dahse HM, Schirmer S, Gropp K, Enghardt T, Wallich R, Hälbich S, Mihlan M, Schlötzer-Schrehardt U, Zipfel PF, Skerka C (2009) Factor H-related protein 1 (CFHR-1) inhibits complement C5 convertase activity and terminal complex formation. Blood 114:2439–2447. doi:10.1182/blood-2009-02-205641

Hellwage J, Meri T, Heikkilä T, Alitalo A, Panelius J, Lahdenne P, Seppälä IJ, Meri S (2001) The complement regulator factor H binds to the surface protein OspE of *Borrelia burgdorferi*. J Biol Chem 276:8427–8435

Hellwage J, Jokiranta TS, Friese MA, Wolk TU, Kampen E, Zipfel PF, Meri S (2002) Complement C3b/C3d and cell surface polyanions are recognized by overlapping binding sites on the most carboxyl-terminal domain of complement factor H. J Immunol 169:6935–6944

Herzberger P, Siegel C, Skerka C, Fingerle V, Schulte-Spechtel U, van Dam A, Wilske B, Brade V, Zipfel PF, Wallich R, Kraiczy P (2007) Human pathogenic *Borrelia spielmanii* sp. nov. resists complement-mediated killing by direct binding of immune regulators factor H and factor H-like protein 1. Infect Immun 75:4817–4825

Hessing M, Vlooswijk RA, Hackeng TM, Kanters D, Bouma BN (1990) The localization of heparin-binding fragments on human C4b-binding protein. J Immunol 144:204–208

Hovis KM, McDowell JV, Griffin L, Marconi RT (2004) Identification and characterization of a linear-plasmid-encoded factor H-binding protein (FhbA) of the relapsing fever spirochete *Borrelia hermsii*. J Bacteriol 186:2612–2618

Hovis KM, Tran E, Sundy CM, Buckles E, McDowell JV, Marconi RT (2006a) Selective binding of *Borrelia burgdorferi* OspE paralogs to factor H and serum proteins from diverse animals: possible expansion of the role of OspE in Lyme disease pathogenesis. Infect Immun 74:1967–1972

Hovis KM, Jones JP, Sadlon T, Raval G, Gordon DL, Marconi RT (2006b) Molecular analyses of the interaction of *Borrelia hermsii* FhbA with the complement regulatory proteins factor H and factor H-like protein 1. Infect Immun 74:2007–2014

Isenman DE (1983) Conformational changes accompanying proteolytic cleavage of human complement protein C3b by the regulatory enzyme factor I and its cofactor H. Spectroscopic and enzymological studies. J Biol Chem 258:4238–4244

Ishihara K, Miura T, Kuramitsu HK, Okuda K (1996) Characterization of the *Treponema denticola* prtP gene encoding a prolyl-phenylalanine-specific protease (dentilisin). Infect Immun 64:5178–5186

Johnson RC, Muschel LH (1965) antileptospiral activity of normal serum. J Bacteriol 89:1625–1626

Johnson RC, Harris VG (1967) Antileptospiral activity of serum. II. Leptospiral virulence factor. J Bacteriol 93:513–519

Jokiranta TS, Cheng ZZ, Seeberger H, Jòzsi M, Heinen S, Noris M, Remuzzi G, Ormsby R, Gordon DL, Meri S, Hellwage J, Zipfel PF (2005) Binding of complement factor H to endothelial cells is mediated by the carboxy-terminal glycosaminoglycan binding site. Am J Pathol 167:1173–1181

Kraiczy P (2016) Hide and seek: how Lyme disease spirochetes overcome complement attack. Front Immunol 7:385

Kraiczy P, Stevenson B (2013) Complement regulator-acquiring surface proteins of *Borrelia burgdorferi*: Structure, function and regulation of gene expression. Ticks Tick Borne Dis 4:26–34. doi:10.1016/j.ttbdis.2012.10.039

Kraiczy P, Skerka C, Kirschfink M, Brade V, Zipfel PF (2001a) Immune evasion of *Borrelia burgdorferi* by acquisition of human complement regulators FHL-1/reconectin and Factor H. Eur J Immunol 31:1674–1684

Kraiczy P, Skerka C, Brade V, Zipfel PF (2001b) Further characterization of complement regulator-acquiring surface proteins of *Borrelia burgdorferi*. Infect Immun 69:7800–7809

Kraiczy P, Hellwage J, Skerka C, Becker H, Kirschfink M, Simon MM, Brade V, Zipfel PF, Wallich R (2004) Complement resistance of *Borrelia burgdorferi* correlates with the expression of BbCRASP-1, a novel linear plasmid-encoded surface protein that interacts with human factor H and FHL-1 and is unrelated to Erp proteins. J Biol Chem 279:2421–2429

Kühn S, Skerka C, Zipfel PF (1995) Mapping the complement regulatory domains in the human factor H-like protein 1 and in factor H1. J Immunol 155:5663–5670

Lachmann PJ (2009) The amplification loop of the complement pathways. Adv Immunol 104:115–149. doi:10.1016/S0065-2776(08)04004-2

Lewis ER, Marcsisin RA, Campeau Miller SA, Hue F, Phillips A, Aucoin DP, Barbour AG (2014) Fibronectin-binding protein of *Borrelia hermsii* expressed in the blood of mice with relapsing fever. Infect Immun 82:2520–2531. doi:10.1128/IAI.01582-14

Macedo ACL, Isaac L (2016) Systemic lupus erythematosus and deficiencies of early components of the complement classical pathway. Front Immunol 7:55. doi:10.3389/fimmu.2016.00055 (eCollection 2016)

Marr N, Luu RA, Fernandez RC (2007) *Bordetella pertussis* binds human C1 esterase inhibitor during the virulent phase, to evade complement-mediated killing. J Infect Dis 195:585–588

Marr N, Shah NR, Lee R, Kim EJ, Fernandez RC (2011) *Bordetella pertussis* autotransporter Vag8 binds human C1 esterase inhibitor and confers serum resistance. PLoS ONE 6:e20585. doi:10.1371/journal.pone.0020585

Matsunaga J, Barocchi MA, Croda J, Young TA, Sanchez Y, Siqueira I, Bolin CA, Reis MG, Riley LW, Haake DA, Ko AI (2003) Pathogenic *Leptospira* species express surface-exposed proteins belonging to the bacterial immunoglobulin superfamily. Mol Microbiol 49:929–945

McDowell JV, Wolfgang J, Tran E, Metts MS, Hamilton D, Marconi RT (2003) Comprehensive analysis of the factor h binding capabilities of *borrelia* species associated with lyme disease: delineation of two distinct classes of factor H binding proteins. Infect Immun 71:3597–3602

McDowell JV, Lankford J, Stamm L, Sadlon T, Gordon DL, Marconi RT (2005) Demonstration of factor H-like protein 1 binding to *Treponema denticola*, a pathogen associated with periodontal disease in humans. Infect Immun 73:7126–7132

McDowell JV, Frederick J, Stamm L, Marconi RT (2007) Identification of the gene encoding the FhbB protein of *Treponema denticola*, a highly unique factor H-like protein 1 binding protein. Infect Immun 75:1050–1054

McRae JL, Duthy TG, Griggs KM, Ormsby RJ, Cowan PJ, Cromer BA, McKinstry WJ, Parker MW, Murphy BF, Gordon DL (2005) Human factor H-related protein 5 has cofactor activity, inhibits C3 convertase activity, binds heparin and C-reactive protein, and associates with lipoprotein. J Immunol 174:6250–6256

Medjeral-Thomas N, Pickering Mathew C (2016) The complement factor H-related proteins. Immunol Reviews 214:191–201. doi:10.1111/imr.12477

Mejia P, Diez-Silva M, Kamena F, Lu F, Fernandes SM, Seeberger PH, Davis AE 3rd, Mitchell JR (2016) Human C1INHibitor suppresses malaria parasite invasion and cytoadhesion via binding to parasite glycosylphosphatidylinositol and host cell receptors. J Infect Dis 213(1):80–89. doi:10.1093/infdis/jiv439

Meri S (2016) Self-nonself discrimination by the complement system. FEBS Lett 590:2418–2434. doi:10.1002/1873-3468.12284

Meri T, Murgia R, Stefanel P, Meri S, Cinco M (2005) Regulation of complement activation at the C3-level by serum resistant leptospires. Microb Pathog 39:139–147

Meri T, Cutler SJ, Blom AM, Meri S, Jokiranta TS (2006) Relapsing fever spirochetes *Borrelia recurrentis* and *B. duttonii* acquire complement regulators C4b-binding protein and factor H. Infect Immun 74:4157–4163

Metts MS, McDowell JV, Theisen M, Hansen PR, Marconi RT (2003) Analysis of the OspE determinants involved in binding of factor H and OspE-targeting antibodies elicited during *Borrelia burgdorferi* infection in mice. Infect Immun 71:3587–3596

Miller DP, Bell JK, McDowell JV, Conrad DH, Burgner JW, Héroux A, Marconi RT (2012) Structure of factor H-binding protein B (FhbB) of the periopathogen, *Treponema denticola*: insights into progression of periodontal disease. J Biol Chem 287:12715–12722. doi:10.1074/jbc.M112.339721

Miller DP, McDowell JV, Rhodes DV, Allard A, Caimano M, Bell JK, Marconi RT (2013) Sequence divergence in the *Treponema denticola* FhbB protein and its impact on factor H binding. Mol Oral Microbiol 28:316–330. doi:10.1111/omi.12027

Miller DP, McDowell JV, Bell JK, Goetting-Minesky MP, Fenno JC, Marconi RT (2014) Analysis of the complement sensitivity of oral treponemes and the potential influence of FH binding, FH cleavage and dentilisin activity on the pathogenesis of periodontal disease. Mol Oral Microbiol 29:194–207. doi:10.1111/omi.12054

Miller DP, Oliver LD Jr, Tegels BK, Reed LA, O'Bier NS, Kurniyati K, Faust LA, Lawson CK, Allard AM, Caimano MJ, Marconi RT (2016) The *Treponema denticola* FhbB protein is a dominant early antigen that elicits FhbB variant-specific antibodies that block factor H binding and cleavage by Dentilisin. Infect Immun 84:2051–2058. doi:10.1128/IAI.01542-15

Morgan BP, Harris CL (1999) Complement regulatory proteins. Harcourt Brace & Company Publishers, Academic Press, London

Müller-Eberhard HJ (1985) The killer molecule of complement. J Invest Dermatol 85:47–52

Pangburn MK (1986) Differences between the binding sites of the complement regulatory proteins DAF, CR1, and factor H on C3 convertases. J Immunol 136:2216–2221

Pangburn MK, Müller-Eberhard HJ (1983) Initiation of the alternative complement pathway due to spontaneous hydrolysis of the thioester of C3. Ann N Y Acad Sci 421:291–298

Pangburn MK, Schreiber RD, Müller-Eberhard HJ (1977) Human complement C3b inactivator: isolation, characterization, and demonstration of an absolute requirement for the serum protein beta1H for cleavage of C3b and C4b in solution. J Exp Med 146:257–270

Pangburn MK, Atkinson MA, Meri S (1991) Localization of the heparin-binding site on complement factor H. J Biol Chem 266:16847–16853

Paster BJ, Boches SK, Galvin JL, Ericson RE, Lau CN, Levanos VA, Sahasrabudhe A, Dewhirst FE (2001) Bacterial diversity in human subgingival plaque. J Bacteriol 183:3770–3783

Pausa M, Pellis V, Cinco M, Giulianini PG, Presani G, Perticarari S, Murgia R, Tedesco F (2003) Serum-resistant strains of *Borrelia burgdorferi* evade complement-mediated killing by expressing a CD59-like complement inhibitory molecule. J Immunol 170:3214–3222

Pietikäinen J, Meri T, Blom AM, Meri S (2010) Binding of the complement inhibitor C4b-binding protein to Lyme disease Borreliae. Mol Immunol 47:1299–1305. doi:10.1016/j.molimm.2009.11.028

Podack ER, Esser AF, Biesecker G, Müller-Eberhard HJ (1980) Membrane attack complex of complement: a structural analysis of its assembly. J Exp Med 151:301–313

Preissner KT, Podack ER, Müller-Eberhard HJ (1985) The membrane attack complex of complement: relation to C7 to the metastable membrane binding site of the intermediate complex C5b-7. J Immunol 135:445–451

Radolf JD, Caimano MJ, Stevenson B, Hu LT (2012) Of ticks, mice and men: understanding the dual-host lifestyle of Lyme disease spirochaetes. Nat Rev Microbiol 10:87–99. doi:10.1038/nrmicro2714

Ramamoorthi N, Narasimhan S, Pal U, Bao F, Yang XF, Fish D, Anguita J, Norgard MV, Kantor FS, Anderson JF, Koski RA, Fikrig E (2005) The Lyme disease agent exploits a tick protein to infect the mammalian host. Nature 436:573–577

Ricklin D, Hajishengallis G, Yang K, Lambris JD (2010) Complement: a key system for immune surveillance and homeostasis. Nat Immunol 9:785–797. doi:10.1038/ni.1923

Rossmann E, Kraiczy P, Herzberger P, Skerka C, Kirschfink M, Simon MM, Zipfel PF, Wallich R (2007) Dual binding specificity of a *Borrelia hermsii*-associated complement regulator-acquiring surface protein for factor H and plasminogen discloses a putative virulence factor of relapsing fever spirochetes. J Immunol 178:7292–7301

Salvesen GS, Catanese JJ, Kress LF, Travis J (1985) Primary structure of the reactive site of human C1-inhibitor. J Biol Chem 260:2432–2436

Schenkein HA, Genco RJ (1977) Gingival fluid and serum in periodontal diseases. I. Quantitative study of immunoglobulins, complement components, and other plasma proteins. J Periodontol 48:772–777

Schott M, Grosskinsky S, Brenner C, Kraiczy P, Wallich R (2010) Molecular characterization of the interaction of *Borrelia parkeri* and *Borrelia turicatae* with human complement regulators. Infect Immun 78:2199–2208. doi:10.1128/IAI.00089-10

Schuijt TJ, Hovius JW, van Burgel ND, Ramamoorthi N, Fikrig E, van Dam AP (2008) The tick salivary protein Salp15 inhibits the killing of serum-sensitive *Borrelia burgdorferi* sensu lato isolates. Infect Immun 76:2888–2894. doi:10.1128/IAI.00232-08

Schuijt TJ, Coumou J, Narasimhan S, Dai J, Deponte K, Wouters D, Brouwer M, Oei A, Roelofs JJ, van Dam AP, van der Poll T, Van't Veer C, Hovius JW, Fikrig E (2011) A tick mannose-binding lectin inhibitor interferes with the vertebrate complement cascade to enhance transmission of the lyme disease agent. Cell Host Microbe 10:136–146. doi:10.1016/j.chom.2011.06.010

Siegel C, Hallström T, Skerka C, Eberhardt H, Uzonyi B, Beckhaus T, Karas M, Wallich R, Stevenson B, Zipfel PF, Kraiczy P (2010) Complement factor H-related proteins CFHR2 and CFHR5 represent novel ligands for the infection-associated CRASP proteins of *Borrelia burgdorferi*. PLoS ONE 5:e13519. doi:10.1371/journal.pone.0013519

Sim RB, Arland GJ, Colomb MG (1979) C1 inhibitor depend dissociation on human complement component C1 bound to the immune complex. Biochem J 179:449–457

Siqueira GH, Atzingen MV, de Souza GO, Vasconcellos SA, Nascimento AL (2016) *Leptospira interrogans* Lsa23 protein recruits plasminogen, factor H and C4BP from normal human serum and mediates C3b and C4b degradation. Microbiology 162:295–308. doi:10.1099/mic.0.000217

Sjöberg AP, Trouw LA, McGrath FD, Hack CE, Blom AM (2006) Regulation of complement activation by C-reactive protein: targeting of the inhibitory activity of C4b-binding protein. J Immunol 176:7612–7620

Skerka C, Chen Q, Fremeaux-Bacchi V, Roumenina LT (2013) Complement factor H related proteins (CFRHs). Mol Immunol 56:170–180. doi:10.1016/j.molimm.2013.06.001

Socransky SS, Haffajee AD, Cugini MA, Smith C, Kent RL Jr (1998) Microbial complexes in subgingival plaque. J Clin Periodontol 25:134–144

Souza NM, Vieira ML, Alves IJ, de Morais ZM, Vasconcellos SA, Nascimento AL (2012) Lsa30, a novel adhesin of *Leptospira interrogans* binds human plasminogen and the complement regulator C4bp. Microb Pathog 53:125–134. doi:10.1016/j.micpath.2012.06.001

Spitzer D, Mitchell LM, Atkinson JP, Hourcade DE (2007) Properdin can initiate complement activation by binding specific target surfaces and providing a platform for de novo convertase assembly. J Immunol 179:2600–2608

Stevenson B, Choy HA, Pinne M, Rotondi ML, Miller MC, Demoll E, Kraiczy P, Cooley AE, Creamer TP, Suchard MA, Brissette CA, Verma A, Haake DA (2007) *Leptospira interrogans* endostatin-like outer membrane proteins bind host fibronectin, laminin and regulators of complement. PLoS ONE 2:e1188

Straight DL, McKee PA (1983) Intrinsic fluorescence analyses of structure-function relationships during alpha 2-macroglobulin-protease reactions. Ann N Y Acad Sci 421:125–133

Tschopp J, Masson D, Stanley KK (1986) Structural/functional similarity between proteins involved in complement-mediated- and cytotoxic T-lymphocyte-mediated cytolysis. Nature 322:831–834

Tyson K, Elkins C, Patterson H, Fikrig E, de Silva A (2007) Biochemical and functional characterization of Salp20, an *Ixodes scapularis* tick salivary protein that inhibits the complement pathway. Insect Mol Biol 16:469–479

Tyson KR, Elkins C, de Silva AM (2008) A novel mechanism of complement inhibition unmasked by a tick salivary protein that binds to properdin. J Immunol 180:3964–3968

van Dam AP, Oei A, Jaspars R, Fijen C, Wilske B, Spanjaard L, Dankert J (1997) Complement-mediated serum sensitivity among spirochetes that cause Lyme disease. Infect Immun 65:1228–1236

Verma A, Hellwage J, Artiushin S, Zipfel PF, Kraiczy P, Timoney JF, Stevenson B (2006) LfhA, a novel factor H-binding protein of Leptospira interrogans. Infect Immun 74:2659–2666

Volanakis JE (1982) Complement activation by C-reactive protein complexes. Ann N Y Acad Sci 389:235–250

von Lackum K, Miller JC, Bykowski T, Riley SP, Woodman ME, Brade V, Kraiczy P, Stevenson B, Wallich R (2005) *Borrelia burgdorferi* regulates expression of complement regulator-acquiring surface protein 1 during the mammal-tick infection cycle. Infect Immun 73:7398–7405

Walport MJ (2001) Complement. First of two parts. N Engl J Med 344(14):1058–1066

Weiler JM, Daha MR, Austen KF, Fearon DT (1976) Control of the amplification convertase of complement by the plasma protein beta1H. Proc Natl Acad Sci U S A 73:3268–3272

Whaley K, Ruddy S (1976) Modulation of C3b hemolytic activity by a plasma protein distinct from C3b inactivator. Science 193:1011–1013

Wolff DG, Castiblanco-Valencia MM, Abe CM, Monaris D, Morais ZM, Souza GO, Vasconcellos SA, Isaac L, Abreu PA, Barbosa AS (2013) Interaction of *Leptospira* elongation factor Tu with plasminogen and complement factor H: a metabolic leptospiral protein with moonlighting activities. PLoS ONE 8:e81818. doi:10.1371/journal.pone.0081818

Yamazaki T, Miyamoto M, Yamada S, Okuda K, Ishihara K (2006) Surface protease of *Treponema denticola* hydrolyzes C3 and influences function of polymorphonuclear leukocytes. Microbes Infect 8:1758–1763

Ziccardi RJ (1982) A new role for C-1-inhibitor in homeostasis: control of activation of the first component of human complement. J Immunol 128:2505–2508

Ziccardi RJ (1985) Demonstration of the interaction of native C1 with monomeric immunoglobulins and C1 inhibitor. J Immunol 134:2559–2563

Spirochetal Lipoproteins in Pathogenesis and Immunity

David A. Haake and Wolfram R. Zückert

Abstract Lipoproteins are lipid-modified proteins that dominate the spirochetal proteome. While found in all bacteria, spirochetal lipoproteins have unique features and play critical roles in spirochete biology. For this reason, considerable effort has been devoted to determining how the lipoproteome is generated. Essential features of the structural elements of lipoproteins are now understood with greater clarity, enabling greater confidence in identification of lipoproteins from genomic sequences. The journey from the ribosome to the outer membrane, and in some cases, to the cellular surface has been defined, including secretion, lipidation, sorting, and export across the outer membrane. Given their abundance and importance, it is not surprising that spirochetes have developed a number of strategies for regulating the spatiotemporal expression of lipoproteins. In some cases, lipoprotein expression is tied to various environmental cues, while in other cases, it is linked to growth rate. This regulation enables spirochetes to express certain lipoproteins at high levels in one phase of the spirochete lifecycle, while dramatically downregulating the same lipoproteins in other phases. The mammalian host has developed specialized mechanisms for recognizing lipoproteins and triggering an immune response. Evasion of that immune response is essential for spirochete persistence. For this reason, spirochetes have developed mechanisms for altering lipoproteins. Lipoproteins recognized by antibodies formed during infection are key serodiagnostic antigens. In addition, lipoprotein vaccines have been developed for generating an immune response to control or prevent a spirochete

---

D. A. Haake (✉)
Division of Infectious Diseases, VA Greater Los Angeles Healthcare System, Los Angeles, CA 90073, USA
e-mail: dhaake@ucla.edu

D. A. Haake
Departments of Medicine, Urology, and Microbiology, Immunology, and Molecular Genetics, The David Geffen School of Medicine at UCLA, Los Angeles, CA 90095, USA

W. R. Zückert (✉)
Department of Microbiology, Molecular Genetics and Immunology, University of Kansas School of Medicine, Kansas City KS 66160, USA
e-mail: wzueckert@kumc.edu

Current Topics in Microbiology and Immunology (2018) 415:239–272
DOI 10.1007/82_2017_78

Published Online: 02 December 2017

infection. This chapter summarizes our current understanding of lipoproteins in interactions of spirochetes with their hosts.

Abbreviation list

ELISA	Enzyme linked immunosorbent assay
BAM	Beta-barrel assembly machinery
PTMs	Post-translational modifications
Osp	Outer surface protein
Lol	Lipoprotein outer membrane localization
LcpA	Leptospiral complement regulator-acquiring protein A
CRASP	Complement regulator-acquiring surface protein
DbpA	Decorin-binding protein A
EMJH	Ellinghausen–McCullough–Johnson–Harris medium
MALDI-TOF	Matrix-assisted laser desorption/ionization time of flight
MudPIT	Multidimensional protein identification technology
TLR	Toll-like receptor
T2SS	Type 2 secretion system

Contents

1 Introduction

Lipoproteins are members of a large family of bacterial membrane proteins that are essential in many bacteria and especially important in spirochetes. Lipoproteins are some of the most abundant spirochetal proteins; for example, the borrelial Osp proteins and leptospiral LipL32 dominate the total protein profile of those spirochetes. In the post-genomic era, we now know that a relatively large percentage of spirochetal genes encode lipoproteins; in spirochetes, the percentage of lipoprotein-encoding genes ranges from 4 to 8% (Setubal et al. 2006), while in *Escherichia coli* that percentage is about 2% (Narita et al. 2004). Bacterial lipoproteins undergo acyl modification during their export and maturation process, resulting in attachment of fatty acids to a cysteine that becomes the amino-terminal residue after removal of the signal peptide. Lipid modification dramatically changes the biophysical characteristics of lipoproteins, rendering them amphipathic, with hydrophobic and hydrophilic domains. Extraction techniques, such as Triton X-114 phase partitioning, take advantage of their unique amphiphilicity to separate lipoproteins from soluble and insoluble components of the bacterial cell (Radolf et al. 1988). The amino-terminal fatty acids serve as anchors to the phospholipid bilayer of spirochetal membranes, decorating the surfaces of the inner and outer membranes.

The approach to studying lipoproteins has changed dramatically in the post-genomic era. Before genome sequences were available, cloning of a lipoprotein-encoding gene involved the labor- and time-intensive process of amino acid sequencing of proteolytically digested proteins (the amino-terminus of lipoproteins is blocked to Edman degradation sequencing) followed by fishing genes out a phage library using degenerate probes based on that amino acid sequence (Shang et al. 1996). Now, an MALDI-TOF analysis leads directly to the gene encoding the lipoprotein of interest. Although the post-genomic era has led to an explosion of knowledge about lipoproteins, we still have much to learn about the spirochetal rules that determine whether a protein becomes a lipoprotein. As in other bacteria, lipid modification requires a "lipobox" in the carboxy-terminal region of the signal peptide. However, comparative analysis of sequences of lipoprotein homologs has revealed that aside from the absolute requirement of a cysteine, spirochete lipobox rules appear to be considerably relaxed compared to other bacteria (Fouts et al. 2016). This means that there remains uncertainty about precisely how to predict what open reading frames do or do not encode lipoproteins.

We are beginning to understand how the cellular destination of spirochetal lipoproteins is determined. The focus on defining, understanding, and disrupting the host–pathogen interface has generally yielded more information on surface-localized lipoproteins than on those found in the periplasmic space, but this bias is slowly dissipating as systematic studies of spirochetal (lipo) proteomes are being completed. Amino-terminal residues following the modified cysteine, generally part of an intrinsically disordered "tether" domain, appear crucial for sorting. No sorting motif such as the "+2 rule" in gram-negative bacteria has emerged, but tethers of surface lipoproteins appear to protect from premature folding in the periplasm. The biological roles of spirochetal homologs of lipoprotein secretion pathway components identified in other systems, such as T2SS and Lol, are pending detailed investigation.

For spirochetes, expression of lipoproteins is a double-edged sword; lipoproteins serve a variety of functions from adhesion to defense against host innate defense mechanisms. At the same time, lipoproteins signal the host innate immune system of their presence through TLR1/2 signaling, given that acyl modification of amino-terminal cysteines is unique to bacterial lipoproteins.

This review covers spirochetal lipoprotein lipidation and export, sorting and secretion, regulation of expression, the host immune response and pathogen escape mechanisms, serodiagnosis, and vaccines. Other chapters in this volume cover some of these lipoprotein topics in greater detail, including the *T. pallidum* outer membrane (Radolf and Kumar), *Borrelia burgdorferi* expression regulation (Stevenson and Seshu), and the interaction of *Leptospira* with the innate immune system (Werts).

2 Generating the Lipoproteome: Export and Lipidation

2.1 Export and Lipidation Pathways

Lipoproteins acquire covalent modification by lipids during export out of the cytoplasm. For lipid modification to occur, lipoproteins are synthesized with a signal peptide near their amino-terminus. Lipoprotein-specific signal peptides are defined by an alpha-helical hydrophobic domain flanked by a short positively charged region on its amino-terminal side and a short "lipobox" ending in cysteine on its carboxy-terminal side. The signal peptide enables the nascent prolipoprotein to interact on the cytoplasmic side of the inner membrane with the SecY secretory complex. Upon translocation across the inner membrane, prolipoproteins are converted to lipoproteins by a series of three enzymes. The first enzyme, prolipoprotein diacylglyceryl transferase (Lgt), recognizes the lipobox to attach two lipid molecules to the cysteine residue's thiol residue. The lipobox is then recognized by a second enzyme, a lipoprotein-specific signal peptidase (Lsp; a.k.a. signal II peptidase) that removes the signal peptide by cleaving the peptide bond on the amino side of the cysteine. The third enzyme, lipoprotein N-acyl transferase (Lnt), aminolinks a third lipid molecule to the cysteine.

In spirochetes, the sequence of events leading to lipidation of lipoproteins likely mirrors that in gram-negative bacteria such as *E. coli*. However, there is growing evidence that there are important differences in the process between *E. coli* and spirochetes. The first line of evidence comes from comparison of lipoprotein lipobox sequences. In *E. coli* lipoproteins, the −1 position relative to cysteine is highly conserved, with only the small nonpolar amino acids Ala and Gly occupying this position in virtually all cases. In contrast, comparison of experimentally verified lipoprotein sequences across spirochete species demonstrated that somewhat larger amino acids, such as Asn and Ser, regularly occur at the −1 position (Setubal et al. 2006). Surprisingly, when sequences of experimentally verified lipoproteins such as LipL21 and Loa22 were compared across pathogenic, intermediate, and saprophytic *Leptospira* spp., it was discovered that bulky aromatic amino acid residues can also occur in the −1 position (Fouts et al. 2016). In other cases, even charged residues were found to occur at this position. To accommodate this unprecedented level of lipobox sequence plasticity, the spirochetal enzymes that recognize the lipobox appear to be significantly different from their homologs in gram-negative bacteria. In the case of prolipoprotein diacylglyceryl transferase, leptospiral Lgt lacks the signature sequence shared by most bacterial Lgt sequences. Likewise, leptospiral Lsp has an insert in the second periplasmic loop, suggesting changes in its active site to accommodate the variability in leptospiral lipobox sequences. Additional work is needed to define more thoroughly the level of variability of lipobox sequences and lipidation enzymes in other spirochetes.

2.2 *Bioinformatics: Lipoprotein Prediction*

Compared to most other bacteria, lipoprotein-encoding genes occur with relatively high frequency in spirochetal genomes. This, together with the relative ease of genome sequencing compared to the effort involved in experimental validation of lipidation (see Sect. 2.4), calls for accurate algorithms for spirochetal lipoprotein prediction. Algorithms that are designed for non-spirochetes are likely to be inaccurate. For example, the Psort lipoprotein prediction tool, which is based on relatively conserved lipobox sequences of gram-negative bacteria, has a low ($\leq$33%) sensitivity for spirochetal lipoproteins. The SpLip algorithm was created to provide a more accurate tool for lipoprotein prediction in spirochetes (Setubal et al. 2006). SpLip was based on sequences from 28 experimentally verified spirochetal lipoproteins and significantly relaxed the lipobox requirements for the −1 position relative to cysteine. Nevertheless, sequences with bulky aromatic or charged amino acids at the −1 position were specifically excluded. In light of the new information from the leptospiral genome sequencing project, it now appears that even the SpLip algorithm is too conservative (Fouts et al. 2016). As a result, the SpLip survey of spirochetal genomes probably misses a number of genuine lipoproteins. Signal peptides followed by a cysteine that are SpLip negative can be considered to currently be in "lipoprotein limbo." One approach is to apply the LipoP algorithm, which utilizes hidden Markov model statistical methods and recognizes some lipoprotein sequences that SpLip misses. While bioinformatic tools will never replace experimental validation of lipidation (see Sect. 2.3), improved bioinformatic approaches to identification of candidate spirochetal lipoproteins are both needed and feasible given the explosion of sequence data available to guide their development.

2.3 Analysis of Lipidation

Partitioning of a predicted lipoprotein with a nonionic detergent phase can be considered an indirect indicator for association with the membrane and thus lipidation, but two standard approaches have been used to more specifically confirm protein lipidation. The first approach uses globomycin, first described in 1978 as a peptide antibiotic leading to bacterial spheroplast formation (Inukai et al. 1978) and recently shown to sterically block the active site of the lipoprotein-specific signal peptidase II LspA as a noncleavable peptide substrate (Vogeley et al. 2016). This leads to a lipoprotein-specific processing defect resulting in higher molecular mass lipoprotein species, as first observed for *B. burgdorferi* OspC (Carter et al. 1994). A second approach takes advantage of palmitate being the predominant fatty acid covalently bound to *B. burgdorferi* and *T. pallidum* (Belisle et al. 1994) by metabolically labeling lipoproteins with radioactive tritiated palmitate and detecting them by autoradiography (Brandt et al. 1990); lipidation of specific lipoproteins can be assessed via pulldowns using epitope tags or immunoprecipitation.

2.4 Additional Post-translational Modifications

Mass spectrometry analysis has found a surprisingly high level of post-translational modifications (PTMs) of leptospiral proteins, including phosphorylation, acetylation, and methylation (Cao et al. 2010). Methylation of both Glx (Glu/Gln) and basic (Arg/Lys) amino acids was documented. A total of 155 methylated proteins were identified, including LipL32 and several other known surface lipoproteins. In the case of LipL32, acetylation or tri-methylation of lysine residues within multiple peptides was observed, including peptides corresponding to regions of LipL32 previously identified as epitopes. As discussed in Sect. 5.2.3, LipL32 PTMs may represent a novel immune evasion mechanism.

3 Operationalizing the Lipoproteome: Sorting and Secretion

3.1 Periplasmic and Surface "Hangouts"

Once spirochetal lipoproteins are processed and fully acylated, they are sent to one or more of three different destinations to assume their membrane-peripheral biological functions: (i) they can remain in the inner membrane, (ii) be transported across the periplasm to the inner leaflet of the outer membrane, or (iii) be secreted through the outer membrane to the bacterial surface.

Experimental localization of lipoproteins has been based on a combination of cellular fractionation and assessments of surface exposure. Often genus-specific, but sometimes generalizable, methods for fractionating spirochetal membranes have been developed based on differential membrane fragility, density or detergent solubility. Incubation of *Borrelia* cells in hypotonic citrate or sucrose buffers leads to the release of outer membrane vesicles that can then be separated from inner membrane-containing protoplasmic cylinders and intact cells by ultracentrifugation in a sucrose gradient (Radolf et al. 1995; Skare et al. 1995). Alternatively, the nonionic detergents Triton X-100 and Triton X-114 preferentially, if not selectively, solubilize the outer membrane of some spirochetes (see below). Assessment of surface exposure can be performed by a variety of methods that may vary in their sensitivity: surface immunoprecipitation, surface biotinylation, surface proteolysis, surface immunofluorescence, or immunoelectron microscopy. The former three methods are better for lipoproteins expressed in low abundance. Specificity of each method needs to be confirmed by the use of proper controls or secondary approaches.

The above approaches have been applied to localize borrelial and leptospiral lipoproteins to either the outer or periplasmic leaflets of the outer membrane or the periplasmic leaflet of the inner membrane (Fig. 1a). Building on a set of independent studies of 49 *B. burgdorferi* lipoproteins, a recent comprehensive study localized the *B. burgdorferi* lipoproteome using an epitope-tagged expression library of SpLip-predicted lipoproteins and quantitative mass spectrometry. Eighty-six of the 125 analyzed lipoproteins localized to the surface. Of the remaining 39 periplasmic lipoproteins, 31 were found in the inner membrane (Dowdell et al. 2017). The significant contribution of lipoproteins to the vector- and host-pathogen interface of *B. burgdorferi* may reflect the unique LPS-deficient envelope architecture of *Borrelia* and the bacterium's exposure to a variety of environments during its enzootic cycle.

Several candidate leptospiral surface lipoproteins have been proposed. Initial studies to identify surface-exposed leptospiral proteins by surface immunoprecipitation identified three proteins, that were later identified as the porin OmpL1 and lipoproteins LipL41 and LipL46 (Haake et al. 1991). LipL41, LipL46, and several other lipoproteins were subsequently localized to the outer membrane by differential extraction with Triton X-114 (Shang et al. 1996). In contrast, as shown in Fig. 1a, the lipoprotein LipL31 was found by Triton insolubility and membrane fractionation studies to be restricted to the inner membrane (Haake and Matsunaga 2002). LipL41 was confirmed to be surface-exposed by immunoprecipitation with specific antibody added to intact bacteria (Shang et al. 1996). These experiments included as a control, LipL36 (Haake et al. 1998), an inner leaflet outer membrane lipoprotein that is accessible to antibody binding only after outer membrane disruption with Triton X-100 (Shang et al. 1996). The most prominent lipoprotein in the leptospiral outer membrane is LipL32 (Haake et al. 2000). LipL32 was initially thought to be surface-exposed based on immunoelectron microscopy and surface biotinylation studies (Cullen et al. 2005). However, subsequent surface proteolysis and immunofluorescence experiments showed that LipL32 is largely, if not entirely,

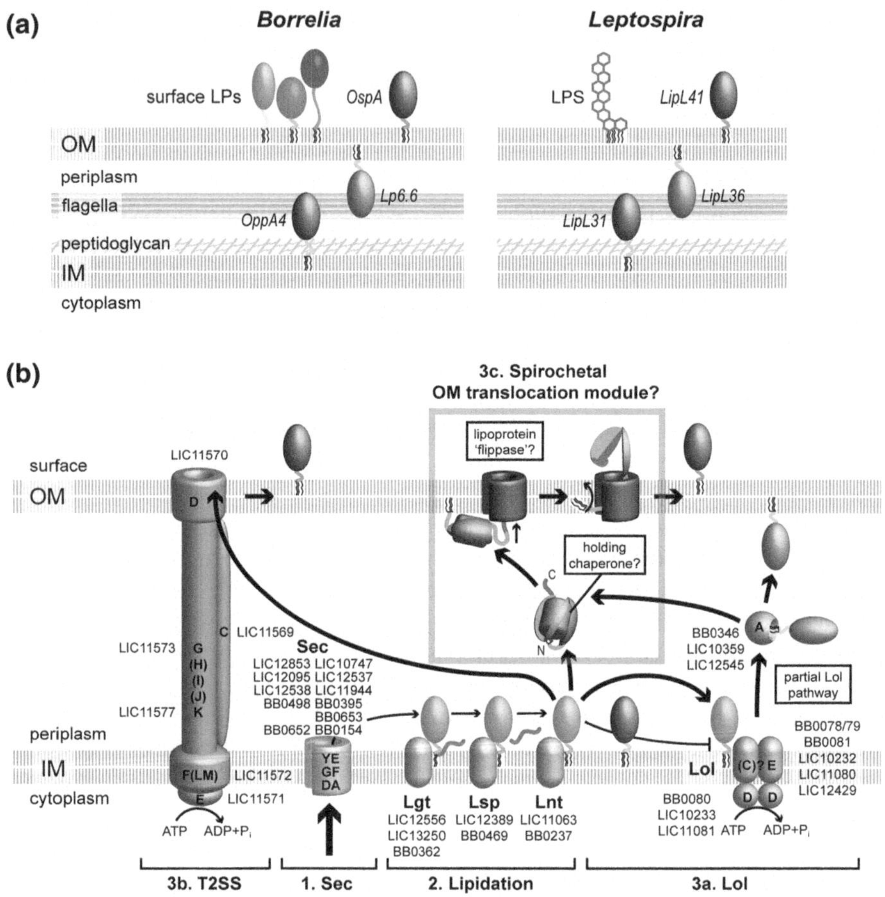

Fig. 1 Envelope structure and lipoprotein secretion pathway models for *Borrelia* and *Leptospira*. **a**. *Envelope structure of* Borrelia *and Leptospira.* Both spirochetal genera are diderm organisms, with an inner (cytoplasmic) membrane, an outer membrane, and a periplasmic space harboring a thin peptidoglycan cell wall and multiple flagella required for motility. *Borrelia* and *Leptospira* lipoproteins have been found in all subcellular compartments, i.e., in the inner membrane, in the periplasmic leaflet of the outer membrane and on the surface; model lipoproteins (e.g., OspA and LipL31) are indicated in narrow italic print. One distinguishing envelope feature is that *Borrelia* exclusively express abundant and serotype-defining surface lipoproteins, while *Leptospira*, similar to gram-negative bacteria, also display serotype-defining lipopolysaccharide (LPS) molecules. **b**. *Model of potential lipoprotein secretion pathways in* Borrelia and *Leptospira*. Proposed lipoprotein secretion pathways are indicated by arrows. After Sec-mediated translocation through the IM (1), proteins lipidated on the periplasmic face of the IM (2) can be targeted to either the Lol pathway (3**a**), a T2SS (3**b**), or a spirochetal OM translocation module that has been proposed based on studies in Borrelia spirochetes (3**c**, see text). *B. burgdorferi* type strain B31 and *Leptospira interrogans* serovar Copenhageni homologs of known components are indicated by their ORF number (e.g., BB01154 and LIC11944 for SecA). Apparently, missing components of protein complexes (e.g., the T2SS GspH homolog) are listed in parentheses. Model substrate proteins are indicated in narrow italic print

a subsurface protein (Pinne and Haake 2013). The experience with LipL32 again underscores the importance of using multiple membrane localization methods and tests for surface exposure to determine the subcellular location of a lipoprotein.

3.2 *Functional Role of N-Terminal "Tether" Peptides*

A review of current PDB database entries shows that spirochetal lipoproteins can form a wide variety of functional folds. Surface lipoproteins dominate among the over 20 available *Borrelia* lipoprotein structures. Demonstrating the breadth of lipoprotein structures, the extensively studied *B. burgdorferi* surface lipoprotein OspA exhibits a unique open beta-sheet (Li and Lawson 1995), whereas the genus-wide *Borrelia* OspC/Vsp1 surface lipoprotein family folds into dimers of tight alpha-helical bundles that together display a membrane-distal dome region with twofold symmetry (Kumaran et al. 2001; Zückert et al. 2001). LipL32, the only leptospiral lipoprotein that has been structurally defined, assumes a compact "jelly roll" fold (Vivian et al. 2009). One striking common feature of lipoproteins, however, is that their mature N-termini are intrinsically disordered, providing a flexible linker or "tether" between the N-terminal fatty acid anchor and the functional fold. Observed/predicted tether lengths for *Borrelia* and *Leptospira* lipoproteins are generally around 10–15 amino acids, with some dramatic outliers such as *B. burgdorferi* BBA66 showing an N-terminal disordered domain of 170 amino acids (Brangulis et al. 2014). In all likelihood, variations in tether lengths ultimately help position lipoproteins for optimal function within the multilayered rainforest of the surface lipoproteome (Bunikis and Barbour 1999). For comparison, a standard tether peptide in fully extended conformation would be dimensionally similar to the cross section of a lipid bilayer (5–6 nm), whereas the BBA66 tether would allow for the projection of the functional fold of the protein up to 60 nm away from the bacterial OM. However, there is already experimental evidence that *B. burgdorferi* surface lipoprotein tether peptides are necessary and sufficient for proper secretion. Rather than containing discernible primary amino acid sorting motifs such as the "+2/+3/+4" sorting motif for periplasmic lipoprotein sorting in gram-negative bacteria, surface lipoprotein tethers help to maintain the proteins in an at least partially unfolded conformation that remains compatible with translocation through the outer membrane (Chen et al. 2011; Chen and Zückert 2011; Dowdell et al. 2017; Kumru et al. 2010, 2011; Schulze and Zückert 2006; Schulze et al. 2010).

3.3 *Lipoprotein Secretion Pathway*

Figure 1b shows a working model of lipoprotein secretion in the two model spirochetes *L. interrogans* and *B. burgdorferi* that is informed by (i) the presence or absence of homologs of known lipoprotein pathway components identified in other diderm bacteria, and (ii) by experimental data obtained from the abovementioned *B. burgdorferi* lipoprotein secretion studies. As expected, both spirochetal genomes contain homologs for the general secretory pathway's Sec complex and the lipidation machinery. Homologs of a partial Lol pathway that is responsible for sorting periplasmic lipoproteins in gram-negative bacteria [reviewed in Okuda and Tokuda (2011) and Zückert (2014)] are present as well, with apparent gene duplications in *Leptospira*. LolC, LolD, and LolE form an ABC transporter-like complex in the inner membrane that uses ATP hydrolysis to release OM lipoproteins to the periplasmic chaperone LolA. 1:1 stoichiometric lipoprotein-LolA complexes then cross the aqueous periplasm and release their lipoprotein cargo to the OM. In a subset of diderm bacteria, the beta- and gamma-proteobacteria, the process of OM lipoprotein insertion is facilitated by the periplasmic outer membrane lipoprotein LolB, but LolB homologs are missing in spirochetes. At the same time, the *Leptospira*, but not the *Borrelia* genome encodes a Type II secretion system that in theory could mediate secretion of lipoproteins to the bacterial surface as in *Klebsiella* (Pugsley 1993). The current data on *B. burgdorferi* surface lipoprotein secretion are compatible with the interaction of surface lipoprotein tethers with a periplasmic holding chaperone that then hands off its cargo to an outer membrane lipoprotein flippase complex. Studies with conditional dominant negative or knockout mutants have indicated that *B. burgdorferi* LolD and LolA homologs are essential for bacterial growth (K. M. Bridges, S. Chen, J. L. Kueker, J. Liu, J. S. Blevins, and W. R. Zückert, unpublished), but their precise roles in lipoprotein secretion are yet to be determined. The *B. burgdorferi* beta barrel assembly machinery (BAM) protein BamA was shown to be at least indirectly involved in this process (Lenhart and Akins 2010).

4 Spatiotemporal Expression

4.1 Expression Patterns in Different Biological Systems

Spirochetes exhibit a variety of different environmental "lifestyles" and vary widely in terms of spatiotemporal expression of lipoproteins. Some spirochetes, such as *Treponema pallidum*, are obligate parasites of the mammalian host, have a small 1.1 Mb genome and relatively few gene regulation mechanisms. *Borrelia* species, alternate between tick or louse vectors and mammalian hosts, have more gene regulation mechanisms, and a genome size of 1.3–1.5 Mb in the case of *B. burgdorferi*. In contrast, pathogenic *Leptospira* species alternate between the commensal, free-living state and parasitic invasion of the mammalian host. The ability to make this transition is reflected in the large 4.6 Mb leptospiral genome and large numbers of signal transduction proteins (Galperin 2005).

4.2 *Environmental Cues*

4.2.1 *Borrelia* in Complex Vector and Host Environments

During transmission from tick to mammal, *B. burgdorferi* undergoes a transformation in terms of which lipoproteins are expressed. In the unfed, resting tick, the major lipoproteins expressed by *B. burgdorferi* are outer surface protein A (OspA) and OspB. OspA promotes survival in the tick by mediating adherence to tick midgut epithelium by binding the TROSPA receptor (Pal et al. 2004) and shielding borrelial surface proteins (Bunikis and Barbour 1999). When the tick attaches to mammalian skin and takes a blood meal, *borreliae* in the tick midgut begin to grow and divide rapidly. As they do, they stop expressing OspA/B and begin to express OspC. OspC-expressing *B. burgdorferi* migrates to the salivary gland in preparation for infection of the mammalian host. OspC expression is essential for establishment of infection in the mammalian host (Grimm et al. 2004; Stewart et al. 2006; Tilly et al. 2006), and the so far identified underlying mechanisms turn out to be immunomodulatory in nature: OspC was recently shown to bind complement factor C4b to interfere with the innate activation of the classical and lectin complement cascade (Caine et al. 2017), which helps explain prior observations that OspC blocks phagocytosis by macrophages (Carrasco et al. 2015). In addition, OspC was shown earlier to bind the tick salivary gland protein Salp15 (Ramamoorthi et al. 2005), thereby inhibiting T cell activation (Juncadella and Anguita 2009).

This transformation from OspA/B expression in the unfed tick to OspC expression in the fed tick can be reproduced in the laboratory by transferring *B. burgdorferi* cultivated at 23 °C into fresh medium at 34 °C (Schwan et al. 1995). Initially, it was thought that these changes in lipoprotein expression were temperature regulated. However, a subsequent study found that similar effects on lipoprotein expression could be obtained simply by increasing the amount of serum in the medium or by increasing the concentration of growth medium while holding the temperature of the culture at 34 °C (Jutras et al. 2013). The type of lipoproteins expressed correlated not with temperature, but instead with the bacterial growth rate. While not an environmental cue, per se, the growth rate is a response to the change in temperature and availability of nutrients that takes place when the tick takes a blood meal.

In addition to OspA/B and OspC, transcriptional analyses have shown that a large number of genes, many of them encoding lipoproteins, are either turned off or turned on during the transition from the unfed to fed tick phase (Iyer et al. 2015). Some of the regulatory mechanisms responsible for this transformation are now understood. Expression of OspC and a number of other lipoproteins directly or indirectly involves the alternative sigma factor RpoS; induction of transcription after temperature upshift requires OspC (Caimano et al. 2007). Likewise, reduced transcription of OspA/B and a different set of lipoproteins after temperature upshift is also OspC dependent. Given the importance of this transition for the

B. burgdorferi life cycle, it is not surprising that a complex regulatory cascade controls RpoS expression (Samuels 2011).

For more detailed and expansive information on regulatory mechanisms in *Borrelia*, see the chapter by Stevenson and Seshu in this volume.

4.2.2 *Leptospira* Response to Changes in Osmolarity

Osmolarity is a key environmental signal that controls the expression of leptospiral lipoproteins. In the process of examining the interaction of leptospires with cells in tissue culture, it was observed that the addition of tissue culture medium to leptospiral culture medium induced expression of leptospiral immunoglobulin-like (Lig) domain lipoproteins LigA and LigB (Matsunaga et al. 2005). The sodium chloride component of tissue culture medium was primarily responsible for these effects. Addition of sodium chloride, potassium chloride, or sodium sulfate to low-osmolarity leptospiral culture medium to the level of osmolarity found in the mammalian host ($\sim$300 mOsm/L) induced expression of both cell-associated LigA and LigB, and release of LigA into the culture supernatant. These results indicate that Lig protein expression is dramatically upregulated when the osmolarity of the leptospiral environment is shifted to that found in mammalian host tissues. Osmotic induction of Lig expression also resulted in enhanced release of LigA and increased surface exposure of LigB, as determined by surface immunofluorescence.

Osmolarity had additional effects on leptospiral lipoproteins and outer membrane components. First, there was complete loss of LipL36 expression, a result confirmed by immunoblot with LipL36 antiserum. LipL36 is an outer membrane lipoprotein that is expressed at high levels in cultivated leptospires, but is not expressed during infection of the mammalian host (Barnett et al. 1999; Haake et al. 1998). Second, there was a significant reduction in LPS content. LPS has been shown to be reduced in leptospires isolated from the liver of infected guinea pigs (Nally et al. 2005). In addition, RT-PCR and immunoblotting show osmo-induction of the secreted sphingomyelinase, Sph2, which is also known to be expressed during infection (Artiushin et al. 2004). These observations suggest that the effects induced by osmolarity parallel the changes in the leptospiral outer membrane that occur within the mammalian host.

The availability of whole genome sequences has made it possible to construct transcriptional microarrays composed of all known open reading frames. A transcriptional microarray study performed with RNA isolated from *L. interrogans* serovar Copenhageni strain L1-130 compared overnight incubation with and without added salt (Matsunaga et al. 2007). The results demonstrated that expression of a select group of leptospiral transcripts is reproducibly upregulated in response to osmolarity. Table 1 lists some of the 28 genes encoding lipoproteins with a >2.5-fold increase in transcript level (**numbers in bold**). All four lipoprotein genes in the *ligB* operon (LIC10461, LIC10462, LIC10463, and LIC10464, see Table 1) are among the top 8 upregulated genes. These results are further confirmed by prior RT-PCR and/or immunoblot studies showing that expression of LIC10464

Table 1 Comparison of transcriptional effects on *L. interrogans* genes

LIC#[1]	LA#[2]	Temp[3]	Osmo[4]	Iron[5]	DMC[6]	Notes
LIC10054	LA0061	0.90	1.10	0.91	**3.84**	Bacterial septal ring protein, RlpA
LIC10371	LA0423	1.11	**2.03**	0.91	**4.61**	Beta-propeller repeat lipoprotein
LIC10373	LA0426	1.08	1.57	0.98	**5.37**	Beta-propeller repeat lipoprotein
LIC10461	LA3781	**1.88**	**7.03**	0.91	**3.54**	Lig operon lipoprotein
LIC10462	LA3780	**1.63**	**4.56**	1.11	**5.08**	Lig operon lipoprotein
LIC10463	LA3779	1.10	**4.95**	0.94	**3.31**	Lig operon lipoprotein
LIC10464/5	LA3778	**1.63**	**4.94**	1.07	**4.57**	LigA/B
LIC10713	LA3469	0.99	1.21	**4.6**	1.68	Iron-regulated lipoprotein
LIC10968	LA3145	**1.44**	**5.03**	1.03	**3.44**	Hypothetical lipoprotein
LIC11030	LA3064	1.09	1.18	1.04	**11.18**	Hypothetical lipoprotein
LIC11058	LA3017	1.13	1.09	1.28	**6.27**	LemA
LIC11082	LA2980	0.93	1.03	1.14	**3.15**	Stage II sporulation protein
LIC11167	LA2877	**0.66**	1.18	0.77	**2.24**	Hypothetical lipoprotein
LIC11695	LA2241	0.85	**3.86**	**4.6**	**4.06**	TonB-dependent operon lipoprotein
LIC11886	LA2023	**0.56**	**0.36**	0.81	**0.66**	Hypothetical lipoprotein
LIC11904	LA2003	1.17	**0.32**	1.10	1.01	Hypothetical lipoprotein
LIC11966	LA1939	1.03	**0.40**	1.39	**2.42**	Hypothetical lipoprotein
LIC12049/50	LA1759	0.95	0.83	**0.13**	0.85	Hypothetical lipoprotein
LIC12099	LIC1691	1.20	**4.15**	0.94	**19.5**	LipL53
LIC12208	LA1569	**1.95**	1.45	1.03	**3.64**	Beta-propeller repeat lipoprotein
LIC12209	LA1567/8	**2.33**	**3.24**	0.90	**7.76**	Beta-propeller repeat lipoprotein
LIC12231	LA1535	1.05	1.04	**0.14**	1.14	Hypothetical lipoprotein
LIC12631	LA1029	**2.06**	**6.56**	1.05	**13.92**	Sph2
LIC12632	LA1027	**2.08**	1.17	0.88	**4.02**	Sph1
LIC12760	LA0872	1.03	0.88	1.10	**49.03**	ColA
LIC13060	LA0492	1.15	**0.34**	0.76	**0.48**	LipL36

(continued)

Table 1 (continued)

LIC#[1]	LA#[2]	Temp[3]	Osmo[4]	Iron[5]	DMC[6]	Notes
LIC13066	LA3834	1.17	**3.56**	0.78	**5.34**	Beta-propeller repeat lipoprotein
LIC13086	LA3867	**9.76**	**2.78**	**2.08**	**0.36**	Hypothetical lipoprotein
LIC13151	LA3946	1.02	**0.40**	1.05	**0.49**	Hypothetical lipoprotein
LIC13198	LA4004	0.86	1.21	0.83	**3.57**	Sph3
LIC13355	LA4203	0.81	1.03	1.19	**3.72**	Hypothetical lipoprotein
LIC20153	LB194	0.99	0.85	**10.1**	**0.34**	Hypothetical lipoprotein
LIC20172	LB216	0.81	**2.89**	0.92	0.71	LruC

Bold font indicates a significant fold change

[1] *L. interrogans* serovar Copenhageni locus tag

[2] *L. interrogans* serovar Lai locus tag

[3] Fold change in transcript level in response to upshift in temperature from 30 to 37 °C (Lo et al. 2006)

[4] Fold change in transcript level in response to shift in osmolarity from that found in EMJH culture medium to that found in the mammalian host (Matsunaga et al. 2007)

[5] Fold change in transcript level in response to iron limitation (Lo et al. 2010)

[6] Fold change in transcript level in response to dialysis membrane chambers implanted into rat peritoneal cavities (Caimano et al. 2014)

(LigA/B) and LIC12631 (Sph2) is induced by osmolarity (Artiushin et al. 2004). Lastly, expression of LipL36, a lipoprotein known to be downregulated during infection of the mammalian host (2, 3), was the second strongest osmo-repressed gene (out of 3518 loci examined).

Genes encoding lipoproteins and other exported proteins are overrepresented among those whose expression is induced by osmolarity. 19/28 (including 10 of the top 10) of the loci with >2.5 fold increase in transcript level in response to salt encoded exported proteins, as determined by the presence of a single hydrophobic transmembrane helix (signal peptide) at the amino-terminus of the sequence. Thirty-two percent (9/28) of these upregulated proteins are predicted to be lipoproteins by the SpLip algorithm, compared to only 4.7% (164/3518) of lipoprotein in the leptospiral genome generally. These results are a strong validation that osmolarity is a key signal for leptospiral adaptation to the host environment.

4.2.3 *Leptospira* in Dialysis Membrane Chambers

Transcriptional microarrays have been replaced with RNA sequencing (RNAseq) approaches utilizing high-throughput sequencing. RNAseq was employed to examine the effects of the host environment on *L. interrogans* gene expression (Caimano et al. 2014). The investigators used a novel system of inoculating leptospiral cells into growth medium at a concentration of 1×10^4 organisms/ml in dialysis membrane chambers (DMCs) followed by implantation into rat peritoneal cavities. The dialysis membranes used in these experiments had an 8000 molecular weight cutoff, allowing electrolytes and small molecules to freely diffuse across the membrane. At 8 days after implantation, leptospires in the DMCs had grown to a density of $\sim 7 \times 10^7$ organisms/ml and were observed to be vigorously motile. RNA was obtained for high-throughput sequencing, and the number of transcripts for each gene was compared to RNA from organisms grown to a similar density in vitro.

Expression of many lipoprotein genes was affected, both positively and negatively, by growth in DMCs, with a surprisingly high level of correlation between the effects of osmolarity and the host environment (Table 1 and Fig. 2). For example, LipL53 and the Lig operon proteins, some of the most strongly upregulated lipoproteins in DMCs, were also strongly upregulated by osmolarity. This was also true of a number of lipoprotein genes encoding beta-propeller repeat proteins, which are implicated in host–pathogen interactions. Likewise, LipL36 and LIC13151, two of the most strongly downregulated lipoproteins in DMCs, were also strongly downregulated by osmolarity. This suggests that of all of the potential host factors regulating leptospiral gene expression, osmolarity is a dominant factor. Not surprisingly, there were a number of important exceptions to this pattern. For example, there was no effect of osmolarity on *colA*, encoding a collagenase-like enzyme, the gene most strongly induced in DMCs, suggesting that factors other than osmolarity are at play.

4.2.4 *Leptospira* Response to Temperature

Temperature is a key environmental cue for many bacterial pathogens, including spirochetes, signaling transition to the mammalian host. For this reason, it is not surprising that many leptospiral genes, including those encoding lipoproteins, are regulated by thermoregulatory mechanisms (Lo et al. 2006). Among these are the five genes in the Lig operon, though as described in Sect. 4.3, the temperature has distinct transcriptional and post-transcriptional effects on Lig protein expression. As shown in Table 1, a number of lipoprotein genes are positively regulated by both temperature and osmolarity, including the Lig operon genes, *lic13086*, and the *lic12209* gene encoding a beta-propeller repeat lipoprotein. Interestingly, LIC13086 is strongly downregulated in DMCs, suggesting that another, as yet unidentified factor controls its regulation in the host environment.

4.2.5 Iron Limitation

Iron limitation is an important host defense against bacterial pathogens. Unlike *B. burgdorferi*, pathogenic leptospires are iron-dependent and have evolved a number of mechanisms for scavenging iron during infection of the mammalian host

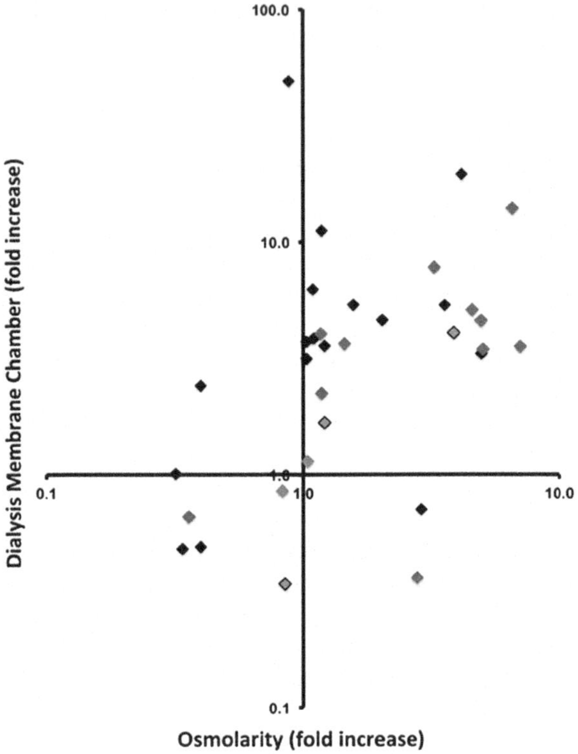

Fig. 2 Comparison of effects of osmolarity and dialysis membrane chamber conditions on transcript levels. Genes encoding leptospiral lipoproteins are plotted on a log-log graph using the relative change in transcript levels resulting from overnight shifts in osmolarity from that found in leptospiral culture medium to that found in the mammalian host (x axis) and in dialysis membrane chambers implanted into rat peritoneal cavities (y axis). Genes with transcript levels that were increased or decreased by an overnight shift in temperature from 30 to 37 °C are indicated by **red** and **blue** diamonds, respectively. Genes with transcript levels that were increased or decreased by iron limiting conditions are indicated by **orange** and **green** diamonds, respectively. The gene with transcript levels that were increased both by temperature and iron limitation is indicated by a **purple** diamond

(Louvel et al. 2006; Sritharan et al. 2006). The effects of low iron on *L. interrogans* gene transcription were examined in a transcriptional array study (Lo et al. 2010). Generally speaking, there was little effect of iron limitation on lipoproteins regulated by osmolarity, temperature, or in DMCs (Table 1). Of the *L. interrogans* lipoprotein genes strongly regulated either positively (LIC10713, LIC11695, and LIC20153) or negatively (LIC12049/50, LIC12231), LIC11695 was also upregulated by osmolarity and in DMCs. Consistent with the effects of iron limitation on expression, LIC11695 is in the same operon as LIC11694, which encodes a TonB-dependent protein that has been implicated in siderophore uptake: Knockout of the *L. biflexa* homolog of LIC11694 resulted in impaired ability to use desferrioxamine as an iron source (Louvel et al. 2006).

4.3 Gene Regulation Mechanisms

Temperature and osmolarity independently increase LigA and LigB expression (Fig. 3a). Interestingly, the effects of temperature are both transcriptional and post-transcriptional (Matsunaga et al. 2013). Post-transcriptional effects were suggested by the fact that the 20- and 14-fold temperature-induced increases in LigA and LigB expression, respectively, were out of proportion to the increases in transcript levels (Fig. 3b). Analysis of the long 5' untranslated region of both the *ligA* and *ligB* transcripts revealed two hairpin loop structures, the second of which includes and obscures their ribosome binding sites and start codons. When the hairpin loop structure is present, the ribosome binding site is prevented from binding to the ribosome and initiation of translation is blocked (Fig. 3c). In *E. coli*, a *lig-bgaB* translational fusion was found to be thermoregulated, demonstrating the

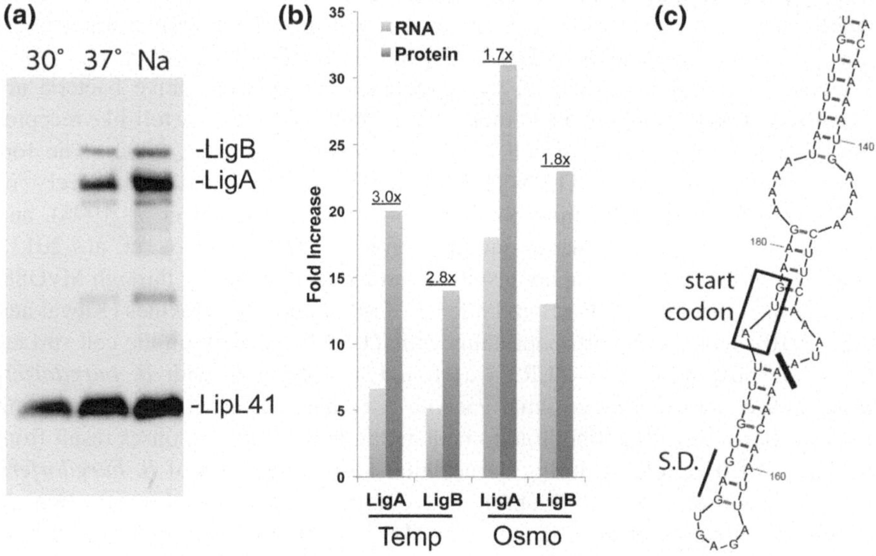

Fig. 3 Regulation of LigA and LigB by temperature and osmolarity. **a** Cultures of *L. interrogans* were incubated at 30 °C, 37 °C, or 30 °C in leptospiral culture medium supplemented with 50 mM sodium chloride. **b** The fold differences of LigA and LigB protein and transcript levels at a growth temperature of 37 °C relative to those at 30 °C (Temp) and in culture medium supplemented with 50 mM sodium chloride relative to those in unsupplemented culture medium (Osmo). Fold increases in protein levels relative to transcript levels are shown. C. Sequence and predicated RNA structure of the 5' end of *ligA* and *ligB* showing the location of the transcriptional start site (heavy bar), Shine Dalgarno binding site (S.D.), and AUG start codon

ability of the *lig* sequences to exert temperature-dependent post-transcriptional control. Mutations on the left or right stem of structure 2 partially relieved inhibition of β-galactosidase expression; inhibition was regained when the mutations were combined to restore base pairing, providing evidence that base-paired RNA is a component of the inhibitory element. These results are consistent with a model in which the hairpin loop structure functions as a thermolabile "thermometer." It should be kept in mind that trans-acting factors may also have an important role in melting the inhibitory hairpin loop structure.

5 Host Immune Response and Pathogen Adaptation

5.1 Immune Pathways

Much of the pathology associated with diseases caused by invasive spirochetes is triggered by host inflammatory responses. For example, Lyme disease is characterized by inflammatory processes including arthritis, carditis, and cranial neuritis. The inflammatory responses associated with spirochetal diseases are triggered at least in part by lipoproteins. Rash is characteristic of both Lyme disease and syphilis, and the dermatopathology can be reproduced by intradermal injection of either synthetic lipopeptides, Tp47, OspA, or OspB (Norgard et al. 1995). Triacylated lipoproteins produced by spirochetes and gram-negative bacteria are recognized as pathogen-associated molecular patterns (PAMPs) by toll-like receptor TLR1/2 heterodimers on host cell surfaces (Alexopoulou et al. 2002). The formation of the lipoprotein-TLR1-TLR2 complex is facilitated by delivery of lipoproteins to TLR1/2 by lipid-binding protein CD14 (Sellati et al. 1998), and possibly also lipopolysaccharide-binding protein (LBP) (Ranoa et al. 2013). Formation of the ternary complex results in intracellular signaling through MyD88, resulting in activation of NF-κB and release of inflammatory cytokines (Kawai and Akira 2010). Interactions of lipoproteins with TLR2 occur both on the cell surface and in the phagolysosome; TLR2 was found to colocalize with *B. burgdorferi* during uptake into the phagosomal vacuole (Cervantes et al. 2011). This finding indicates that some of the lipoprotein-mediated inflammatory responses result from the release of lipoproteins during intracellular killing and lysis of *B. burgdorferi*. TLR2-mediated responses to lipoproteins are important for control of infection; *B. burgdorferi* infection of mice deficient in TLR2 results in much higher spirochete tissue burdens (Wooten et al. 2002). Although TLR2-deficient mice are unable to respond to lipoproteins, they have much more severe arthritis than wild-type mice (Wang et al. 2004), indicating that spirochetal components other than lipoproteins also play important roles in inflammation. In fact, TLR2-deficient macrophages exposed to live *B. burgdorferi* have only a moderate reduction in TNFα production compared to normal macrophages (Salazar et al. 2009).

5.2 *Pathogen Immune Evasion Mechanisms*

5.2.1 Complement Evasion

Many of the mechanisms by which spirochetes resist complement-mediated killing involve surface lipoprotein interactions with complement regulatory pathways (Zipfel et al. 2013). An important mechanism of complement resistance involves binding to factor H and factor H-like protein 1 (FHL1), which inhibits the alternative pathway of complement by preventing binding of Factor B to C3b. *B. burgdorferi* has a family of at least five complement regulator-acquiring surface proteins (CRASPs) (Kraiczy and Stevenson 2013), all of which are predicted to be surface lipoproteins based on surface proteolysis studies (Dowdell et al. 2017). CspA (BBA68), which binds factor H and FHL1, is expressed during the initial stages of mammalian infection. CspA's importance is indicated by the finding that CspA-deficient *B. burgdorferi* strain B31 is complement sensitive (Brooks et al. 2005). In addition, *B. garinii* expressing CspA acquires a serum-resistant phenotype, indicating that CspA is sufficient to medicate complement resistance (Brooks et al. 2005). After the establishment of infection in the mammalian host, CspA expression is downregulated and complement defense is taken up by CspZ, ErpA, ErpC, and ErpP, which binding factor H and various factor H related proteins (Kraiczy and Stevenson 2013).

The classical pathway of complement can be inhibited by binding to C1 esterase inhibitor (C1-INH) and/or C4 binding protein (C4BP), which interferes with the function of the C3 convertase. *B. burgdorferi* and the relapsing fever spirochetes *Borrelia recurrentis* and *Borrelia duttonii* express surface lipoproteins that inhibit the classical pathway by binding C1-INH and C4BP (Grosskinsky et al. 2010; Meri et al. 2006; Pietikainen et al. 2010). BBK32 is a fibronectin- and glycosaminoglycan-binding lipoprotein that mediates interactions of intravascular *B. burgdorferi* with endothelial surfaces (Norman et al. 2008). BBK32 binding is unique in that it has a disordered region that undergoes a conformational change to assume a beta-sheet structure to adapt to the structure of its binding partner (Kim et al. 2004). For this reason, it is not surprising that BBK32 is able to interact with a variety of different host proteins. Recently, it was demonstrated that BBK32 binds the first component of complement (C1) and blocks the classical complement cascade by preventing activation of C1 (Garcia et al. 2016). As described above, OspC now needs to be added to the list of proteins blocking the classical complement pathway due to its binding of C4b (Caine et al. 2017).

Several *L. interrogans* surface lipoproteins have been shown to bind complement regulatory proteins (Fraga et al. 2016). These include LigB, leptospiral complement regulator-acquiring protein A (LcpA, LIC11947) and leptospiral surface adhesion 23 (Lsa23, LIC11360), and leptospiral endostatin-like protein A (LenA, LfhA, Lsa24, and LIC12906). Evidence has been presented indicating that all four of these leptospiral proteins bind Factor H (Choy 2012; da Silva et al. 2015; Siqueira et al. 2016; Verma et al. 2006). LcpA, Lsa23, and LigB also bind C4BP

(Barbosa et al. 2010; Choy 2012; Siqueira et al. 2016). Finally, LcpA has been shown to inhibit polymerization of C9, which is required to form the membrane attack complex (da Silva et al. 2015).

5.2.2 Antigenic Variation

Relapsing fever *Borrelia* spp. persist in the bloodstream by periodically switching expression of the dominant surface lipoprotein, referred to as the variable major protein (VMP). During infection, organisms increase to densities as high as 10^6–10^7 ml^{-1} in the bloodstream. As the host immune response clears these organisms from the bloodstream, a serologically distinct variant emerges expressing a new surface lipoprotein. This new escape variant becomes the dominant organism in the bloodstream until the host immune response clears it. Up to 13 relapses have been reported in tick-borne relapsing fever, although the average number of relapses is 3 (Southern and Sanford 1969). The mechanism of antigenic variation involves a recombinational event between the expression locus and one of up to 59 different silent loci, in the case of *B. hermsii*, encoding variant surface lipoprotein genes (Barbour et al. 2006). Switching is facilitated by homology between the regions upstream and downstream of the expression locus and the silent loci. The degree of homology determines the likelihood of recombination with a particular silent locus, such that a semi-predictable pattern of serotype variants tends to emerge with each relapse (Barbour et al. 2006). Analogous systems of antigenic variation occur in trypanosomiasis and falciparum malaria.

Like the relapsing fever *Borrelia* spp., *B. burgdorferi* also undergoes antigenic variation of a surface lipoprotein, VlsE, as a mechanism of persistence in the mammalian host. However, the manner of variation seen in *B. burgdorferi* is significantly different from that seen in relapsing fever. In contrast to relapsing fever where a single antigenic variant emerges periodically, the VlsE sequence undergoes continuous variation during mammalian infection by *B. burgdorferi*. After a few weeks of infection of mice, numerous VlsE variants are present simultaneously, even in the same skin biopsy. The mechanism of antigenic variation is also different. As in relapsing fever *Borrelia* spp., there is an expression locus and a number of silent "cassettes." However, instead of the entire *vlsE* locus being replaced by one of these silent cassettes, a gene conversion mechanism occurs in which segments of the *vlsE* gene in the expression locus are replaced by sequences from the silent cassettes (Norris 2014). This results in *vlsE* sequence variants with changes clustered in six different variable regions. Random, segmental recombination results in a *vlsE* gene in the expression locus that contains sequences from several different silent cassettes. *vls* sequences have been found in all Lyme *Borrelia* organisms, and *vlsE*–deficient mutants are cleared by immunocompetent mice, indicating that antigenic variation is essential for immune evasion (Rogovskyy and Bankhead 2013).

5.2.3 Lipoprotein Methylation

There is provocative preliminary evidence for LipL32 post-translational methylation as an immune evasion mechanism. LipL32 is methylated at charged lysine and acidic residues, which is consistent with the appearance of LipL32 on two-dimensional gels, which frequently show a number of LipL32 isoforms (Nally et al. 2007). Interestingly, LipL32 from infection-derived organisms was found to be much more extensively methylated than LipL32 from cultivated organisms (Witchell et al. 2014). This increased methylation included lysine and aspartate residues at known LipL32 epitopes. Synthetic peptides corresponding to these LipL32 epitopes were prepared with and without methylation at amino acids shown to be methylated during infection. Surprisingly, there was 2.8-fold weaker reactivity of convalescent human leptospirosis sera with the methylated peptide LipL32-1 M compared with the unmethylated version (LipL32-1). These data suggest that LipL32 methylation during infection produces epitopes that are poorly recognized by the immune system during infection. Because the amino and ester linkages of methyl groups to basic and acidic amino acids, respectively, are acid labile, uptake of LipL32 into acidic endosomes of antigen-presenting cells would result in demethylation. As a result, the LipL32 immune response primarily recognizes unmethylated epitopes but not methylated epitopes of live organisms. This would be a novel "cloaking" mechanism for masking epitopes from the mammalian immune system.

6 Serodiagnostics and Vaccines

6.1 Serodiagnostics

Traditionally, serodiagnosis of Lyme disease is performed with a two-tier strategy involving ELISA followed by reflex testing by Western blot if the ELISA is positive. In this case, both the ELISA and Western Blot involve a crude mixture of *B. burgdorferi* antigens. The greater the number of positive bands on the Western blot, the greater the chance that the patient has been exposed to *B. burgdorferi*. A number of these Western blot antigens are lipoproteins including p35 (VlsE), p24 (OspC), p39 (BmpA lipoprotein), p31 (OspA lipoprotein), and p17 (DbpA lipoprotein). More recently, studies are evaluating the serodiagnostic role of ELISA based on reactivity with the carboxy-terminal six amino acids (C6 peptide) of the variable surface lipoprotein VlsE (Branda et al. 2011). When sera from patients with early Lyme disease were tested by VlsE C6 ELISA alone versus crude antigen ELISA followed by VlsE C6 ELISA, both approaches had a similar sensitivity (64% vs. 61%). Both ELISA approaches were more sensitive than crude antigen ELISA followed by Western blot (48%). In patients with late Lyme disease, all three approaches had 100% sensitivity. Crude antigen ELISA followed by VlsE

C6 ELISA had equivalent specificity as ELISA followed by Western blot (99.5% for each). Both of these two-tier strategies had higher specificity than VlsE C6 ELISA alone (98.4%). As a result of the fact that most individuals are seronegative, this reduced specificity resulted in a reduced positive predictive value of the VlsE C6 ELISA of 43% compared to that of crude antigen ELISA followed by VlsE C6 ELISA (70%) or ELISA followed by Western blot (66%).

Leptospiral lipoproteins have been identified as useful serodiagnostic antigens in a number of studies. Early studies of whole leptospiral immunoblots probed with sera from leptospirosis patients resulted in reactivity with leptospiral lipopolysaccharide (LPS), primarily by IgM antibodies, and a number of protein bands (Chapman et al. 1991). A later effort to identify the reactive proteins by one- and two-dimensional whole leptospiral immunoblots included treatment with sodium periodate to remove LPS carbohydrates and reduce background reactivity (Guerreiro et al. 2001). This effort resulted in identification of LipL21, LipL32, and LipL41 as immunoreactive antigens. A follow-up ELISA study found that 56% of sera from patients in the acute phase and 94% of sera from patients in the convalescent phase of leptospirosis contained higher levels of antibodies to LipL32 than control sera from individuals from endemic and non-endemic regions (Flannery et al. 2001). A number of other studies have documented the usefulness of these lipoproteins as serodiagnostic antigens (Bomfim et al. 2005; Chen et al. 2013; Natarajaseenivasan et al. 2008; Ye et al. 2014).

The immunoblot approach described in the previous paragraph involving whole leptospires was limited by the fact that the only proteins that can be recognized are those expressed by cultivated organisms. For this reason, we pursued the hypothesis that leptospiral antigens expressed selectively during infection could be identified by screening an expression library of phage containing leptospiral DNA fragments with sera from leptospirosis patients (Matsunaga et al. 2003). A number of phage clones were recovered that contained DNA encoding immunoglobulin-like domains corresponding to the LigA, LigB, and LigC lipoproteins. The later discovery that the Lig proteins are highly regulated (as described above) immunodominant lipoproteins confirmed our original hypothesis that such proteins could be identified by screening an expression library with clinical sera. The Lig proteins appear to be expressed early in infection based on a subsequent study showing immunoreactivity with the Lig proteins by sera during the acute phase of infection (Croda et al. 2007).

To search more comprehensively for leptospiral serodiagnostic antigens, essentially all leptospiral proteins encoded in the *L. interrogans* serovar Copenhageni genome were expressed by in vitro transcription translation and printed in a proteome array. The leptospiral proteome array was screened with sera from 80 patients in the acute phase and 80 patients in the convalescent phase of leptospirosis and compared with control sera from individuals from endemic and non-endemic regions (Lessa-Aquino et al. 2013). LipL32 and the Lig proteins had the highest reactivity with sera from acute and convalescent leptospirosis patients. In addition, lipoproteins Loa22 and LipL31 were selectively recognized by sera from acute leptospirosis patients. A subsequent immunoproteome study was performed in a similar way except that this time, the study compared antigens recognized by patients with mild or severe leptospirosis, as defined by whether or not patients had been admitted to the hospital (Lessa-Aquino et al. 2017). In both mild and severe leptospirosis, the Lig proteins were the most sensitive and specific serodiagnostic antigens. When used together, LigA domains 8–13 and LigA/B domains 1–6 were 94% sensitive and 100% specific in severe leptospirosis patients.

6.2 Vaccines

Surface-exposed lipoproteins that play key roles in bacterial virulence and that can serve as targets of a protective immune response are potential vaccine candidates. In the case of *B. burgdorferi*, immunization with the lipoprotein OspA is highly effective at preventing infection in the experimental mouse model either by inoculation of culture grown organisms (Fikrig et al. 1990; 1992a; Probert and LeFebvre 1994) or by tick transmission (Fikrig et al. 1992b). The mechanism of action of OspA antibodies was shown to be blocking of transmission from the tick to the mammalian host (de Silva et al. 1996), which is consistent with tick phase-specific expression of OspA (see Sect. 4.2). Lipid modification of recombinant OspA was shown to be required for immunogenicity (Erdile et al. 1993), indicating that the lipid component functions as an internal adjuvant thru activation of TLR1/2 (Yoder et al. 2003). OspA seemed like the ideal recombinant, subunit vaccine, as its sequence is highly conserved, providing cross-protection from a variety of *B. burgdorferi* strains (Probert et al. 1997). OspA immunization was found to be safe and effective in the nonhuman primate model of *B. burgdorferi* infection (Philipp et al. 1997). In 1998, the FDA approved the LYMErix recombinant OspA vaccine for the prevention of Lyme disease, which was marketed by SmithKline Beecham. The LYMErix was subsequently withdrawn in 2002 due to a variety of problems including safety concerns related to the hypothesis that a short region of OspA could stimulate autoimmunity to human lymphocyte function-associated antigen-1 (hLFA-1) (Poland 2011). Subsequently, a novel lipidated, hexavalent OspA vaccine, designated VLA15, was developed consisting of the carboxy-terminal half of six OspA serotypes from four *Borrelia* species, with the region of *B. burgdorferi* OspA with partial homology to hLFA-1 replaced by the corresponding region from *B. afzelii* (Comstedt et al. 2014). Recently, the FDA approved a phase I safety trial of VLA15.

OspB is similar to OspA both in sequence and structure (Becker et al. 2005) and can substitute for OspA during tick colonization (Tilly et al. 2016). Although it does not bind TROSPA, OspB is important for adherence to the tick gut and survival in the tick (Neelakanta et al. 2007). Interestingly, OspB is the target of bactericidal antibodies that are able to kill *B. burgdorferi* in the absence of complement (Sadziene et al. 1994). Immunization with recombinant OspB is protective against some strains of *B. burgdorferi* (Probert and LeFebvre 1994), but not in others expressing truncated forms of OspB (Fikrig et al. 1993).

Immunization with surface lipoprotein OspC is protective in mouse models of *B. burgdorferi* protection both by needle inoculation (Probert and LeFebvre 1994) and challenge with infected ticks (Gilmore et al. 1996). OspC exists as a dimer and some protective epitopes may be specific to the OspC dimer; disulfide-linked dimeric OspC was a more effective immunogen than monomeric OspC and was found to protect mice from infection either by subcutaneous or tick challenge (Edmondson et al. 2017). Although OspC sequences are highly variable and cross-protection is limited (Probert et al. 1997), there is a relatively conserved ten-amino-acid epitope near the OspC carboxy terminus that can induce bactericidal IgM antibodies (Ikushima et al. 2000; Jobe et al. 2003; Lovrich et al. 2005).

Decorin-binding protein A (DbpA) is a surface lipoprotein protein consisting of a bundle of five helices held together by a hydrophobic core (Wang 2012). *B. burgdorferi* mutants lacking both *dbpA* and *dbpB* are attenuated for infection in mice challenged by needle inoculation but not by tick challenge (Blevins et al. 2008). Similarly, immunization with recombinant DbpA protects mice from challenge by needle inoculation (Hanson et al. 1998) but not by tick-transmitted infection (Hagman et al. 2000). When combined with DbpA and OspC, immunization with fibronectin binding protein Bbk32 was effective against challenge by needle inoculation and tick challenge (Brown et al. 2005).

A number of leptospiral lipoproteins have been tested for their protective ability, including LipL32 (Branger et al. 2005; Humphryes et al. 2014; Lucas et al. 2011), LipL41 (Haake et al. 1999; Lin et al. 2011), LigA (Coutinho et al. 2011; Faisal et al. 2008, 2009; Hartwig et al. 2014; Koizumi and Watanabe 2004; Lourdault et al. 2014; Lucas et al. 2011; Monaris et al. 2015; Palaniappan et al. 2006; Silva et al. 2007), LigB (Cao et al. 2011; Yan et al. 2009), and the combination of LigA and LigB (Forster et al. 2013). Of these, immunization with LigA domains 7–13 has been the most consistently successful recombinant lipoprotein for protection of hamsters from lethal challenge. A key to successful protection with LigA is immunization with the protein in its native conformation. This can be achieved either by expressing LigA in *E. coli* as a soluble protein (Coutinho et al. 2011) or by refolding the protein from a denatured form. Multiple immunoglobulin-like domains appear to be required, as protection has been observed either with LigA domains 10–12 or 11–13 (Coutinho et al. 2011). There are several problems with LigA as a commercial vaccine. One problem is that LigA is not expressed by all pathogenic leptospires; the genome sequence of the highly virulent Lai serovar does not include the *ligA* gene (Ren et al. 2003). Another problem is that there are at least two LigA sequence variants among pathogenic leptospires; LigA domains 11–13 of some Canicola and Pomona serovars are significantly different from those of other leptospires (McBride et al. 2009). The third and most significant problem is that while there is excellent protection from lethal infection, LigA-immunized hamsters challenged with *L. interrogans* had kidney infection (Coutinho et al. 2011). Prevention of renal colonization is a major requirement of any commercial leptospirosis vaccine intended for animals. Recently, sterilizing immunity was achieved by immunization with a recombinant protein encompassing domains 1–6 and part of domain 7 of LigB (Conrad et al. 2017), which is expressed by all leptospiral pathogens.

7 Summary and Conclusions

Lipoproteins are relatively more abundant in spirochetes than in other bacteria and play essential roles in spirochete structure and function and in host–pathogen interactions. Spirochete export and lipidation pathways are similar to those in gram-negative bacteria, but with some unique features. Sequence-based lipoprotein prediction algorithms must take into account the increased variability at the −1 position relative to the invariant cysteine. Export to the outer leaflet of the outer membrane depends on an N-terminal tether region that prevents premature periplasmic folding, likely by binding to a holding chaperone that ushers nascent lipoproteins to a putative outer membrane lipoprotein flippase. The spatiotemporal expression of many lipoproteins is finely tuned, based on a multitude of environmental cues, and ever more complex transcriptional and post-transcriptional gene regulation mechanisms are emerging.

Lipoproteins are frequently targeted by the host immune response, in part driven by recognition of lipidation by the TLR1/2 complex. Responses to lipoproteins are important in triggering the host inflammatory responses and spirochetal clearance. Because persistence is a key part of spirochetal infections, elaborate lipoprotein variation mechanisms have evolved to evade the host immune response. Lipoproteins are some of the most important serodiagnostic and vaccine antigens.

References

Alexopoulou L, Thomas V, Schnare M, Lobet Y, Anguita J, Schoen RT, Medzhitov R, Fikrig E, Flavell RA (2002) Hyporesponsiveness to vaccination with *B. burgdorferi* OspA in humans and in TLR1- and TLR2-deficient mice. Nat Med 8:878–884

Artiushin S, Timoney JF, Nally J, Verma A (2004) Host-inducible immunogenic sphingomyelinase-like protein, Lk73.5, of *L. interrogans*. Infect Immun 72:742–749

Barbosa AS, Monaris D, Silva LB, Morais ZM, Vasconcellos SA, Cianciarullo AM, Isaac L, Abreu PA (2010) Functional characterization of LcpA, a surface-exposed protein of *Leptospira* spp. that binds the human complement regulator C4BP. Infect Immun 78:3207–3216

Barbour AG, Dai Q, Restrepo BI, Stoenner HG, Frank SA (2006) Pathogen escape from host immunity by a genome program for antigenic variation. Proc Natl Acad Sci U S A 103:18290–18295

Barnett JK, Barnett D, Bolin CA, Summers TA, Wagar EA, Cheville NF, Hartskeerl RA, Haake DA (1999) Expression and distribution of leptospiral outer membrane components during renal infection of hamsters. Infect Immun 67:853–861

Becker M, Bunikis J, Lade BD, Dunn JJ, Barbour AG, Lawson CL (2005) Structural investigation of *B. burgdorferi* OspB, a bactericidal Fab target. J Biol Chem 280:17363–17370

Belisle J, Brandt M, Radolf J, Norgard M (1994) Fatty acids of *T. pallidum* and *B. burgdorferi* lipoproteins. J Bacteriol 176:2151–2157

Blevins JS, Hagman KE, Norgard MV (2008) Assessment of decorin-binding protein A to the infectivity of *B. burgdorferi* in the murine models of needle and tick infection. BMC Microbiol 8:82

Bomfim MR, Ko A, Koury MC (2005) Evaluation of the recombinant LipL32 in enzyme-linked immunosorbent assay for the serodiagnosis of bovine leptospirosis. Vet Microbiol 109:89–94

Branda JA, Linskey K, Kim YA, Steere AC, Ferraro MJ (2011) Two-tiered antibody testing for Lyme disease with use of 2 enzyme immunoassays, a whole-cell sonicate enzyme immunoassay followed by a VlsE C6 peptide enzyme immunoassay. Clin Infect Dis 53:541–547

Brandt M, Riley B, Radolf J, Norgard M (1990) Immunogenic integral membrane proteins of *B. burgdorferi* are lipoproteins. Infect Immun 58:983–991

Branger C, Chatrenet B, Gauvrit A, Aviat F, Aubert A, Bach JM, Andre-Fontaine G (2005) Protection against *L. interrogans* sensu lato challenge by DNA immunization with the gene encoding hemolysin-associated protein 1. Infect Immun 73:4062–4069

Brangulis K, Petrovskis I, Kazaks A, Tars K, Ranka R (2014) Crystal structure of the infectious phenotype-associated outer surface protein BBA66 from the Lyme disease agent *B. burgdorferi*. Ticks Tick Borne Dis 5:63–68

Brooks CS, Vuppala SR, Jett AM, Alitalo A, Meri S, Akins DR (2005) Complement regulator-acquiring surface protein 1 imparts resistance to human serum in *B. burgdorferi*. J Immunol 175:3299–3308

Brown EL, Kim JH, Reisenbichler ES, Höök M (2005) Multicomponent Lyme vaccine: three is not a crowd. Vaccine 23:3687–96

Bunikis J, Barbour A (1999) Access of antibody or trypsin to an integral outer membrane protein (P66) of *B. burgdorferi* is hindered by Osp lipoproteins. Infect Immun 67:2874–2883

Caimano MJ, Iyer R, Eggers CH, Gonzalez C, Morton EA, Gilbert MA, Schwartz I, Radolf JD (2007) Analysis of the RpoS regulon in *B. burgdorferi* in response to mammalian host signals provides insight into RpoS function during the enzootic cycle. Mol Microbiol 65:1193–1217

Caimano MJ, Sivasankaran SK, Allard A, Hurley D, Hokamp K, Grassmann AA, Hinton JC, Nally JE (2014) A model system for studying the transcriptomic and physiological changes associated with mammalian host-adaptation by *L. interrogans* serovar Copenhageni. PLoS Pathog 10:e1004004

Caine JA, Lin YP, Kessler JR, Sato H, Leong JM, Coburn J (2017) *B. burgdorferi* outer surface protein C (OspC) binds complement component C4b and confers bloodstream survival. Cell Microbiol. https://doi.org/10.1111/cmi.12786

Cao XJ, Dai J, Xu H, Nie S, Chang X, Hu BY, Sheng QH, Wang LS, Ning ZB, Li YX, Guo XK, Zhao GP, Zeng R (2010) High-coverage proteome analysis reveals the first insight of protein modification systems in the pathogenic spirochete *L. interrogans*. Cell Res 20:197–210

Cao Y, Faisal SM, Yan W, Chang YC, McDonough SP, Zhang N, Akey BL, Chang YF (2011) Evaluation of novel fusion proteins derived from extracellular matrix binding domains of LigB as vaccine candidates against leptospirosis in a hamster model. Vaccine 29:7379–7386

Carrasco SE, Troxell B, Yang Y, Brandt SL, Li H, Sandusky GE, Condon KW, Serezani CH, Yang XF (2015) Outer surface protein OspC is an antiphagocytic factor that protects *B. burgdorferi* from phagocytosis by macrophages. Infect Immun 83:4848–4860

Carter C, Bergström S, Norris S, Barbour A (1994) A family of surface-exposed proteins of 20 kilodaltons in the genus Borrelia. Infect Immun 62:2792–2799

Cervantes JL, Dunham-Ems SM, La Vake CJ, Petzke MM, Sahay B, Sellati TJ, Radolf JD, Salazar JC (2011) Phagosomal signaling by *B. burgdorferi* in human monocytes involves Toll-like receptor (TLR) 2 and TLR8 cooperativity and TLR8-mediated induction of IFN-beta. Proc Natl Acad Sci U S A 108:3683–3688

Chapman AJ, Everard CO, Faine S, Adler B (1991) Antigens recognized by the human immune response to severe leptospirosis in Barbados. Epidemiol Infect 107:143–155

Chen S, Zückert W (2011) Probing the *B. burgdorferi* surface lipoprotein secretion pathway using a conditionally folding protein domain. J Bacteriol 193:6724–6732

Chen S, Kumru O, Zückert W (2011) Determination of borrelia surface lipoprotein anchor topology by surface proteolysis. J Bacteriol 193:6379–6383

Chen HW, Zhang Z, Halsey ES, Guevara C, Canal E, Hall E, Maves R, Tilley DH, Kochel TJ, Ching WM (2013) Detection of *Leptospira*-specific antibodies using a recombinant antigen-based enzyme-linked immunosorbent assay. Am J Trop Med Hyg 89:1088–1094

Choy HA (2012) Multiple activities of LigB potentiate virulence of *L. interrogans*: inhibition of alternative and classical pathways of complement. PLoS ONE 7:e41566

Comstedt P, Hanner M, Schuler W, Meinke A, Lundberg U (2014) Design and development of a novel vaccine for protection against Lyme borreliosis. PLoS ONE 9:e113294

Conrad NL, Cruz McBride FW, Souza JD, Silveira MM, Felix S, Mendonca KS, Santos CS, Athanazio DA, Medeiros MA, Reis MG, Dellagostin OA, McBride AJ (2017) LigB subunit vaccine confers sterile immunity against challenge in the hamster model of leptospirosis. PLoS Negl Trop Dis 11:e0005441

Coutinho ML, Choy HA, Haake D (2011) A LigA three-domain region protects hamsters from lethal infection by *L. interrogans*. PLoS Neg Trop Dis 5:e1422

Croda J, Ramos JG, Matsunaga J, Queiroz A, Homma A, Riley LW, Haake DA, Reis MG, Ko AI (2007) *Leptospira* immunoglobulin-like proteins as a serodiagnostic marker for acute leptospirosis. J Clin Microbiol 45:1528–1534

Cullen PA, Xu X, Matsunaga J, Sanchez Y, Ko AI, Haake DA, Adler B (2005) Surfaceome of *Leptospira* spp. Infect Immun 73:4853–4863

da Silva LB, Miragaia Ldos S, Breda LC, Abe CM, Schmidt MC, Moro AM, Monaris D, Conde JN, Jozsi M, Isaac L, Abreu PA, Barbosa AS (2015) Pathogenic *Leptospira* species acquire factor H and vitronectin via the surface protein LcpA. Infect Immun 83:888–897

de Silva AM, Telford SRr, Brunet LR, Telford SRr, Barthold SW, Fikrig E (1996) *B. burgdorferi* OspA is an arthropod-specific transmission-blocking Lyme disease vaccine. J Exp Med 183:271–275

Dowdell AS, Murphy MD, Azodi C, Swanson SK, Florens L, Chen S, Zückert WR (2017) Comprehensive spatial analysis of the *B. burgdorferi* lipoproteome reveals a compartmentalization bias toward the bacterial surface. J Bacteriol 199:e00658–16

Edmondson DG, Prabhakaran S, Norris SJ, Ullmann AJ, Piesman J, Dolan M, Probst C, Radzimski C, Stocker W, Komorowski L (2017) Enhanced protective immunogenicity of homodimeric *B. burgdorferi* outer surface protein C. Clin Vaccine Immunol 24

Erdile LF, Brandt MA, Warakomski DJ, Westrack GJ, Sadziene A, Barbour AG, Mays JP (1993) Role of attached lipid in immunogenicity of *B. burgdorferi* OspA. Infect Immun 61:81–90

Faisal SM, Yan W, Chen CS, Palaniappan RU, McDonough SP, Chang YF (2008) Evaluation of protective immunity of *Leptospira* immunoglobulin like protein A (LigA) DNA vaccine against challenge in hamsters. Vaccine 26:277–287

Faisal SM, Yan W, McDonough SP, Chang YF (2009) *Leptospira* immunoglobulin-like protein A variable region (LigAvar) incorporated in liposomes and PLGA microspheres produces a robust immune response correlating to protective immunity. Vaccine 27:378–387

Fikrig E, Barthold SW, Kantor FS, Flavell RA (1990) Protection of mice against the Lyme disease agent by immunizing with recombinant OspA. Science 250:553–556

Fikrig E, Barthold SW, Kantor FS, Flavell RA (1992a) Long-term protection of mice from Lyme disease by vaccination with OspA. Infect Immun 60:773–777

Fikrig E, Telford SR, Barthold SW, Kantor FS, Spielman A, Flavell RA (1992b) Elimination of *B. burgdorferi* from vector ticks feeding on OspA-immunized mice. Proc Natl Acad Sci U S A 89:5418–5421

Fikrig E, Tao H, Kantor FS, Barthold SW, Flavell RA (1993) Evasion of protective immunity by *B. burgdorferi* by truncation of outer surface protein B. Proc Natl Acad Sci U S A 90:4092–4096

Flannery B, Costa D, Carvalho FP, Guerreiro H, Matsunaga J, Da Silva ED, Ferreira AG, Riley LW, Reis MG, Haake DA, Ko AI (2001) Evaluation of recombinant *Leptospira* antigen-based enzyme-linked immunosorbent assays for the serodiagnosis of leptospirosis. J Clin Microbiol 39:3303–3310

Forster KM, Hartwig DD, Seixas FK, Bacelo KL, Amaral M, Hartleben CP, Dellagostin OA (2013) A conserved region of leptospiral immunoglobulin-like A and B proteins as a DNA vaccine elicits a prophylactic immune response against leptospirosis. Clin Vaccine Immunol 20:725–731

Fouts DE, Matthias MA, Adhikarla H, Adler B, Amorim-Santos L, Berg DE, Bulach D, Buschiazzo A, Chang YF, Galloway RL, Haake DA, Haft DH, Hartskeerl R, Ko AI, Levett PN, Matsunaga J, Mechaly AE, Monk JM, Nascimento AL, Nelson KE, Palsson B, Peacock SJ, Picardeau M, Ricaldi JN, Thaipandungpanit J, Wunder EA Jr, Yang XF, Zhang JJ, Vinetz JM (2016) What makes a bacterial species pathogenic?: comparative genomic analysis of the genus *Leptospira*. PLoS Negl Trop Dis 10:e0004403

Fraga TR, Isaac L, Barbosa AS (2016) Complement evasion by pathogenic *Leptospira*. Front Immunol 7:623

Galperin MY (2005) A census of membrane-bound and intracellular signal transduction proteins in bacteria: bacterial IQ, extroverts and introverts. BMC Microbiol 5:35

Garcia BL, Zhi H, Wager B, Hook M, Skare JT (2016) *B. burgdorferi* BBK32 Inhibits the classical pathway by blocking activation of the C1 complement complex. PLoS Pathog 12:e1005404

Gilmore RD Jr, Kappel KJ, Dolan MC, Burkot TR, Johnson BJ (1996) Outer surface protein C (OspC), but not P39, is a protective immunogen against a tick-transmitted *B. burgdorferi* challenge: evidence for a conformational protective epitope in OspC. Infect Immun 64:2234–2239

Grimm D, Tilly K, Byram R, Stewart PE, Krum JG, Bueschel DM, Schwan TG, Policastro PF, Elias AF, Rosa PA (2004) Outer-surface protein C of the Lyme disease spirochete: a protein induced in ticks for infection of mammals. Proc Natl Acad Sci U S A 101:3142–3147

Grosskinsky S, Schott M, Brenner C, Cutler SJ, Simon MM, Wallich R (2010) Human complement regulators C4b-binding protein and C1 esterase inhibitor interact with a novel outer surface protein of *B. recurrentis*. PLoS Negl Trop Dis 4:e698

Guerreiro H, Croda J, Flannery B, Mazel M, Matsunaga J, Galvao Reis M, Levett PN, Ko AI, Haake DA (2001) Leptospiral proteins recognized during the humoral immune response to leptospirosis in humans. Infect Immun 69:4958–4968

Haake DA, Matsunaga J (2002) Characterization of the leptospiral outer membrane and description of three novel leptospiral membrane proteins. Infect Immun 70:4936–4945

Haake DA, Walker EM, Blanco DR, Bolin CA, Miller MN, Lovett MA (1991) Changes in the surface of *L. interrogans* serovar grippotyphosa during in vitro cultivation. Infect Immun 59:1131–1140

Haake DA, Martinich C, Summers TA, Shang ES, Pruetz JD, McCoy AM, Mazel MK, Bolin CA (1998) Characterization of leptospiral outer membrane lipoprotein LipL36: downregulation associated with late-log-phase growth and mammalian infection. Infect Immun 66:1579–1587

Haake DA, Mazel MK, McCoy AM, Milward F, Chao G, Matsunaga J, Wagar EA (1999) Leptospiral outer membrane proteins OmpL1 and LipL41 exhibit synergistic immunoprotection. Infect Immun 67:6572–6582

Haake DA, Chao G, Zuerner RL, Barnett JK, Barnett D, Mazel M, Matsunaga J, Levett PN, Bolin CA (2000) The leptospiral major outer membrane protein LipL32 is a lipoprotein expressed during mammalian infection. Infect Immun 68:2276–2285

Hagman KE, Yang X, Wikel SK, Schoeler GB, Caimano MJ, Radolf JD, Norgard MV (2000) Decorin-binding protein A (DbpA) of *B. burgdorferi* is not protective when immunized mice are challenged via tick infestation and correlates with the lack of DbpA expression by *B. burgdorferi* in ticks. Infect Immun 68:4759–4764

Hanson MS, Cassatt DR, Guo BP, Patel NK, McCarthy MP, Dorward DW, Hook M (1998) Active and passive immunity against *B. burgdorferi* decorin binding protein A (DbpA) protects against infection. Infect Immun 66:2143–2153

Hartwig DD, Bacelo KL, Oliveira PD, Oliveira TL, Seixas FK, Amaral MG, Hartleben CP, McBride AJ, Dellagostin OA (2014) Mannosylated LigANI produced in *Pichia pastoris* protects hamsters against leptospirosis. Curr Microbiol 68:524–530

Humphryes PC, Weeks ME, AbuOun M, Thomson G, Nunez A, Coldham NG (2014) Vaccination with leptospiral outer membrane lipoprotein LipL32 reduces kidney invasion of *L. interrogans* serovar canicola in hamsters. Clin Vaccine Immunol 21:546–551

Ikushima M, Matsui K, Yamada F, Kawahashi S, Nishikawa SK (2000) Specific immune response to a synthetic peptide derived from outer surface protein C of *B. burgdorferi* predicts protective borreliacidal antibodies. FEMS Immunol Med Microbiol 29:15–21

Inukai M, Takeuchi M, Shimizu K, Arai M (1978) Mechanism of action of globomycin. J Antibiot (Tokyo) 31:1203–1205

Iyer R, Caimano MJ, Luthra A, Axline D Jr, Corona A, Iacobas DA, Radolf JD, Schwartz I (2015) Stage-specific global alterations in the transcriptomes of Lyme disease spirochetes during tick feeding and following mammalian host adaptation. Mol Microbiol 95:509–538

Jobe DA, Lovrich SD, Schell RF, Callister SM (2003) C-terminal region of outer surface protein C binds borreliacidal antibodies in sera from patients with Lyme disease. Clin Diagn Lab Immunol 10:573–578

Juncadella IJ, Anguita J (2009) The immunosuppresive tick salivary protein, Salp15. Adv Exp Med Biol 666:121–131

Jutras BL, Chenail AM, Stevenson B (2013) Changes in bacterial growth rate govern expression of the *B. burgdorferi* OspC and Erp infection-associated surface proteins. J Bacteriol 195:757–764

Kawai T, Akira S (2010) The role of pattern-recognition receptors in innate immunity: update on Toll-like receptors. Nature Immunol 11:373–384

Kim JH, Singvall J, Schwarz-Linek U, Johnson BJ, Potts JR, Hook M (2004) BBK32, a fibronectin binding MSCRAMM from *B. burgdorferi*, contains a disordered region that undergoes a conformational change on ligand binding. J Biol Chem 279:41706–41714

Koizumi N, Watanabe H (2004) Leptospiral immunoglobulin-like proteins elicit protective immunity. Vaccine 22:1545–1552

Kraiczy P, Stevenson B (2013) Complement regulator-acquiring surface proteins of *B. burgdorferi*: structure, function and regulation of gene expression. Ticks Tick Borne Dis 4:26–34

Kumaran D, Eswaramoorthy S, Luft B, Koide S, Dunn J, Lawson C, Swaminathan S (2001) Crystal structure of outer surface protein C (OspC) from the Lyme disease spirochete, *B. burgdorferi*. EMBO J 20:971–978

Kumru O, Schulze R, Slusser J, Zückert W (2010) Development and validation of a FACS-based lipoprotein localization screen in the Lyme disease spirochete *B. burgdorferi*. BMC Microbiol 10:277

Kumru O, Schulze R, Rodnin M, Ladokhin A, Zückert W (2011) Surface localization determinants of Borrelia OspC/Vsp family lipoproteins. J Bacteriol 193:2814–2825

Lenhart T, Akins D (2010) *B. burgdorferi* locus BB0795 encodes a BamA orthologue required for growth and efficient localization of outer membrane proteins. Mol Microbiol 75:692–709

Lessa-Aquino C, Borges Rodrigues C, Pablo J, Sasaki R, Jasinskas A, Liang L, Wunder EA Jr, Ribeiro GS, Vigil A, Galler R, Molina D, Liang X, Reis MG, Ko AI, Medeiros MA, Felgner PL (2013) Identification of seroreactive proteins of *L. interrogans* serovar copenhageni using a high-density protein microarray approach. PLoS Negl Trop Dis 7:e2499

Lessa-Aquino C, Lindow JC, Randall A, Wunder E, Pablo J, Nakajima R, Jasinskas A, Cruz JS, Damiao AO, Nery N, Ribeiro GS, Costa F, Hagan JE, Reis MG, Ko AI, Medeiros MA, Felgner PL (2017) Distinct antibody responses of patients with mild and severe leptospirosis determined by whole proteome microarray analysis. PLoS Negl Trop Dis 11:e0005349

Li H, Lawson CL (1995) Crystallization and preliminary X-Ray analysis of *B. burgdorferi* Outer surface protein a (Ospa) complexed with a murine monoclonal antibody Fab fragment. J Struct Biol 115:335–337

Lin X, Sun A, Ruan P, Zhang Z, Yan J (2011) Characterization of conserved combined T and B cell epitopes in *L. interrogans* major outer membrane proteins OmpL1 and LipL41. BMC Microbiol 11:21

Lo M, Bulach DM, Powell DR, Haake DA, Matsunaga J, Paustian ML, Zuerner RL, Adler B (2006) Effects of temperature on gene expression patterns in *L. interrogans* serovar Lai as assessed by whole-genome microarrays. Infect Immun 74:5848–5859

Lo M, Murray GL, Khoo CA, Haake DA, Zuerner RL, Adler B (2010) Transcriptional response of *L. interrogans* to iron limitation and characterization of a PerR homolog. Infect Immun 78:4850–4859

Lourdault K, Wang LC, Vieira A, Matsunaga J, Melo R, Lewis MS, Haake DA, Gomes-Solecki M (2014) Oral immunization with *E. coli* expressing a lipidated form of LigA protects hamsters against challenge with *L. interrogans* serovar copenhageni. Infect Immun 82:893–902

Louvel H, Bommezzadri S, Zidane N, Boursaux-Eude C, Creno S, Magnier A, Rouy Z, Medigue C, Saint Girons I, Bouchier C, Picardeau M (2006) Comparative and functional genomic analyses of iron transport and regulation in *Leptospira* spp. J Bacteriol 188:7893–7904

Lovrich SD, Jobe DA, Schell RF, Callister SM (2005) Borreliacidal OspC antibodies specific for a highly conserved epitope are immunodominant in human lyme disease and do not occur in mice or hamsters. Clin Diagn Lab Immunol 12:746–751

Lucas DS, Cullen PA, Lo M, Srikram A, Sermswan RW, Adler B (2011) Recombinant LipL32 and LigA from *Leptospira* are unable to stimulate protective immunity against leptospirosis in the hamster model. Vaccine 29:3413–3418

Matsunaga J, Barocchi MA, Croda J, Young TA, Sanchez Y, Siqueira I, Bolin CA, Reis MG, Riley LW, Haake DA, Ko AI (2003) Pathogenic *Leptospira* species express surface-exposed proteins belonging to the bacterial immunoglobulin superfamily. Mol Microbiol 49:929–945

Matsunaga J, Sanchez Y, Xu X, Haake DA (2005) Osmolarity, a key environmental signal controlling expression of leptospiral proteins LigA and LigB and the extracellular release of LigA. Infect Immun 73:70–78

Matsunaga J, Lo M, Bulach DM, Zuerner RL, Adler B, Haake DA (2007) Response of *L. interrogans* to physiologic osmolarity: relevance in signaling the environment-to-host transition. Infect Immun 75:2864–2874

Matsunaga J, Schlax PJ, Haake DA (2013) Role for cis-acting RNA sequences in the temperature-dependent expression of the multiadhesive lig proteins in *L. interrogans*. J Bacteriol 195:5092–5101

McBride AJ, Cerqueira GM, Suchard MA, Moreira AN, Zuerner RL, Reis MG, Haake DA, Ko AI, Dellagostin OA (2009) Genetic diversity of the leptospiral immunoglobulin-like (Lig) genes in pathogenic *Leptospira* spp. Infect Genet Evol 9:196–205

Meri T, Cutler SJ, Blom AM, Meri S, Jokiranta TS (2006) Relapsing fever spirochetes *B. recurrentis* and *B. duttonii* acquire complement regulators C4b-binding protein and factor H. Infect Immun 74:4157–4163

Monaris D, Sbrogio-Almeida ME, Dib CC, Canhamero TA, Souza GO, Vasconcellos SA, Ferreira LC, Abreu PA (2015) Protective immunity and reduced renal colonization induced by vaccines containing recombinant *L. interrogans* outer membrane proteins and flagellin adjuvant. Clin Vaccine Immunol 22:965–973

Nally JE, Chow E, Fishbein MC, Blanco DR, Lovett MA (2005) Changes in lipopolysaccharide O antigen distinguish acute versus chronic *L. interrogans* infections. Infect Immun 73:3251–3260

Nally JE, Whitelegge JP, SBBlanco DR, Lovett MA (2007) Characterization of the outer membrane proteome of *L. interrogans* expressed during acute lethal infection. Infect Immun 75:766–773

Narita S, Matsuyama S, Tokuda H (2004) Lipoprotein trafficking in *E. coli*. Arch Microbiol 182:1–6

Natarajaseenivasan K, Vijayachari P, Sharma S, Sugunan AP, Selvin J, Sehgal SC (2008) Serodiagnosis of severe leptospirosis: evaluation of ELISA based on the recombinant OmpL1 or LipL41 antigens of *L. interrogans* serovar autumnalis. Ann Trop Med Parasitol 102:699–708

Neelakanta G, Li X, Pal U, Liu X, Beck DS, DePonte K, Fish D, Kantor FS, Fikrig E (2007) Outer surface protein B is critical for *B. burgdorferi* adherence and survival within Ixodes ticks. PLoS Pathog 3:e33

Norgard MV, Riley BS, Richardson JA, Radolf JD (1995) Dermal inflammation elicited by synthetic analogs of *T. pallidum* and *B. burgdorferi* lipoproteins. Infect Immun 63:1507–1515

Norman MU, Moriarty TJ, Dresser AR, Millen B, Kubes P, Chaconas G (2008) Molecular mechanisms involved in vascular interactions of the Lyme disease pathogen in a living host. PLoS Pathog 4:e1000169

Norris SJ (2014) *vls* Antigenic variation systems of Lyme disease Borrelia: eluding host immunity through both random, segmental gene conversion and framework heterogeneity. Microbiol Spectr 2(6):MDNA3-0038-2014

Okuda S, Tokuda H (2011) Lipoprotein sorting in bacteria. Annu Rev Microbiol 65:239–259

Pal U, Li X, Wang T, Montgomery RR, Ramamoorthi N, Desilva AM, Bao F, Yang X, Pypaert M, Pradhan D, Kantor FS, Telford S, Anderson JF, Fikrig E (2004) TROSPA, an Ixodes scapularis receptor for *B. burgdorferi*. Cell 119:457–468

Palaniappan RU, McDonough SP, Divers TJ, Chen CS, Pan MJ, Matsumoto M, Chang YF (2006) Immunoprotection of recombinant leptospiral immunoglobulin-like protein A against *L. interrogans* serovar Pomona infection. Infect Immun 74:1745–1750

Philipp MT, Lobet Y, Bohm RP Jr, Roberts ED, Dennis VA, Gu Y, Lowrie RC Jr, Desmons P, Duray PH, England JD, Hauser P, Piesman J, Xu K (1997) The outer surface protein A (OspA) vaccine against Lyme disease: efficacy in the rhesus monkey. Vaccine 15:1872–1887

Pietikainen J, Meri T, Blom AM, Meri S (2010) Binding of the complement inhibitor C4b-binding protein to Lyme disease Borreliae. Mol Immunol 47:1299–1305

Pinne M, Haake DA (2013) LipL32 Is a Subsurface Lipoprotein of *L. interrogans*: presentation of new data and reevaluation of previous studies. PLoS ONE 8:e51025

Poland GA (2011) Vaccines against Lyme disease: what happened and what lessons can we learn? Clin Infect Dis 52(Suppl 3):s253–s258

Probert WS, LeFebvre RB (1994) Protection of C3H/HeN mice from challenge with *B. burgdorferi* through active immunization with OspA, OspB, or OspC, but not with OspD or the 83-kilodalton antigen. Infect Immun 62:1920–1926

Probert WS, Crawford M, Cadiz RB, LeFebvre RB (1997) Immunization with outer surface protein (Osp) A, but not OspC, provides cross-protection of mice challenged with North American isolates of *B. burgdorferi*. J Infect Dis 175:400–405

Pugsley A (1993) The complete general secretory pathway in gram-negative bacteria. Microbiol Rev 57:50–108

Radolf JD, Chamberlain NR, Clausell A, Norgard MV (1988) Identification and localization of integral membrane proteins of virulent *T. pallidum* ssp. *pallidum* by phase partitioning with the nonionic detergent triton X-114. Infect Immun 56:490–498

Radolf J, Goldberg M, Bourell K, Baker S, Jones J, Norgard M (1995) Characterization of outer membranes isolated from *B. burgdorferi*, the Lyme disease spirochete. Infect Immun 63:2154–2163

Ramamoorthi N, Narasimhan S, Pal U, Bao F, Yang XF, Fish D, Anguita J, Norgard MV, Kantor FS, Anderson JF, Koski RA, Fikrig E (2005) The Lyme disease agent exploits a tick protein to infect the mammalian host. Nature 436:573–577

Ranoa DR, Kelley SL, Tapping RI (2013) Human lipopolysaccharide-binding protein (LBP) and CD14 independently deliver triacylated lipoproteins to Toll-like receptor 1 (TLR1) and TLR2 and enhance formation of the ternary signaling complex. J Biol Chem 288:9729–9741

Ren SX, Fu G, Jiang XG, Zeng R, Miao YG, Xu H, Zhang YX, Xiong H, Lu G, Lu LF, Jiang HQ, Jia J, Tu YF, Jiang JX, Gu WY, Zhang YQ, Cai Z, Sheng HH, Yin HF, Zhang Y, Zhu GF, Wan M, Huang HL, Qian Z, Wang SY, Ma W, Yao ZJ, Shen Y, Qiang BQ, Xia QC, Guo XK, Danchin A, Saint Girons I, Somerville RL, Wen YM, Shi MH, Chen Z, Xu JG, Zhao GP (2003) Unique physiological and pathogenic features of *L. interrogans* revealed by whole-genome sequencing. Nature 422:888–893

Rogovskyy AS, Bankhead T (2013) Variable VlsE is critical for host reinfection by the Lyme disease spirochete. PLoS ONE 8:e61226

Sadziene A, Jonsson M, Bergström S, Bright RK, Kennedy RC, Barbour AG (1994) A bactericidal antibody to *B. burgdorferi* is directed against a variable region of the OspB protein. Infect Immun 62:2037–2045

Salazar JC, Duhnam-Ems S, La Vake C, Cruz AR, Moore MW, Caimano MJ, Velez-Climent L, Shupe J, Krueger W, Radolf JD (2009) Activation of human monocytes by live *B. burgdorferi* generates TLR2-dependent and -independent responses which include induction of IFN-beta. PLoS Pathog 5:e1000444

Samuels DS (2011) Gene regulation in *B. burgdorferi*. Annu Rev Microbiol 65:479–499

Schulze R, Zückert W (2006) *B. burgdorferi* lipoproteins are secreted to the outer surface by default. Mol Microbiol 59:1473–1484

Schulze R, Chen S, Kumru O, Zückert W (2010) Translocation of *B. burgdorferi* surface lipoprotein OspA through the outer membrane requires an unfolded conformation and can initiate at the C-terminus. Mol Microbiol 76:1266–1278

Schwan TG, Piesman J, Golde WT, Dolan MC, Rosa PA (1995) Induction of an outer surface protein on *B. burgdorferi* during tick feeding. Proc Natl Acad Sci U S A 92:2909–2913

Sellati TJ, Bouis DA, Kitchens RL, Darveau RP, Pugin J, Ulevitch RJ, Gangloff SC, Goyert SM, Norgard MV, Radolf JD (1998) *T. pallidum* and *B. burgdorferi* lipoproteins and synthetic lipopeptides activate monocytic cells via a CD14-dependent pathway distinct from that used by lipopolysaccharide. J Immunol 160:5455–5464

Setubal JC, Reis M, Matsunaga J, Haake DA (2006) Lipoprotein computational prediction in spirochaetal genomes. Microbiol 152:113–121

Shang ES, Summers TA, Haake DA (1996) Molecular cloning and sequence analysis of the gene encoding LipL41, a surface-exposed lipoprotein of pathogenic *Leptospira* species. Infect Immun 64:2322–2330

Silva EF, Medeiros MA, McBride AJ, Matsunaga J, Esteves GS, Ramos JB, Santos CS, Croda J, Homma A, Dellagostin OA, Haake DA, Reis MG, Ko AI (2007) The terminal portion of leptospiral immunoglobulin-like protein LigA confers protective immunity against lethal infection in the hamster model of leptospirosis. Vaccine 25:6277–6286

Siqueira GH, Atzingen MV, de Souza GO, Vasconcellos SA, Nascimento AL (2016) *L. interrogans* Lsa23 protein recruits plasminogen, factor H and C4BP from normal human serum and mediates C3b and C4b degradation. Microbiol 162:295–308

Skare J, Shang E, Foley D, Blanco D, Champion C, Mirzabekov T, Sokolov Y, Kagan B, Miller J, Lovett M (1995) Virulent strain associated outer membrane proteins of *B. burgdorferi*. J Clin Invest 96:2380–2392

Southern PM, Sanford JP (1969) Relapsing fever. A Clinic Microbiol Rev Med. 48:129–149

Sritharan M, Asuthkar S, Sridhar V (2006) Understanding iron acquisition by pathogenic leptospires: a review. Indian J Med Microbiol 24:311–316

Stewart PE, Wang X, Bueschel DM, Clifton DR, Grimm D, Tilly K, Carroll JA, Weis JJ, Rosa PA (2006) Delineating the requirement for the *B. burgdorferi* virulence factor OspC in the mammalian host. Infect Immun 74:3547–3553

Tilly K, Krum JG, Bestor A, Jewett MW, Grimm D, Bueschel D, Byram R, Dorward D, Vanraden MJ, Stewart P, Rosa P (2006) *B. burgdorferi* OspC protein required exclusively in a crucial early stage of mammalian infection. Infect Immun 74:3554–3564

Tilly K, Bestor A, Rosa PA (2016) Functional equivalence of OspA and OspB, but not OspC, in tick colonization by *B. burgdorferi*. Infect Immun 84:1565–1573

Verma A, Hellwage J, Artiushin S, Zipfel PF, Kraiczy P, Timoney JF, Stevenson B (2006) LfhA, a novel factor H-binding protein of *L. interrogans*. Infect Immun 74:2659–2666

Vivian J, Beddoe T, McAlister A, Wilce M, Zaker-Tabrizi L, Troy S, Byres E, Hoke D, Cullen P, Lo M, Murray G, Adler B, Rossjohn J (2009) Crystal structure of LipL32, the most abundant surface protein of pathogenic *Leptospira* spp. J Mol Biol 387:1229–1238

Vogeley L, El Arnaout T, Bailey J, Stansfeld PJ, Boland C, Caffrey M (2016) Structural basis of lipoprotein signal peptidase II action and inhibition by the antibiotic globomycin. Science 351:876–880

Wang X (2012) Solution structure of decorin-binding protein A from *B. burgdorferi*. Biochem 51:8353–8362

Wang G, Ma Y, Buyuk A, McClain S, Weis JJ, Schwartz I (2004) Impaired host defense to infection and Toll-like receptor 2-independent killing of *B. burgdorferi* clinical isolates in TLR2-deficient C3H/HeJ mice. FEMS Microbiol Lett 231:219–225

Witchell TD, Eshghi A, Nally JE, Hof R, Boulanger MJ, Wunder EA Jr, Ko AI, Haake DA, Cameron CE (2014) Post-translational modification of LipL32 during *L. interrogans* infection. PLoS Negl Trop Dis 8:e3280

Wooten RM, Ma Y, Yoder RA, Brown JP, Weis JH, Zachary JF, Kirschning CJ, Weis JJ (2002) Toll-like receptor 2 plays a pivotal role in host defense and inflammatory response to *B. burgdorferi*. Vector Borne Zoonotic Dis 2:275–278

Yan W, Faisal SM, McDonough SP, Divers TJ, Barr SC, Chang CF, Pan MJ, Chang YF (2009) Immunogenicity and protective efficacy of recombinant *Leptospira* immunoglobulin-like protein B (rLigB) in a hamster challenge model. Microbes Infect 11:230–237

Ye C, Yan W, McDonough PL, McDonough SP, Mohamed H, Divers TJ, Chang YF, Yang Z (2014) Serodiagnosis of equine leptospirosis by enzyme-linked immunosorbent assay using four recombinant protein markers. Clin Vaccine Immunol 21:478–483

Yoder A, Wang X, Ma Y, Philipp MT, Heilbrun M, Weis JH, Kirschning CJ, Wooten RM, Weis JJ (2003) Tripalmitoyl-S-glyceryl-cysteine-dependent OspA vaccination of toll-like receptor 2-deficient mice results in effective protection from *B. burgdorferi* challenge. Infect Immun 71:3894–3900

Zipfel PF, Hallstrom T, Riesbeck K (2013) Human complement control and complement evasion by pathogenic microbes–tipping the balance. Mol Immunol 56:152–160

Zückert WR (2014) Secretion of bacterial lipoproteins: through the cytoplasmic membrane, the periplasm and beyond. Biochim Biophys Acta 1843:1509–1516

Zückert W, Kerentseva T, Lawson C, Barbour A (2001) Structural conservation of neurotropism-associated VspA within the variable Borrelia Vsp-OspC lipoprotein family. J Biol Chem 276:457–463

Colonic Spirochetes: What Has Genomics Taught Us?

David John Hampson and Penghao Wang

Abstract The 'colonic' spirochetes assigned to the genus *Brachyspira* are slow-growing anaerobic bacteria. The genus includes both pathogenic and non-pathogenic species, and these variously colonise the large intestines of different species of birds and animals, including humans. Scientific understanding of the physiology and molecular biology of *Brachyspira* spp. remains very limited compared with that of other pathogenic spirochetes, and there are few descriptions of successful genetic manipulations undertaken to investigate gene function. An important boost to knowledge occurred in 2009 when, for the first time, the whole genome sequence of a *Brachyspira* strain (*Brachyspira hyodysenteriae* strain WA1) was obtained. The genomics analysis provided a significant increase in knowledge: for example, a previously unknown ~36 Kb plasmid was discovered and metabolic pathways were constructed. The study also revealed likely acquisition of genes involved in transport and central metabolic functions from other enteric bacterial species. Four subsequent publications have provided a similarly detailed analysis of other *Brachyspira* genomes, but of these only two included more than one strain of a species (20 strains of *B. hyodysenteriae* in one and three strains of *B. pilosicoli* in the other). Since then, more *Brachyspira* genomes have been made publicly available, with the sequences of at least one representative of each of the nine officially recognised species deposited at public genome repositories. All species have a single circular chromosome varying in size from ~2.5 to 3.3 Mb, with a C + G content of around 27%. In this chapter, we summarise the current knowledge and present a preliminary comparative genomic analysis conducted on 56 strains covering the official *Brachyspira* species. Besides providing detailed genetic maps of the bacteria, this analysis has revealed gene island rearrangements,

---

D.J. Hampson (✉) · P. Wang (✉)
School of Veterinary and Life Sciences, Murdoch University, Perth, WA 6150, Australia
e-mail: d.hampson@murdoch.edu.au

P. Wang
e-mail: P.Wang@murdoch.edu.au

Current Topics in Microbiology and Immunology (2018) 415:273–294
DOI 10.1007/82_2017_48

Published Online: 07 September 2017

putative phenotypes (including antimicrobial drug resistance) and genetic mutation mechanisms that enable brachyspires to evolve and respond to stress. The application of Next-Generation Sequencing (NGS) to generate genomic data from many more *Brachyspira* species and strains increasing will improve our understanding of these enigmatic spirochetes.

Contents

1 Introduction

This chapter presents an overview of the anaerobic spirochetes belonging to the genus *Brachyspira*, emphasising new knowledge that is emerging from the recent availability of whole genome sequences, and indicating future directions for research. It provides some technical background to the methods used in the preliminary pan-genomic analysis presented here, including limitations and suggestions for future studies. Such studies will improve understanding of *Brachyspira* genomics and develop practical applications based on this information.

1.1 The Genus Brachyspira

The genus *Brachyspira* currently includes nine officially named species and several proposed spirochete species. When they were originally described, *Brachyspira hyodysenteriae* and *Brachyspira innocens* were assigned to the genus *Treponema*:

subsequently they were transferred to the genus *Serpula* and then to *Serpulina*. Finally these and other species were transferred to the genus *Brachyspira*, to join the previously described *B. aalborgi* (Ochiai et al. 1997). These species are all anaerobes that can tolerate exposure to oxygen. The different species colonise the large intestines of various species of animals and birds (Table 1). Depending on the spirochete species, growth can take 3 days to 3 weeks to appear as a low flat haze on the surface of agar. Due to their slow growth and potential presence of other bacterial species, specialised selective agar is required for effective isolation from clinical samples. Subsequently the purified isolates can be grown in pre-reduced anaerobic broth.

Table 1 *Brachyspira* species, strength of hemolysis on blood agar, host species and disease associations

Species	Haemolysis	Main host species	Disease	References
B. hyodysenteriae	Strong[a]	Pigs	Swine dysentery	Burrough (2017)
B. hampsonii	Strong	Pigs, waterbirds	Swine dysentery	Mirajkar et al. (2016a)
B. suanatina	Strong	Ducks, pigs	Swine dysentery	Mushtaq et al. (2015)
B. pilosicoli	Weak	Many species, including humans	Intestinal spirochetosis	Trott et al. (1996b)
B. intermedia	Weak	Chickens, pigs	Avian intestinal spirochetosis; mild colitis?	Stanton et al. (1997)
B. innocens	Weak	Pigs, chickens, rats	Not recorded	Kinyon and Harris (1979)
B. murdochii	Weak	Pigs, chickens, rats	Mild colitis in pigs	Stanton et al. (1997)
B. alvinipulli	Weak	Chickens	Avian intestinal spirochetosis	Stanton et al. (1998)
B. aalborgi	Weak	Humans and other primates	Intestinal spirochetosis	Hovind-Hougen et al. (1982)
"*B. canis*"	Weak	Dogs	Not recorded	Duhamel et al. (1998)
"*B. pulli*"	Weak	Chickens	Not recorded	Stephens and Hampson (1999)
"*B. corvi*"	Weak	Corvid birds	Not recorded	Jansson et al. (2008)
Others	Weak	Rodents	Not recorded	Backhans et al. (2010)

[a]Weakly hemolytic strains have been recorded

1.2 Diseases Caused by Brachyspira *Species*

1.2.1 Swine Dysentery

Swine dysentery (SD) is the most economically significant and best-described disease resulting from infection with *Brachyspira* species. SD manifests as a mucohemorrhagic colitis that occurs mainly in growing and fattening pigs (typically between 12 and 20 weeks of age), although it can occur in other age groups. Spirochetes are found adjacent to the colonic luminal epithelium and in the colonic crypts, where they induce an outpouring of mucus and cause local damage to the epithelium, resulting in bleeding and secondary infection.

B. hyodysenteriae is the species that was originally described as the etiological agent of SD and remains the most commonly reported cause (Burrough 2017), although in recent years the newly described species *B. hampsonii* and *B. suanatina* also have been shown to be able to cause SD (Mushtaq et al. 2015; Mirajkar et al. 2016a). Strains of all three of these species induce a strong beta-hemolysis on blood agar, whilst all the other six officially named *Brachyspira* species are weakly hemolytic. Atypical weakly hemolytic strains of *B. hyodysenteriae* with reduced virulence have been reported (Mahu et al. 2016; La et al. 2016b).

1.2.2 Other Diseases

The weakly hemolytic *Brachyspira* species include those that are considered to be harmless commensals (*Brachyspira innocens*), and others that are variably associated with mild enteric disease in different species. The nomenclature used for these conditions in different animal species can be confusing, and is outlined here.

In human beings, the condition called "Intestinal Spirochetosis" is characterised by attachment of spirochetes by one cell end to the apical epithelium of the colon and rectum, forming a dense "false brush border" of attached spirochete cells. Two species can cause this condition: *B. aalborgi* (Hovind-Hougen et al. 1982) and *B. pilosicoli* (Trott et al. 1996b). Of these, *B. pilosicoli* is the most strongly associated with disease (Mikosza and Hampson 2001), with manifestations including mild colitis, rectal bleeding, chronic diarrhea and failure to thrive in infants (Brooke et al. 2006).

In pigs, *B. pilosicoli* is the cause of a condition called "Porcine Intestinal Spirochetosis" (PIS) or "Porcine Colonic Spirochetosis" (Trott et al. 1996b). As in humans, spirochetes may form a false brush border covering the apical epithelium in parts of the colon and rectum. This condition is widespread in growing pigs, and is associated with mild colitis, diarrhea and poor growth rates. *B. pilosicoli* also colonises dogs, horses and other species, where it may cause diarrhea. In many avian species, *B. pilosioli* is an agent of a condition called "Avian Intestinal Spirochetosis" (AIS), again with end-on attachment of spirochetes, mild colitis and increased fecal water content. AIS occurs naturally in adult poultry, where besides diarrhea it can cause reduced egg production. Other *Brachyspira* species also associated with forms of AIS include *B. intermedia* and *B. alvinipulli* (McLaren et al. 1997), of which *B. intermedia* is the most common and widespread species.

In pigs, the term "Brachyspiral colitis" has been used for a mild colitis caused by weakly hemolytic species other than *B. pilosicoli*, particularly *B. intermedia* and/or *B. murdochii*. Both the latter species have occasionally been associated with mild localised colitis and diarrhea (Weissenböck et al. 2005; Komarek et al. 2009; Jensen et al. 2010), although they also are commonly found in healthy pigs. Whether or not disease occurs may depend on the specific strain attributes, the microenvironment in the colon and on the extent of the colonisation.

2 *Brachyspira* Genomes

A number of *Brachyspira* genomes now have been sequenced and reported. A list of the number of genomes of the nine officially named *Brachyspira* species that are available at the time of writing is presented in Table 2. Only *B. hyodysenteriae*, *B. hampsonii, B. suanatina* and *B. pilosicoli* are represented by more than one strain. Unfortunately, this is very limiting for comparative genomic analysis both at the systematic level and when making specific between-species comparisons.

Brachyspira species DNA has a low GC content, in the range 24.6%–28%. Genome sizes for the species vary from ~2.5 to 3.3 million base pairs (Mbp), each with >2300 protein coding sequences. Most species share close similarities in their

Table 2 *Brachyspira* species for which complete or near complete genome sequences are available in NCBI or another public site

Species	Unique genomes	Genome sizes (Mb)	G + C content (%)	References
B. hyodysenteriae	20[a]	2.99–3.17	~27.1	Bellgard et al. (2009), Black et al. (2015), Mirajkar et al. (2016b, c)
B. hampsonii	5	~3.2	~27.4	Mirajkar et al. (2016a)
B. suanatina	2	3.26	27.0	Mushtaq et al. (2015)
B. pilosicoli	4	2.56–2.89	27.44–27.90	Wanchanthuek et al. (2010), Mappley et al. (2012), Lin et al. (2013)
B. intermedia	1	3.30	27.22	Håfström et al. (2011)
B. innocens	1	3.28	27.7	DOE Joint Genome Institute
B. murdochii	1	3.24	27.75	Pati et al. (2010)
B. alvinipulli	1	3.42	26.8	Stanton et al. (1998); DOE Joint Genome Institute
B. aalborgi	1	~2.5	~27.0	Mappley et al. (2012)

[a]Three sequences available for type strain B78^{T}

16S rRNA gene sequences, implying that separation of species occurred relatively recently. *B. hyodysenteriae* strains contain a ~36 Kb plasmid (Bellgard et al. 2009; Black et al. 2015), a 3.2 Kb plasmid is present in *B. intermedia* strain PWS/A^T (Håfström et al. 2011), and putative plasmid and a bacteriophage sequences occur in *B. suanatina* strain AN4859/03 (Mushtaq et al. 2015). *B. hyodysenteriae* contains a prophage-like gene transfer agent (GTA) that randomly packages ~7.5 Kb fragments of host DNA and transfers it to other *B. hyodysenteriae* strains (Humphrey et al. 1997; Matson et al. 2005). Other *Brachyspira* species harbor similar GTA sequences (Stanton et al. 2003; Motro et al. 2009), although it is not known whether they are functional. The GTAs could contribute to the extensive gene rearrangements that occur within and between *Brachyspira* species (Zuerner et al. 2004; Mappley et al. 2012).

The complete or near complete genome sequences of the most common *Brachyspira* spp. have been described and/or deposited with publicly available genome sequence repositories. At the time of writing, the deposited genome sequences include those from 20 strains of *B. hyodysenteriae* (plus three repeats of the type strain B78^T), five strains of *B. hampsonii*, four strains of *B. pilosicoli*, two strains of *B. suanatina*, and one strain each of *B. intermedia*, *B. murdochii* and *B. innocens* (Bellgard et al. 2009; Black et al. 2015; Håfström et al. 2011; Lin et al. 2013; Mappley et al. 2012; Mirajkar et al. 2016a, b; Mushtaq et al. 2015; Pati et al. 2010; Wanchanthuek et al. 2010). The partial sequence of *B. aalborgi* 513^T is available at: www.sanger.ac.uk/resources/downloads/bacteria/metahit/.

2.1 Genomic Features of B. Hyodysenteriae

There have been three publications devoted to analysing the genome sequence of *B. hyodysenteriae.* The first described the genome of strain WA1 (Bellgard et al. 2009), the second included 19 other strains of *B. hyodysenteriae*, in addition to WA1 (Black et al. 2015), and the third publication and corrections made to it (Mirajkar et al. 2016b; Mirajkar et al. 2017) described the sequence of the type strain B78^T, which had been included in the publication by Black et al. (2015), and for which an incomplete genome sequence also had been submitted by the DOE Joint Genome Institute.

The first paper demonstrated that WA1 had a single circular chromosome of ~3 Mb and a plasmid of ~36 Kb which previously had not been identified. The predicted protein coding sequences of WA1 were found to have more similarities to proteins from *Escherichia coli* and to *Clostridium* species than to those from other spirochetes: many of the proteins were associated with transport and metabolism, and it was considered that these are likely to have been acquired by the spirochete during its adaptation to life in the environment of the large intestine. The spirochete also had large numbers of genes associated with chemotaxis and motility, which are presumed to be important for colonisation. This study reconstructed central metabolic pathways for the spirochete, and confirmed that it did not have a complete tricarboxylic acid cycle, but had pathways for glycolysis, gluconeogenesis, a non-oxidative pentose phosphate pathway and a respiratory electron transport chain. ATP generating mechanisms were predicted to be substrate-level phosphorylation reactions.

The second study provided an opportunity to assess the extent of variation in the genomes between multiple (20) strains of *B. hyodysenteriae* from different countries, including several strains that have been reported to have low virulence potential (Black et al. 2015). On an average 88% of the identified genes were found to be "core", being shared by the different strains, 8.6% were ancillary genes and 2.9% were unique and strain-specific. Overall the genomes were conserved compared to many other pathogenic bacterial species, although strain-specific markers also occurred. The results confirmed the clonal population structure of this species. The high degree of conservation in the presence of genes between strains suggests that differences in their gene sequences or in gene expression may provide a better explanation for variations in phenotype than just gene presence or absence.

Immediate practical outcomes of the availability of the genome sequences have included the ability to predicted potential immunogenic outer membrane proteins or lipoproteins for use in vaccines, as part of a "reverse vaccinology" approach (Song et al. 2009); the development of serological assays using surface proteins in recombinant form (Song et al. 2015); and development of new strain typing methods such as multiple-locus variable-number tandem repeats analysis (Hidalgo et al. 2010).

2.2 Genomic Features of **B. Pilosicoli**

There have been three publications relating to the analysis of *B. pilosicoli* genomes. The first described the genome sequence of *B. pilosicoli* strain 95/1000, recovered from a pig (Wanchanthuek et al. 2010), and compared that with the sequence of *B. hyodysenteriae* strain WA1 as well as with the partial sequence of *B. murdochii* strain 56-150^{T} that previously had been deposited in GenBank (Pati et al. 2010). The second study described a comparative analysis of strain 95/1000 and two other *B. pilosicoli* strains, B2904 and WesB, isolated from a chicken and a human being respectively (Mappley et al. 2012): the study included an analysis of phenotypic properties of the strains using a Biolog phenotype Microarray™. The third paper described the sequence of porcine type strain P43/6/78^{T} (Lin et al. 2013).

Wanchanthuek et al. (2010) identified a single circular chromosome of ~2.59 Mb in *B. pilosicoli* 95/1000, without any extra-chromosomal elements. As well as being absent in 95/1000, the ~36 Kb plasmid of *B. hyodysenteriae* WA1 was not found in *B. murdochii* 56-150^{T}. A previously unreported bacteriophage sequence was detected in 95/1000. The three strains of the three species had differences in genome sizes (Table 2), and showed extensive genome rearrangements. Despite this variability, they had many similar features, including in their core metabolic pathways. Strain 95/1000 differed from *B. hyodysenteriae* WA1 in having a glycine reductase complex and genes for aconitase and related enzymes that allow for glutamate synthesis and function of the TCA cycle during oxidative stress. There also were differences in numbers of methyl-accepting chemotaxis genes between the species that may help to explain their different preferences for local niches and colonisation sites in the large intestine.

Mappley et al. (2012) demonstrated that three strains of *B. pilosicoli* varied considerably in their genome sizes (from ~2.6 to 2.9 Mb), and showed major genome rearrangements that correlated mainly with the positions of mobile genetic elements. Differences were found in the occurrence of certain genes, including those encoding glycine reductase complex components and transposases. Overall, 2132 genes were shared by the three strains, with between 112 and 256 other genes being unique to each strain. Another 236, 38 and 19 genes were shared between different pairs of strains, respectively. Using the Cluster of Orthologous Genes (COG) database, the three strains were shown to have similar distributions of features. Screening of the strains for utilisation of 178 carbon compounds showed that their metabolic capabilities were highly conserved. Differences were only found in seven carbon sources, and these correlated with genotypic variation between the strains.

Finally, Lin et al. (2013) described the genome of strain P43/6/78^T, which was ~2.56 Mb in size, and was most similar to the genome of 95/1000. The majority of genes that were unique to P43/6/78^T encoded hypothetical proteins, although some may be involved in O-antigen variation.

2.3 *Genomic Features of* B. Murdochii, B. Intermedia *and Other* Brachyspira *Species*

Other publications describing genomic features of *Brachyspira* species include a short paper by Pati et al. (2010) describing the genome of *B. murdochii* type strain 56-150^T, the paper by Håfström et al. (2011) describing *B. intermedia*-type strain PWS/A^T (and comparative analysis with other published *Brachyspira* species), and a pan-genomic analysis that formed part of the paper by Mappley et al. (2012), and that included unpublished sequences from *B. aalborgi* 513^T, *B. alvinipulli* C1^T and *B. intermedia* HB60. Descriptions of the new species *B. suanatina* (Mushtaq et al. 2015) and *B. hampsonii* (Mirajkar et al. 2016a) also contain whole genome sequence data to support the distinct genetic groupings of these species.

The analysis of the genome of *B. murdochii* 56-150^T by Pati et al. (2010) demonstrated that it had a circular genome of ~3.24 Mb. Approximately 66% of the 2853 protein-encoding genes were assigned a putative function, and their COG functional categories were broadly similar to those of the other *Brachyspira* species. Analysis of the genome sequence of *B. intermedia* strain PWS/A^T by Håfström et al. (2011), which included comparisons with the published sequences of *B. hyodysenteriae* WA1, *B. murdochii* 56-150^T and *B. pilosicoli* 95/1000 identified a circular chromosome of ~3.3 Mb (the largest *Brachyspira* genome then described), with a 3.2 Kb plasmid that was also found in *B. murdochii* 56-150^T. Not surprisingly, the pan-genomic analysis that included PWS/A^T substantially reduced the number of unique genes previously reported for the other three sequenced species/strains (Pati et al. 2010), with 2,184 core genes identified for the four species/strains. The four species shared a similar COGs profile. Evidence was found for gene translocation, as well as for remodeling of bacteriophage sequences and incorporation of genes into them that may contribute to horizontal gene transfer.

In the analysis that was included in the study of *B. pilosicoli* strains by Mappley et al. (2012), the pan-genome was further reduced to genes encoding 1,647 proteins. Novel bacteriophage sequences were detected which are likely to be involved in intra-species and inter-species horizontal gene transfer. The draft genome sequence of *B. suanatina* strain AN4859/03 identified a genome of ~3.2 Mb including a plasmid sequence with 88% identity over 50% of its length with the 36 Kb plasmid of *B. hyodysenteriae* (Mushtaq et al. 2015). The comparative genomic information was used to assist with the official recognition of the new species. In the same way, using digital DNA–DNA hybridisation and average nucleotide identity, the genomic sequences of three strains of *B. hampsonii* (NSH-16, NSH-24 and P280/1) were used to support the official proposal and subsequent approval of the new species and the three recognised genomovars of the species (Mirajkar et al. 2016a). *B. hampsonii* was shown to have a genome size of ~3.2 Mb.

The advent of NGS methods has opened up the potential for rapid and cheap sequencing of the genomes of clinical isolates of *Brachyspira* species. Coupled with improved data about specific gene content and function derived from earlier sequencing studies, this provides a powerful new tool for epidemiological and diagnostic investigations. An example was presented by La et al. (2016b) who sequenced the genomes of 23 German field isolates of *B. hyodysenteriae* and investigated these for 332 genes that previously had been annotated as associated with virulence. It is anticipated that this type of approach will be used increasingly by diagnostic laboratories, with large volumes of clinically relevant data quickly being generated.

3 A Preliminary Pan-Genomic Comparison of *Brachyspira* Species and Strains

3.1 Overview of the Pan-Genomics Comparative Analysis

In this section, we provide a summary of the results of a preliminary pan-genomic analysis of *Brachyspira* species and strains that were available to us at the time of writing. The sequences used were derived from 56 *Brachyspira* strains covering nine different species, namely *B. hyodysenteriae*, *B. hampsonii*, *B. suanatina*,

Table 3 The 39 *Brachyspira* species genomes included in the pan-genomics analysis and the basic genome statistics

Species	Unique genomes	Genome sizes (Mb)	Assembly condition	Source
B. hyodysenteriae	17	2.99–3.17	2 Complete, 15 Near complete	NCBI
	2	2.99–3.17	1 Complete, 1 Near complete	Our study
B. hampsonii	1	~3.2	Complete	NCBI
	5	~3.2	3 Complete, 2 Near complete	Our study
B. suanatina	2	~3.2	Near complete	NCBI
B. pilosicoli	3	~2.6	Complete	NCBI
	1	~2.6	Near complete	Our study
B. intermedia	1	~3.4	Complete	NCBI
	2	~3.4	Near complete	Our study
B. innocens	1	3.2	Complete	NCBI
B. murdochii	1	3.2	Complete	NCBI
	1	3.2	Near complete	Our study
B. alvinipulli	1	3.4	Complete	NCBI
B. aalborgi	1	~2.5	Near complete	Our study

B. pilosicoli, *B. intermedia*, *B. intermedia*, *B. murdochii*, *B. alvinipulli* and *B. aalborgi*. The genomic sequences of these strains were collected from two sources: sequences derived from our own sequencing studies or those of our collaborators, and sequences obtained from online repositories. We applied several quality control criteria and finally focused our analysis on 39 strains that had complete or near complete genome sequences. Some of the other genome assemblies were rather fragmented and large parts of the genomes were not captured: these were not included as they could have rendered the analysis unreliable. Table 3 presents a summary of the *Brachyspira* strains used, and provides some basic information about their origins.

Although this study is preliminary, to our knowledge it is the most comprehensive pan-genomic analysis reported for members of the genus *Brachyspira,* and covers all the officially recognised species. Our analysis provides some useful insights about potential genomic functions and mutational characteristics of *Brachyspira*, which were only revealed by the inclusion of a relatively large number of strains. In addition, this pan-genome analysis provides a guideline to analytical methodology and procedures for future large-scale whole genomics comparative analyses.

Two methods were used to perform the comparative analysis on annotated genomes: the local mapping method and the whole genome mapping method. Genome and plasmid contiguous sequences were locally aligned to reference *B. hyodysenteriae* strain WA1 using the BWA alignment method. The alignments were indexed and sorted by using Samtools software (Li et al. 2009). The whole genome alignment method utilised the progressiveMauve algorithm (Darling et al. 2004, 2010). Downstream analysis was performed mainly in R.

3.2 Genomic Relationships Among **Brachyspira** *Species*

The *Brachyspira* species analysed in this study originated from different sources and displayed different characteristics. As shown in Table 1, some species are found mainly in pigs, for example *B. hyodysenteriae*, while some species such as *B. pilosicoli* are able to infect a wide range of hosts including humans, pigs and birds. Some of the species such as *B. innocens* are not considered to be associated with disease, while others such as *B. hyodysenteriae* and *B. hampsonii* can cause severe disease (i.e. SD). Little is known about how *Brachyspira* species have evolved from a common origin and developed these special features, although mutation is likely to be an important driver in addition to acquisition of new genomic sequences through horizontal gene transfer.

A phylogenetic analysis was performed on the whole genome sequences of the species to evaluate their evolutionary relationships. Due to the gaps presented in the assemblies, for phylogenetic analysis 15 strains with high-quality sequence were selected, and relationships inferred by a two-step procedure. First, the sequences were subjected to multiple sequence alignment using ClustalW (Thompson et al. 1994) and Muscle (Tamura et al. 2011). The alignment procedure was modified using in-house programs to accurately take account of genomic rearrangements. Second, phylogenetic relationships were inferred from the aligned sequences by using the maximum likelihood-based PhyML algorithm version 3.0 (Guindon et al. 2010). The phylogenetic tree was visualised using TreeView version 1.6.6 (Page 1996). The inferred genetic relationships are shown in Fig. 1.

The *Brachyspira* strains that were analysed were classified into four clades, named A–D respectively (Fig. 1). These clades were classified into two groups: the first consisted of clades A, B and C and the second was represented by clade D. The strains in the first group clearly demonstrated that *B. alvinipulli* (clade A), *B. pilosicoli* (clade B) and *B. aalborgi* (clade C) are quite distinct, and are correctly regarded as separate species. All these weakly hemolytic species cause forms of intestinal spirochetosis: the two slightly closer neighbors *B. pilosicoli* and *B. aalborgi* colonise humans and other animals whilst *B. alvinipulli* is associated with poultry. *B. pilosicoli* is able to infect multiple species and has a unique genetic profile compared to all other species, and shows the greatest genetic distance from *B. hyodysenteriae*.

The second group consisted of *B. hyodysenteriae*, *B. murdochii*, *B. suanatina*, *B. intermedia* and *B. hampsonii*. Strains of these species formed 3 clusters based on their genetic relationships: (1) *B. hampsonii*; (2) *B. intermedia* and (3) *B. hyodysenteriae*, *B. suanatina* and *B. murdochii*. It must be emphasised that this analysis

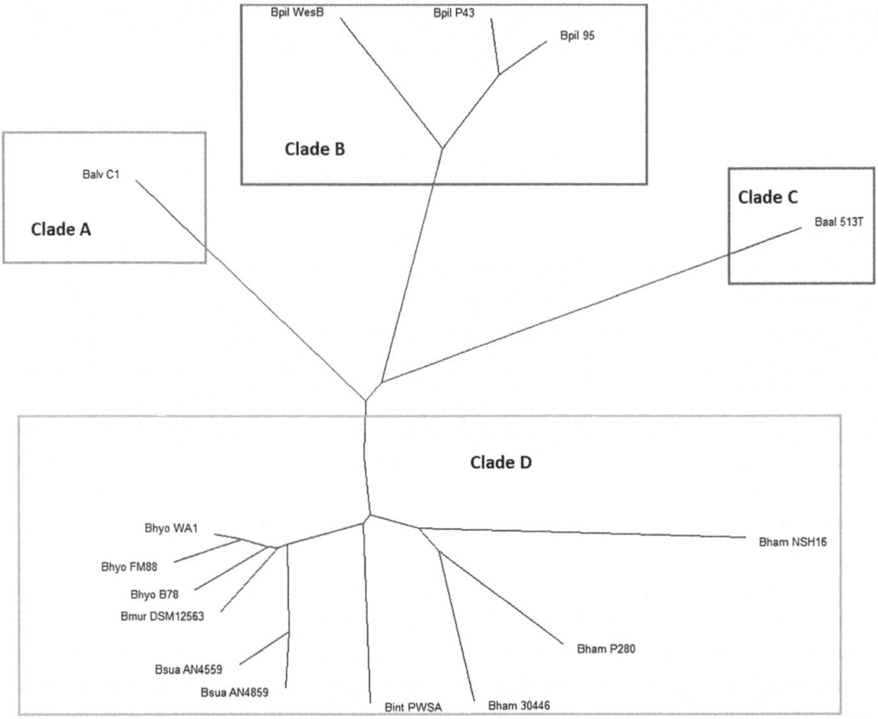

Fig. 1 Inferred phylogenetic relationships among the 15 *Brachyspira* species strains that were analysed

was very much limited by the small number of strains available, and more robust analysis will require the availability of dozens of strains of all these species. All three strongly hemolytic species causing SD fell into clade D, and all strains of *B. hyodysenteriae* were tightly clustered. Their genomic composition seemed to be very conserved. *B. suanatina* strains have only been found in Scandinavia, and to date this species has not been thoroughly studied. Our analysis indicated that the two available *B. suanatina* strains are closely related to *B. hyodysenteriae*, and consequently it is perhaps not surprising that that they can colonise pigs and cause SD.

The available *B. hampsonii* strains were quite diverse, and on the tree were genetically relatively distant from the *B. hyodysenteriae* strains. Like *B. suanatina, B. hampsonii* has been found colonising waterfowl and pigs, and identified as a cause of SD. Interestingly, the *B. hampsonii* strains were more distantly related to *B. hyodysenteriae* than to *B. intermedia*, with the latter species also being found in birds and pigs and associated with disease in poultry. Although the three genomovars of *B. hampsonii* were more distantly related to *B. hyodysenteriae* than was *B. suanatina*, and consequently had a more distant genetic origin, all three species that cause SD are strongly hemolytic. This relationship points to the hemolytic phenotype as an important indicator of the ability of these species to cause SD, rather than the broad genomic similarities between them.

The position of *B. murdochii* DSM12563 (56-150^T) within clade D was particularly interesting. This species is weakly beta-hemolytic and generally is considered to be an intestinal commensal in pigs. However, Jensen et al. (2010) reported a field case of *B. murdochii*-associated catarrhal colitis in a pig, characterised by extensive spirochetal colonisation of the surface epithelium. When eight weaned pigs were challenged with the *B. murdochii* isolate, catarrhal colitis was reproduced in two animals. By applying fluorescent in situ hybridisation using a species-specific oligonucleotide probe targeting 23S rRNA, *B. murdochii* organisms were found in high numbers and were closely associated with the surface epithelium in the pigs with catarrhal colitis. The current pan-genomic analysis indicated that *B. murdochii* DSM12563 was very closely related to strains of *B. hyodysenteriae*, supporting the likelihood that some strains of *B. murdochii* may have the potential to be more virulent than generally has been thought. Clearly, high quality genome sequences from many more strains need to be analysed to confirm the consistency of this finding.

3.3 *Similarities Across* Brachyspira *Species Revealed by Comparative Genomics*

Genome-level comparison of the *Brachyspira* strains showed high sequence similarities among the species, despite the greatly varying quality of the genome assemblies and the widely different geographic origins and host species from which they were recovered. The genomic sequences of *Brachyspira* strains belonging to the same species were all quite conserved. Approximately 30% of the proteins predicted from the genome sequences were highly similar across all strains, despite strong evidence to indicate that many of the species have undergone substantial genome rearrangements. Most strains had relatively similar genome sizes of around 3 Mbp and encoded approximately 3000 proteins. Nearly all species had a high portion of genes encoding proteins responsible for catalytic activity, metabolic activity, transporter activity, antioxidant activity and protein binding.

When the analysis was focused on strains mainly infecting pigs, the sequence similarity became even higher. Around 60% of the genomic sequences of *B. hyodysenteriae*, *B. hampsonii*, *B. murdochii*, *B. intermedia* and *B. suanatina* were conserved and were predicted to have similar biological functions, particularly for surface proteins and proteins with transporter activity. Each *Brachyspira* species was estimated to have different numbers of genes, with an average of around 2500. The *B. hyodysenteriae* strains appeared to consistently encode around 2600 proteins, and this species had the fewest genes. The only two available *B. suanatina* strains, the closest neighbor of *B. hyodysenteriae*, encoded around 2700 proteins on average, slightly more than *B. hyodysenteriae*. On the other hand, *B. hampsonii* and

B. intermedia strains were predicted to encode around 3000 proteins, which was substantially more than the *B. hyodysenteriae* strains. Interestingly, despite having smaller genome sizes, *B. pilosicoli* strains also were predicted to encode almost 3000 proteins, again more than *B. hyodysenteriae*. Given the observation that all strains had similar numbers of unique genes, a possible explanation may be that some species have a higher proportion of duplicated or homologous genes than others. Another explanation may be that the evolution of *B. hyodysenteriae* has involved reduction of genome size by gene loss, a feature that has been seen in other specialised pathogens. This would be consistent with the observed high genetic similarity of the strains, the deduced clonal population structure and the virulence of this species.

The comparative genome analysis also indicated that all *Brachyspira* species have a circular genome. When mapping the complete genome of one species to another, loci were identified where the sequence orientation of certain genes was reversed. It also may explain the high coverage of sequence similarity found for nearly every strain. Large genetic indels were not very common among the *Brachyspira* strains. The indels that were detected all appeared random, and only a few large and unique indels were found across the species.

Predicted proteins involved in motility, hydrolase activities, lipopolysaccharide biosynthesis, cell adhesion and chemotaxis were found to have greater genetic divergence than the core genes. These genes are likely to be linked to colonisation, virulence and even antimicrobial drug resistance in the bacteria. The chemotaxis-related proteins are usually found in the surfaceomes. Other similar proteins detected included several related to McpA and McpC proteins that were observed only in *B. hyodysenteriae*. Interestingly, the *mcpC* gene was not identified in the *B. pilosicoli* strains, whereas *mcpA* was detected. Other proteins in this class were related to histidine kinases and flavin biosynthesis.

3.4 Unique Features of Each Species

Previously Black et al. (2015) analysed and compared the genomes of 20 strains of *B. hyodysenteriae*, the species that has attracted the most research attention. Here we extended that analysis, but used only 19 strains that were selected based on the quality and completeness of their genome assemblies. The overall relationship of the *B. hyodysenteriae* strains is shown in Fig. 2. They demonstrated the highest similarity in their genome composition compared to the other species of *Brachyspira*, with up to 90% of their predicted proteins being conserved for all strains.

The sequenced *B. hyodysenteriae* strains could be classified into two large groups based on their genetic variation, and these appeared to be associated with the geographic locations where the strains were isolated. The first group, represented by strain WA1, contained isolates from Australia and the Asia-Pacific region. Strains in this group shared many genetic characteristics. The strains isolated in Australia,

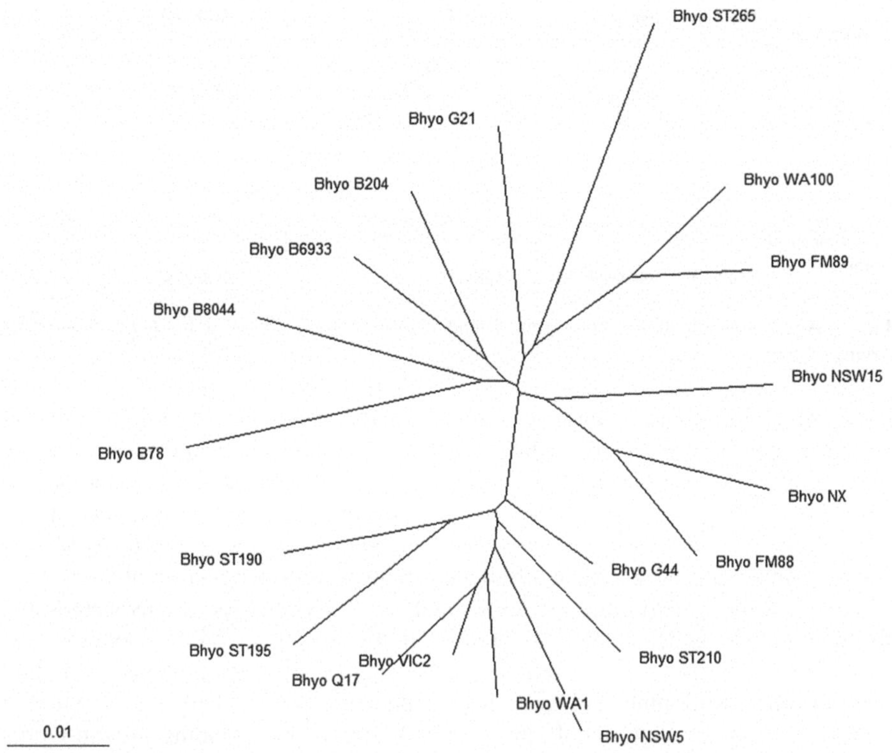

Fig. 2 The inferred relationship among 19 *B. hyodysenteriae* strains using whole genome sequence comparisons

namely WA1 from Western Australia, NSW5 from New South Wales, VIC2 from Victoria and Q17 from Queensland were very conserved. *B. hyodysenteriae* strain WA100 from Western Australia deviated from this tight cluster, but this may be due to the incompleteness of its assembly, as 5% of the genome was estimated to be missing. The second group of strains was more divergent and these mostly were isolated from Europe and North America. This suggests that the species may have evolved in isolation following its introduction into Australia with pigs imported from Europe and/or North American. Importation of live pigs into Australia stopped in the mid-1980s, and La et al. (2016a) have showed that since then Australian strains of *B. hyodysenteriae* have lost plasmid genes that are associated with virulence, and have postulated that this adaptation has allowed these less virulent strains to become widely established across Australia.

The analysis indicated that *B. hyodysenteriae* has a unique large gene island rearrangement that allows the species to adapt, but strains shared a high level of similarity in exome sequences. Comparing the two major *B. hyodysenteriae* groups, the genomes were found to have four major gene island anchor points forming four distinctive large genomic islands. *B. hyodysenteriae* strains, therefore, appear to

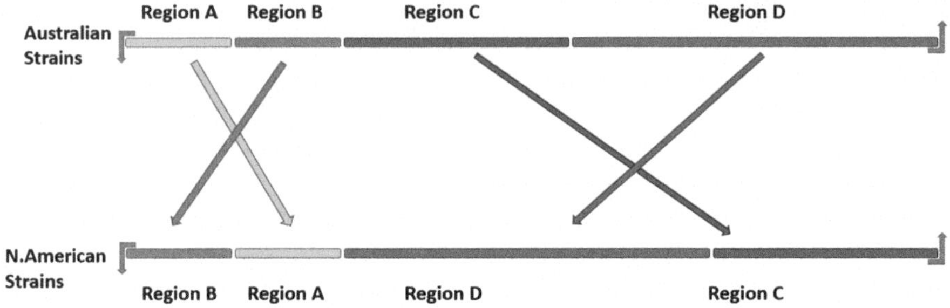

Fig. 3 An illustration of the large gene island rearrangements observed in *B. hyodysenteriae* strains

have the ability to rearrange the circular genome structure into four different potential configurations to achieve different biological phenotypes (Fig. 3). Interestingly the European and North America strains and the Asia-Pacific/ Australian strains belonged to the two most distant possible rearrangement pathways. Smaller scale rearrangements also were observed among the *B. hyodysenteriae* strains, with these potentially being involved in functional regulation.

Amongst the *Brachyspira* species studied, *B. hyodysenteriae* strains were found to have unique genes related to preprotein translocation, ATP generation and transporter activities. For example, these genes encode proteins including chromosome-partitioning ATPase Soj, replicative DNA helicase, preprotein translocase subunit YajC family proteins and Oligopeptide-binding protein OppA precursor. These genes may contribute to the capacity of *B. hyodysenteriae* to undergo genomic rearrangements by providing both unique chromosome-partitioning capacity and the energy that is required for this activity.

Although closely related to *B. hyodysenteriae*, *B. suanatina* has been proposed as a separate *Brachyspira* species. The two strains of *B. suanatina* examined here

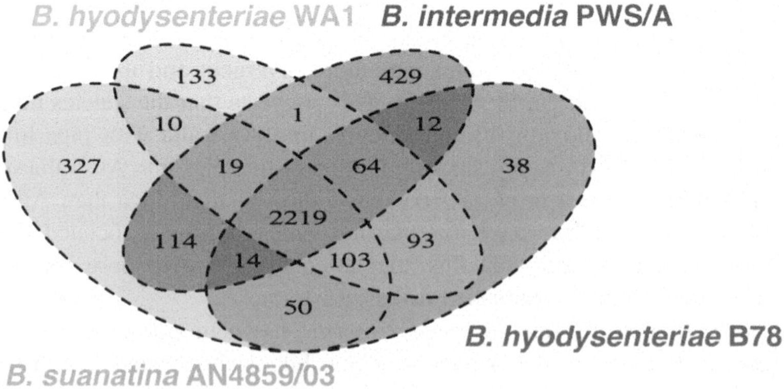

Fig. 4 Venn diagram showing the numbers of common and unique genes comparing two strains of *B. hyodysenteriae* and single strains of *B. suanatina* and *B. intermedia*

had more than 90% similarity in their genome sequences, and the majority of their genes were very conserved. The most prominent genomic divergence found between *B. suanatina* and *B. hyodysenteriae* strains was associated with indels and SNPs rather than large and obvious genomic rearrangements, or different gene content (Fig. 4). It appears that the *B. suanatina* strains have highly similar protein composition and almost no large gene rearrangements compared to the *B. hyodysenteriae* strain WA1, which was used as the reference genome. This finding is consistent with a previous study that recommended that *B. suanatina* should be classified as a separate species, despite having a high sequence similarity with *B. hyodysenteriae* (Mushtaq et al. 2015). The available data suggest that *B. suanatina* relies almost entirely on sequence point mutation to derive new phenotypes or resistance to stress. Nevertheless, it is unfortunate that there are only two *B. suanatina* genome sequences available, and the assemblies are not complete. This deficiency seriously undermines the reliability of the analysis, as these two strains may not be representative of the whole species.

Multilocus sequence typing of indole positive, weakly hemolytic *Brachyspira* strains that were identified as *B. intermedia* has shown that this species is extremely diverse and actually may represent several species (Phillips et al. 2010). In the current analysis, the genomes of the three strains for which data were available were shown to be relatively conserved (although data for two strains was quite fragmented), and they were genetically quite distinct from *B. hyodysenteriae* and *B. hampsonii*. Like *B. hyodysenteriae*, when mapping the genome sequences of the three strains to one another, a large number of genomic rearrangements were observed, indicating that *B. intermedia* species apparently have undergone genomic island rearrangements rather than point mutations in generating the diversity that was found. Whole genome alignment-based analysis suggests that *B. intermedia* and *B. suanatina* may have evolved from a common ancestor relatively recently, evidenced by their estimated genetic distance. Again, the genomes of many more strains need to be examined to strengthen these conclusions.

The *B. hampsonii* strains displayed relatively larger genetic distance between them compared to strains of *B. hyodysenteriae* and *B. suanatina*, and more random mutations seem to have occurred within their genomes. The large differences are consistent with the species being divided into three genomovars (Mirajkar et al. 2016a). Although the assemblies of *B. hampsonii* strains had vastly varying quality and completeness, there was evidence that *B. hampsonii* strains are able to employ a high degree of genome rearrangement. The rearrangement mechanism(s) of *B. hampsonii* strains appears to be more complicated and involve much smaller gene islands than for *B. hyodysenteriae*. The most typical examples of these rearrangements were observed when comparing a pair of strains which were the most similar within the species, and for which a number of possible gene island rearrangements were identified. Based on the data collected, around eight major rearrangement anchors were identified, and one particular region had a much higher frequency of rearrangement events than other regions of the genome. The encoded proteins in these regions were associated mainly with APTase activity coupled to transmembrane movement of substances, peptidase activity, pyrophosphatase activities, phosphate-containing compound metabolic process and protein metabolic process. Many of these proteins have been reported to be related to virulence, and consequently it is conjectured that strains of *B. hampsonii* may vary in their virulence.

The *B. hampsonii* strains also were found to have a large number of mutations. Fast mutation rates may result in more rapid adaptation, an example of which could be related to their ability to infect a wider range of hosts. Some strains shared large number of genes similar to those found in *B. aalborgi,* a species that infects humans. It is possible that in the future more strains or genomovars of *B. hampsonii* will be found given the high rate and complicated mechanisms involved in mutation in this species.

The *B. murdochii* strains were highly conserved in their genomic sequences. The genomes of *B. murdochii* strain $56\text{-}150^T$ obtained from NCBI and another field strain had more than 90% sequence similarity, but both were quite different from the *B. hyodysenteriae* strains. A large consecutive section of the *B. hyodysenteriae* sequence that encodes around 200 proteins was not present in the *B. murdochii* strains. Many of these proteins were annotated as being related to phosphorylation and the sugar phosphotransferase system, such as phosphor kinase and phospho carriers. The phosphotransferase system is a major carbohydrate transport system in bacteria and catalyses the phosphorylation of sugar substrates during their translocation across the cell membrane. Comparison of the genomes of the two *B. murdochii* strains identified some interesting potential genomic mechanisms. Although having very high genome similarity, these strains had unique mutations in genes related to antimicrobial drug resistance, including the multidrug resistance protein MdtC, efflux pump periplasmic linker BepF, outer membrane efflux protein and cysteine-rich secretory protein family proteins. Mutations in these proteins are likely to contribute to variations in antimicrobial drug resistance between the strains. It thus may be hypothesised that *B. murdochii* species can develop resistance to antimicrobial drugs more easily than other *Brachyspira* species.

Apart from *B. aalborgi, B. pilosicoli* had the smallest genome size compared with the other *Brachyspira* species (although the genome size varies between *B. pilosicoli* strains), but *B. pilosicoli* also encoded more proteins. One explanation may be that *B. pilosicoli* has a greater number of gene duplications and homogeneity than other species. A very high rate of genomic rearrangement events was observed between *B. pilosicoli* strains and *B. hyodysenteriae* strains, and this also has been observed between *B. pilosicoli* strains. *B. pilosicoli* and *B. aalborgi* may have almost double the recombination rates of other *Brachyspira* species. In particular, and as previously reported (Mappley et al. 2012), the genome of *B. pilosicoli* strain WesB appears to have undergone many recombination events. *B. pilosicoli* strains BR2001 and P43/6/78 were quite similar, and shared 80% of their proteins. Interestingly, the genomic rearrangement pattern observed in three of the four *B. pilosicoli* strains (95/1000, BR2001 and P43/6/78) closely resembled that observed in the *B. hyodysenteriae* strains. It appears these strains also can move large gene islands to form four potential pathways. This leads to the hypothesis that these large genome rearrangements are employed by members of the species to adapt to their environments, and also suggests that these three strains originally came from a common source that was different from the one from which strain WesB originated. Based on the different genome sizes and organisations we suspect that the *B. pilosicoli* strains also may have differences in their virulence potential.

Despite some similarities, there were clear differences between *B. hyodysenteriae* and *B. pilosicoli*. *B. pilosicoli* strains appeared to have incorporated more mutations and genome repeats than *B. hyodysenteriae,* and as a result may have acquired the ability to colonise a broader range of host species. This is particularly evident in *B. pilosicoli* strain WesB: although isolated from an Australian Aboriginal child, WesB has been shown to be capable of colonising and causing disease in pigs and chickens under experimental conditions (Trott et al. 1995, 1996a). Strains such as this may have greater potential for cross-species transmission than other strains.

4 Future Studies

As previously indicated, it is important that the genome sequences of many more strains of *Brachyspira* species be obtained to strengthen the pan-genomic and species-level comparisons. The use of the third-generation sequencing instruments that significantly increase read lengths should make this task easier. In addition, proteomic and metabolomic analysis should be applied to *Brachyspira* species in a systems biology approach to improve understanding of complex cellular mechanisms. Finally, investigations of gene expression and the development and application of simple methods for genetic manipulation to test and confirm the function of the encoded proteins remain greatly needed. It is evident that future research on *Brachyspira* should be focused on these activities.

Acknowledgements We thank Dr. Tom La of Murdoch University for critically reading the manuscript and Dr. Eric Burrough (Iowa State University) and Dr. Connie Gebhart (University of Minnesota) for providing genomic sequence from some of the *Brachyspira* strains that were included in this preliminary analysis.

References

Backhans A, Johansson K-E, Fellström C (2010) Phenotypic and molecular characterization of *Brachyspira* spp. isolated from wild rodents. Environ Microbiol Report 2:720–727

Bellgard MI, Wanchanthuek P, La T, Ryan K, Moolhuijzen P, Albertyn Z et al (2009) Genome sequence of the pathogenic intestinal spirochete *Brachyspira hyodysenteriae* reveals adaptations to its lifestyle in the porcine large intestine. PLoS ONE 4(3):e4641

Black M, Moolhuijzen P, Barrero R, La T, Phillips N, Hampson DJ et al (2015) Analysis of multiple *Brachyspira hyodysenteriae* genomes confirms that the species is relatively conserved but has potentially important strain variation. PLoS ONE 10(6):e0131050

Brooke CJ, Riley TV, Hampson DJ (2006) Comparison of prevalence and risk factors for faecal carriage of the intestinal spirochaetes *Brachyspira aalborgi* and *Brachyspira pilosicoli* in four Australian populations. Epidemiol Infect 134:627–634

Burrough ER (2017) Swine dysentery. Vet Pathol 54:22–31

Darling AC, Mau B, Blattner FR, Perna NT (2004) Mauve: multiple alignment of conserved genomic sequence with rearrangements. Genome Res 14:1394–1403

Darling AE, Mau B, Perna NT (2010) progressiveMauve: multiple genome alignment with gene gain, loss and rearrangement. PLoS ONE 5(6):e11147

Duhamel GE, Trott DJ, Muniappa N, Mathiesen MR, Tarasiuk K, Lee JI et al (1998) Canine intestinal spirochetes consist of *Serpulina pilosicoli* and a newly identified group provisionally designated "*Serpulina canis*" sp. nov. J Clin Microbiol 36:2264–2270

Guindon S, Dufayard JF, Lefort V, Anisimova M, Hordijk W, Gascuel O (2010) New algorithms and methods to estimate maximum-likelihood phylogenies: assessing the performance of PhyML 3.0. Syst Biol 59:307–321

Håfström T, Jansson DS, Segerman B (2011) Complete genome sequence of *Brachyspira intermedia* reveals unique genomic features in *Brachyspira* species and phage-mediated horizontal gene transfer. BMC Genom 12:395

Hidalgo A, Carvajal A, La T, Naharro G, Rubio P, Phillips ND et al (2010) Multiple-locus variable-number tandem-repeat analysis of the swine dysentery pathogen, *Brachyspira hyodysenteriae*. J Clin Microbiol 48:2859–2865

Hovind-Hougen K, Birch-Andersen A, Henrik-Nielsen R, Orholm M, Pedersen JO, Teglbjaerg PS et al, (1982) Intestinal spirochetosis: morphological characterization and cultivation of the spirochete *Brachyspira aalborgi* gen. nov., sp. nov. J Clin Microbiol 16:1127–1136

Humphrey SB, Stanton TB, Jensen NS, Zuerner RL (1997) Purification and characterization of VSH-1, a generalized transducing bacteriophage of *Serpulina hyodysenteriae*. J Bacteriol 179:323–329

Jansson DS, Fellström C, Johansson KE (2008) Intestinal spirochetes isolated from wild-living jackdaws, hooded crows and rooks (genus *Corvus*): provisionally designated "*Brachyspira corvi*" sp. nov. Anaerobe 14:287–295

Jensen TK, Christensen AS, Boye M (2010) *Brachyspira murdochii* colitis in pigs. Vet Pathol 47:334–338

Kinyon JM, Harris DL (1979) *Treponema innocens*, a new species of intestinal bacteria and emended description of the type strain of *Treponema hyodysenteriae* Harris et al. Int J System Bacteriol 29:102–109

Komarek V, Maderner A, Spergser J, Weissenböck H (2009) Infections with weakly haemolytic *Brachyspira* species in pigs with miscellaneous chronic diseases. Vet Microbiol 134:311–317

La T, Phillips ND, Hampson DJ (2016a) An investigation into the etiological agents of swine dysentery in Australian pig herds. PLoS One 11(12):e0167424

La T, Rohde J, Phillips ND, Hampson DJ (2016b) Comparison of *Brachyspira hyodysenteriae* isolates recovered from pigs in apparently healthy multiplier herds with isolates from herds with swine dysentery. PLoS One 11(8):e0160362. 0362

Li H, Handsaker B, Wysoker A, Fennell T, Ruan J, Homer N et al (2009) The Sequence alignment/map (SAM) format and SAMtools. Bioinformatics 25:2078–2079

Lin C, den Bakker HC, Suzuki H, Lefébure T, Ponnala L, Sun Q, et al. (2013) Complete genome sequence of the porcine strain *Brachyspira pilosicoli* P43/6/78(T.). Genome Announc 1(1)

Mahu M, De Pauw N, Vande Maele L, Verlinden M, Boyen F, Ducatelle R et al (2016) Variation in hemolytic activity of *Brachyspira hyodysenteriae* strains from pigs. Vet Res 47:66

Mappley LJ, Black ML, AbuOun M, Darby AC, Woodward MJ, Parkhill J et al (2012) Comparative genomics of *Brachyspira pilosicoli* strains: genome rearrangements, reductions and correlation of genetic compliment with phenotypic diversity. BMC Genom 13:454

Matson EG, Thompson MG, Humphrey SB, Zuerner RL, Stanton TB (2005) Identification of genes of VSH-1, a prophage-like gene transfer agent of *Brachyspira hyodysenteriae*. J Bacteriol 187:5885–5892

McLaren AJ, Trott DJ, Swayne DE, Oxberry SL, Hampson DJ (1997) Genetic and phenotypic characterization of intestinal spirochetes colonizing chickens, and allocation of known pathogenic isolates to three distinct genetic groups. J Clin Microbiol 35:412–417

Mikosza AS, Hampson DJ (2001) Human intestinal spirochetosis: *Brachyspira aalborgi* and/or *Brachyspira pilosicoli*? Anim Health Res Rev 2:101–110

Mirajkar NS, Phillips ND, La T, Hampson DJ, Gebhart CJ (2016a) Characterization and recognition of *Brachyspira hampsonii* sp. nov., a novel intestinal spirochete that is pathogenic to pigs. J Clin Microbiol 54:2942–2949

Mirajkar NS, Johnson TJ, Gebhart CJ (2016b) Complete genome sequence of *Brachyspira hyodysenteriae* type strain B-78 (ATCC 27164). Genome Announc 4(4)

Mirajkar NS, Johnson TJ, Gebhart CJ, Correction for Mirajkar et al. (2017) Complete genome sequence of *Brachyspira hyodysenteriae* type train B78 (ATCC 27164). Genome Announc 5(3)

Motro Y, La T, Bellgard MI, Dunn DS, Phillips ND, Hampson DJ (2009) Identification of genes associated with prophage-like gene transfer agents in the pathogenic intestinal spirochaetes *Brachyspira hyodysenteriae*, *Brachyspira pilosicoli* and *Brachyspira intermedia*. Vet Microbiol 134:340–345

Mushtaq M, Zubair S, Råsbäck T, Bongcam-Rudloff E, Jansson DS (2015) *Brachyspira suanatina* sp. nov., an enteropathogenic intestinal spirochaete isolated from pigs and mallards: genomic and phenotypic characteristics. BMC Microbiol 15:208

Ochiai S, Adachi Y, Mori K (1997) Unification of the genera *Serpulina* and *Brachyspira*, and proposals of *Brachyspira hyodysenteriae* Comb. Nov., *Brachyspira innocens* Comb. Nov. and *Brachyspira pilosicoli* Comb. Nov. Microbiol Immunol 41:445–452

Page RD (1996) TREEVIEW: An application to display phylogenetic trees on personal computers. Comput Appl Biosci 12:357–358

Pati A, Sikorski J, Gronow S, Munk C, Lapidus A, Copeland A et al (2010) Complete genome sequence of *Brachyspira murdochii* type strain (56-150). Stand Genomic Sci 2:260–269

Phillips ND, La T, Amin MA, Hampson DJ (2010) *Brachyspira intermedia* strain diversity and relationships to other indole-positive *Brachyspira* species. Vet Microbiol 143:246–254

Song Y, La T, Phillips ND, Bellgard MI, Hampson DJ (2009) A reverse vaccinology approach to swine dysentery vaccine development. Vet Microbiol 137:111–119

Song Y, La T, Phillips ND, Hampson DJ (2015) Development of a serological ELISA using a recombinant protein to identify pig herds infected with *Brachyspira hyodysenteriae*. Vet J 206:365–370

Stanton TB, Fournie-Amazouz E, Postic D, Trott DJ, Grimont PAD, Baranton G et al (1997) Recognition of two new species of intestinal spirochetes: *Serpulina intermedia* sp. nov. and *Serpulina murdochii* sp. nov. Int J System Bacteriol 47:1007–1012

Stanton TB, Postic D, Jensen NS (1998) *Serpulina alvinipulli* sp. nov., a new *Serpulina* species that is enteropathogenic for chickens. Int J Syst Bacteriol 48:669–676

Stanton TB, Thompson MG, Humphrey SB, Zuerner RL (2003) Detection of bacteriophage VSH-1 svp38 gene in *Brachyspira* spirochetes. FEMS Microbiol Lett 224:225–229

Stephens CP, Hampson DJ (1999) Prevalence and disease association of intestinal spirochaetes in chickens in eastern Australia. Avian Pathol 28:447–454

Tamura K, Peterson D, Peterson N, Stecher G, Nei M, Kumar S. (2011) MEGA5: molecular evolutionary genetics analysis using maximum likelihood, evolutionary distance, and maximum parsimony methods. Mol Biol Evol 28:2731–2739

Thompson JD, Higgins DG, Gibson TJ (1994) Clustal-W: improving the sensitivity of progressive multiple sequence alignment through sequence weighting, position-specific gap penalties and weight matrix choice. Nucleic Acids Res 22(4673):4680

Trott DJ, McLaren AJ, Hampson DJ (1995) Pathogenicity of human and porcine intestinal spirochetes in day-old specific pathogen free chicks: an animal model of intestinal spirochetosis. Infect Immun 63:3705–3710

Trott DJ, Huxtable CR, Hampson DJ (1996a) Experimental infection of newly weaned pigs with human and porcine strains of *Serpulina pilosicoli*. Infect Immun 64:4648–4654

Trott DJ, Stanton TB, Jensen NS, Duhamel GE, Johnson JL, Hampson DJ (1996b) *Serpulina pilosicoli* sp. nov., the agent of porcine intestinal spirochetosis. Int J System Bacteriol 46:206–215

Wanchanthuek P, Bellgard MI, La T, Ryan K, Moolhuijzen P, Chapman B et al. (2010) The complete genome sequence of the pathogenic intestinal spirochete *Brachyspira pilosicoli* and comparison with other *Brachyspira* genomes. PLoS One 6, 5(7):e11455

Weissenböck H, Maderner A, Herzog AM, Lussy H, Nowotny N (2005) Amplification and sequencing of *Brachyspira* spp. specific portions of nox using paraffin-embedded tissue samples from clinical colitis in Austrian pigs shows frequent solitary presence of *Brachyspira murdochii*. Vet Microbiol 111:67–75

Zuerner RL, Stanton TB, Minion FC, Li C, Charon NW, Trott DJ et al (2004) Genetic variation in *Brachyspira*: chromosomal rearrangements and sequence drift distinguish *B. pilosicoli* from *B. hyodysenteriae*. Anaerobe 10:229–237

Erratum to: Gene Regulation, Two Component Regulatory Systems, and Adaptive Responses in *Treponema Denticola*

Richard T. Marconi

Erratum to: R. T. Marconi, *Current Topics in Microbiology and Immunology*, The Authors, http://doi.org/10.1007/82_2017_66

By mistake this chapter was published open access and with "The Authors" as copyright holder. This has now been corrected.

The updated online version for this chapter can be found under http://doi.org/10.1007/82_2017_66.

R. T. Marconi (✉)
Department of Microbiology and Immunology, Virginia Commonwealth University, Medical Center, 1112 East Clay Street, Room 101 McGuire Hall, 980678, Richmond, VA 23298-0678, USA
e-mail: Richard.marconi@vcuhealth.org

Current Topics in Microbiology and Immunology (2018) 415:E1
DOI 10.1007/82_2018_95

Published Online: 28 June 2018

The manufacturer's authorised representative in the EU is Springer Nature Customer Service Centre GmbH, Europaplatz 3, 69115 Heidelberg, Germany. If you have any concerns regarding our products, please contact ProductSafety@springernature.com

Printed and bound by CPI Group (UK) Ltd, Croydon, CR0 4YY
02/10/2026
02227135-0004